PLATE AND PANEL STRUCTURES OF ISOTROPIC, COMPOSITE AND PIEZOELECTRIC MATERIALS, INCLUDING SANDWICH CONSTRUCTION

SOLID MECHANICS AND ITS APPLICATIONS
Volume 120

Series Editor: G.M.L. GLADWELL
Department of Civil Engineering
University of Waterloo
Waterloo, Ontario, Canada N2L 3GI

Aims and Scope of the Series

The fundamental questions arising in mechanics are: *Why?, How?,* and *How much?* The aim of this series is to provide lucid accounts written bij authoritative researchers giving vision and insight in answering these questions on the subject of mechanics as it relates to solids.

The scope of the series covers the entire spectrum of solid mechanics. Thus it includes the foundation of mechanics; variational formulations; computational mechanics; statics, kinematics and dynamics of rigid and elastic bodies: vibrations of solids and structures; dynamical systems and chaos; the theories of elasticity, plasticity and viscoelasticity; composite materials; rods, beams, shells and membranes; structural control and stability; soils, rocks and geomechanics; fracture; tribology; experimental mechanics; biomechanics and machine design.

The median level of presentation is the first year graduate student. Some texts are monographs defining the current state of the field; others are accessible to final year undergraduates; but essentially the emphasis is on readability and clarity.

For a list of related mechanics titles, see final pages.

Plate and Panel Structures of Isotropic, Composite and Piezoelectric Materials, Including Sandwich Construction

by

JACK R. VINSON
Center for Composite Materials and College of Marine Studies,
Department of Mechanical Engineering,
Spencer Laboratory, University of Delaware,
Newark, Delaware, U.S.A.

A C.I.P. Catalogue record for this book is available from the Library of Congress.

ISBN 978-90-481-6795-1 (PB)
ISBN 978-1-4020-3111-3 (e-book)

Published by Springer,
P.O. Box 17, 3300 AA Dordrecht, The Netherlands.

Sold and distributed in North, Central and South America
by Springer,
101 Philip Drive, Norwell, MA 02061, U.S.A.

In all other countries, sold and distributed
by Springer,
P.O. Box 322, 3300 AH Dordrecht, The Netherlands.

Printed on acid-free paper

Printed in the Netherlands.

This textbook is dedicated to my beautiful wife Midge, who through her encouragement and nurturing over these last two decades, has made the writing of this book possible.

Table of Contents

PREFACE

Plates and panels are primary structural components in many structures from space vehicles, aircraft, automobiles, buildings and homes, bridges decks, ships, and submarines. The ability to design, analyze, optimize and select the proper materials and architecture for plates and panels is a necessity for all structural designers and analysts, whether the adjective in front of the "engineer" on their degree reads aerospace, civil, materials or mechanical.

This text is broken into four parts. The first part deals with the behavior of isotropic plates. Most metals and pure polymeric materials used in structures are isotropic, hence this part covers plates and panels using metallic and polymeric materials.

The second part involves plates and panels of composite materials. Because these fiber reinforced matrix materials can be designed for the particular geometry and loading, they are very often anisotrópic with the properties being functions of how the fibers are aligned, their volume fraction, and of course the fiber and matrix materials used. In general, plate and panel structures involving composite materials will weigh less than a plate or panel of metallic material with the same loads and boundary conditions, as well as being more corrosion resistant. Hence, modern structural engineers must be knowledgeable in the more complicated anisotropic material usage for composite plates and panels.

Sandwich plates and panels offer spectacular advantages over the monocoque constructions treated above. By having suitable face and core materials, isotropic or anisotropic, sandwich plates and panels subjected to bending loads can be 300 times as stiff in bending, with face stresses 1/30 of those using a monocoque construction of a thickness equal to the two faces of the sandwich. Thus, for only the additional weight of the light core material, the spectacular advantages of sandwich construction can be attained. In Part 3, the analyses, design and optimization of isotropic and anisotropic sandwich plates and panels are presented.

In Part 4, the use of piezoelectric materials in beams, plates and panels are treated. Piezoelectric materials are those that when an electrical voltage is applied, the effects are tensile, compressive or shear strains in the material. Conversely, with piezoelectric materials, when loads cause tensile, compressive or shear strains, an electrical voltage is generated. Thus, piezoelectric materials can be used as damage sensors, used to achieve a planned structural response due to an electrical signal, or to increase damping. Piezoelectric materials are often referred to as smart or intelligent materials. The means to describe this behavior and incorporate this behavior into beam, plate and panel construction is the theme of Part 4.

This book is intended for three purposes: as an undergraduate textbook for those students who have taken a mechanics of material course, as a graduate textbook, and as a reference for practicing engineers. It therefore provides the fundamentals of plate and panel behavior. It does not include all of the latest research information nor the complications associated with numerous complex structures – but those structures can be studied and analyzed better using the information provided herein.

Several hundred problems are given at the end of Chapters. Most if not all of these problems are homework and exam problems used by the author over several decades of teaching this material. Appreciation is expressed to Alejandro Rivera, who as the first student to take the course using this text, worked most of the problems at the end of the chapters. These solutions will be the basis of a solutions manual which will be available to professors using this text who contact me.

Special thanks is given to James T. Arters, Research Assistant, who has typed this entire manuscript including all of its many changes and enhancements. Finally, many thanks are given to Dr. Moti Leibowitz who reviewed and offered significant suggestions toward improving Chapter 18, 19 and 20.

CHAPTER 1

EQUATIONS OF LINEAR ELASTICITY IN CARTESIAN COORDINATES

References [1.1-1.6][*] derive in detail the formulation of the governing differential equations of elasticity. Those derivations will not be repeated here, but rather the equations are presented and then utilized to systematically make certain assumptions in the process of deriving the governing equations for rectangular plates and beams.

1.1 Stresses

Consider an elastic body of any general shape. Consider the material to be a *continuum*, ignoring its crystalline structure and its grain boundaries. Also consider the continuum to be *homogeneous*, i.e., no variation of material properties with respect to the spatial coordinates. Then, consider a *material point* anywhere in the interior of the elastic body. If one assigns a Cartesian reference frame with axes x, y and z, shown in Figure 1.1, it is then convenient to assign a rectangular parallelepiped shape to the material point, and label it a *control element* of dimensions dx, dy and dz. The control element is defined to be infinitesimally small compared to the size of the elastic body, yet infinitely large compared to elements of the molecular structure, in order that the material can be considered a continuum.

On the surfaces of the control element there can exist both *normal stresses* (those perpendicular to the plane of the face) and *shear stresses* (those parallel to the plane of the face). On any one face these three stress components comprise a vector, called a *surface traction.*

It is important to note the sign convention and the meaning of the subscripts of these surfaces stresses. For a stress component on a positive face, that is, a face whose outer normal is in the direction of a positive axis, that stress component is positive when it is directed in the direction of that positive axis. Conversely, when a stress is on a negative face of the control element, it is positive when it is directed in the negative axis direction. This procedure is followed in Figure 1.1. Also, the first subscript of any stress component on any face signifies the axis to which the outer normal of that face is parallel. The second subscript refers to the axis to which that stress component is parallel. In the case of normal stresses the subscripts are seen to be repeated and often the two subscripts are shortened to one, i.e. σ_i σ_{ii} where $i = x, y$ or z.

---

[*] Numbers in brackets refer to references given at the end of chapters.

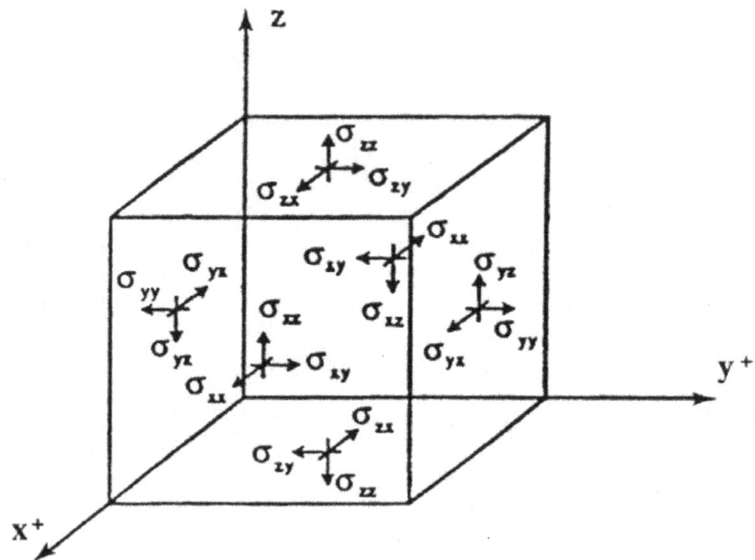

Figure 1.1. Control element in an elastic body showing positive direction of stresses.

1.2 Displacements

The displacements u, v and w are parallel to the x, y and z axes respectively and are positive when in the positive axis direction.

1.3 Strains

Strains in an elastic body are also of two types, extensional and shear. *Extensional strains*, where $i = x, y$ or z, are directed parallel to each of the axes respectively and are a measure of the change in dimension of the control volume in the subscripted direction due to the normal stresses acting on all surfaces of the control volume. Looking at Figure 1.2, one can define *shear strains*.

The shear strain γ_{ij} (where i and $j = x, y$ or z, and $i \neq j$) is a change of angle. As an example shown in Figure 1.2, in the x-y plane, defining γ_{xy} to be

$$\gamma_{xy} = \frac{\pi}{2} - \phi \quad \text{(in radians),} \tag{1.1}$$

then,

$$\varepsilon_{xy} = \frac{1}{2}\gamma_{xy}. \tag{1.2}$$

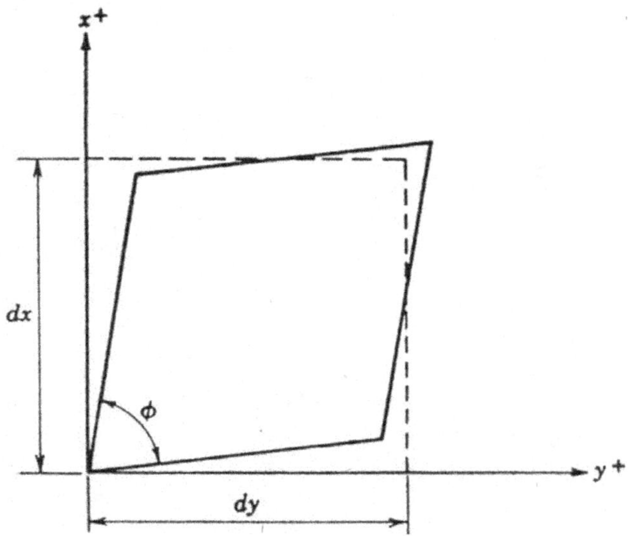

Figure 1.2. Shearing of a control element.

It is important to define the shear strain ε_{xy} to be one half the angle γ_{xy} in order to use tensor notation. However, in many texts and papers the shear strain is defined as γ_{xy}. Care must be taken to insure awareness of which definition is used when reading or utilizing a text or research paper, to obtain correct results in subsequent analysis. Sometimes ε_{ij} is termed tensor strain, and γ_{xy} is referred to as engineering shear strain (not a tensor quantity).

The rules regarding subscripts of strains are identical to those of stresses presented earlier.

1.4 Isotropy and Its Elastic Constants

An isotropic material is one in which the mechanical and physical properties do not vary with orientation. In mathematically modeling an isotropic material, the constant of proportionality between a normal stress and the resulting extensional strain, in the sense of tensile tests is called the modulus of elasticity, *E*.

Similarly, from mechanics of materials, the proportionality between shear stress and the resulting angle γ_{ij} described earlier, in a state of pure shear, is called the shear modulus, *G*.

One final quantity must be defined – the Poisson's ratio, denoted by ν. It is defined as the ratio of the negative of the strain in the *j* direction to the strain in the *i* direction caused by a stress in the *i* direction, σ_{ii}. With this definition it is a positive quantity of magnitude $0 \leq \nu \leq 0.5$, for all isotropic materials.

The well known relationship between the modulus of elasticity, the shear modulus and Poisson's ratio for an isotropic material should be remembered:

$$G = \frac{E}{2(1+\nu)}. \tag{1.3}$$

It must also be remembered that (1.3) can only be used for isotropic materials.

The basic equations of elasticity for a control element of an elastic body in a Cartesian reference frame can now be written. They are written in detail in the following sections and the compact Einsteinian notation of tensor calculus is also provided.

1.5 Equilibrium Equations

A material point within an elastic body can be acted on by two types of forces: *body forces* (F_i) and surface tractions. The former are forces which are proportional to the mass, such as magnetic forces. Because the material is homogeneous, the body forces can be considered to be proportional to the volume. The latter involve stresses caused by neighboring control elements.

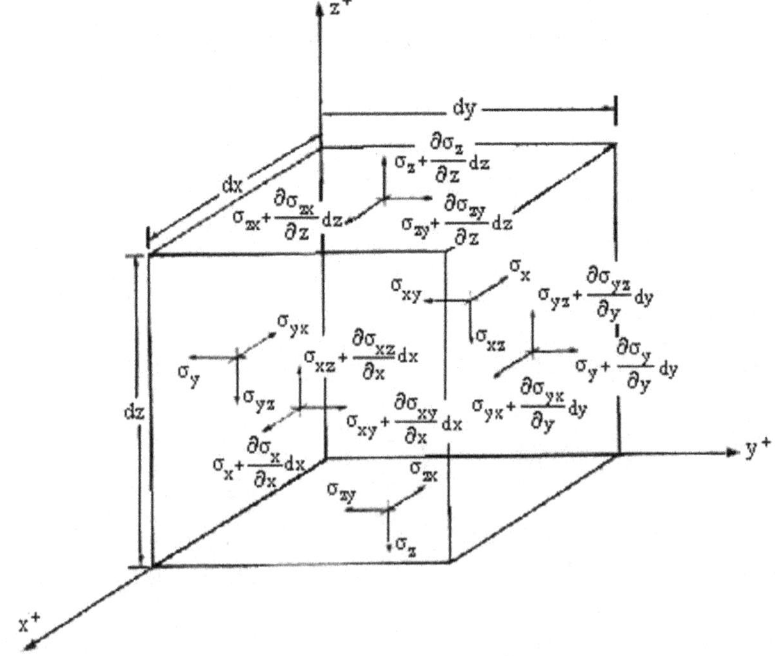

Figure 1.3. Control element showing variation of stresses.

Figure 1.1 is repeated above, but in Figure 1.3, the provision for stresses varying with respect to space is provided. Thus on the back face the stress σ_x is shown, while on the front face that stress value differs because σ_x is a function of x; hence, its value is $\sigma_x + (\partial\sigma_x / \partial x)\mathrm{d}x$. Also shown are the appropriate expressions for the shear stresses.

The body forces per unit volume, $F_i (i = x, y, z)$ are proportional to mass and, as stated before, because the body is homogeneous, are proportional to volume.

The summation of forces in the x direction can be written as

$$\begin{aligned}&\left(\sigma_x + \frac{\partial\sigma_x}{\partial x}\mathrm{d}x\right)\mathrm{d}y\,\mathrm{d}z + \left(\sigma_{yx} + \frac{\partial\sigma_{yx}}{\partial y}\mathrm{d}y\right)\mathrm{d}x\,\mathrm{d}z \\ &+\left(\sigma_{zx} + \frac{\partial\sigma_{zx}}{\partial z}\mathrm{d}z\right)\mathrm{d}x\,\mathrm{d}y - \sigma_x\,\mathrm{d}y\,\mathrm{d}z - \sigma_{yx}\,\mathrm{d}x\,\mathrm{d}z \\ &- \sigma_{zx}\,\mathrm{d}x\,\mathrm{d}y + F_x\,\mathrm{d}x\,\mathrm{d}y\,\mathrm{d}z = 0.\end{aligned} \tag{1.4}$$

After cancellations, every term is multiplied by the volume, which upon division by the volume, results in

$$\frac{\partial\sigma_x}{\partial x} + \frac{\partial\sigma_{yx}}{\partial y} + \frac{\partial\sigma_{zx}}{\partial z} + F_x = 0. \tag{1.5}$$

Likewise, in the y and z direction, the equilibrium equations are:

$$\frac{\partial\sigma_{xy}}{\partial x} + \frac{\partial\sigma_y}{\partial y} + \frac{\partial\sigma_{zy}}{\partial z} + F_y = 0 \tag{1.6}$$

$$\frac{\partial\sigma_{xz}}{\partial x} + \frac{\partial\sigma_{yz}}{\partial y} + \frac{\partial\sigma_z}{\partial z} + F_z = 0. \tag{1.7}$$

In the compact Einsteinian notation, the above three equilibrium equations are written as

$$\sigma_{ki,k} + F_i = 0 \quad (i, k = x, y, z) \tag{1.8}$$

where this is the ith equation, and the repeated subscripts k refer to each term being repeated in x, y and z, and where the comma means partial differentiation with respect to the subsequent subscript.

1.6 Stress-Strain Relations

The relationship between the stresses and strains at a material point in a three dimensional body mathematically describe the way the elastic material behaves. They are often referred to as the constitutive equations and are given below without derivation, because easy reference to many texts on elasticity can be made, such as [1.1 - 1.7].

$$\varepsilon_x = \frac{1}{E}\left[\sigma_x - \nu(\sigma_y + \sigma_z)\right], \qquad \varepsilon_y = \frac{1}{E}\left[\sigma_y - \nu(\sigma_x + \sigma_z)\right] \qquad (1.9), (1.10)$$

$$\varepsilon_z = \frac{1}{E}\left[\sigma_z - \nu(\sigma_x + \sigma_y)\right], \qquad \varepsilon_{xy} = \frac{1}{2G}\sigma_{xy} \qquad (1.11), (1.12)$$

$$\varepsilon_{yz} = \frac{1}{2G}\sigma_{yz}, \qquad \varepsilon_{zx} = \frac{1}{2G}\sigma_{zx} \qquad (1.13), (1.14)$$

From (1.9) the proportionality between the strain ε_x and the stress σ_x is clearly seen. It is also seen that stresses σ_y and σ_z affect the strain σ_x, due to the Poisson's ratio effect.

Similarly, in (1.12) the proportionality between the shear strain ε_{xy} and the shear stress σ_{xy} is clearly seen, the number 'two' being present due to the definition of ε_{xy} given in (1.2).

In the compact Einsteinian notation, the above six equations can be written as

$$\varepsilon_{ij} = a_{ijkl}\sigma_{kl} \qquad (1.15)$$

where a_{ijkl} is the generalized compliance tensor.

1.7 Linear Strain-Displacement Relations

The strain-displacement relations are the kinematic equations relating the displacements that result from an elastic body being strained due to applied loads, or the strains that occur in the material when an elastic body is physically displaced.

$$\varepsilon_x = \frac{\partial u}{\partial x}, \qquad \varepsilon_y = \frac{\partial v}{\partial y} \qquad (1.16), (1.17)$$

$$\varepsilon_z = \frac{\partial w}{\partial z}, \qquad \varepsilon_{xy} = \frac{1}{2}\left(\frac{\partial u}{\partial y} + \frac{\partial v}{\partial x}\right) \qquad (1.18), (1.19)$$

$$\varepsilon_{xz} = \frac{1}{2}\left(\frac{\partial u}{\partial z}+\frac{\partial w}{\partial x}\right), \qquad \varepsilon_{yz} = \frac{1}{2}\left(\frac{\partial v}{\partial z}+\frac{\partial w}{\partial y}\right) \qquad (1.20), (1.21)$$

In compact Einsteinian notation, these six equations are written as:

$$\varepsilon_{ij} = \frac{1}{2}(u_{i,j} + u_{j,i}) \quad (i, j = x, y, z) \qquad (1.22)$$

1.8 Compatibility Equations

The purpose of the compatibility equations is to insure that the displacements of an elastic body are single-valued and continuous. They can be written as:

$$\frac{\partial^2 \varepsilon_{xx}}{\partial y\, \partial z} = \frac{\partial}{\partial x}\left(-\frac{\partial \varepsilon_{yz}}{\partial x}+\frac{\partial \varepsilon_{zx}}{\partial y}+\frac{\partial \varepsilon_{xy}}{\partial z}\right) \qquad (1.23)$$

$$\frac{\partial^2 \varepsilon_{yy}}{\partial z\, \partial x} = \frac{\partial}{\partial y}\left(-\frac{\partial \varepsilon_{zx}}{\partial y}+\frac{\partial \varepsilon_{xy}}{\partial z}+\frac{\partial \varepsilon_{yz}}{\partial x}\right) \qquad (1.24)$$

$$\frac{\partial^2 \varepsilon_{zz}}{\partial x\, \partial y} = \frac{\partial}{\partial z}\left(-\frac{\partial \varepsilon_{xy}}{\partial z}+\frac{\partial \varepsilon_{yz}}{\partial x}+\frac{\partial \varepsilon_{zx}}{\partial y}\right) \qquad (1.25)$$

$$2\frac{\partial^2 \varepsilon_{xy}}{\partial x\, \partial y} = \frac{\partial^2 \varepsilon_{xx}}{\partial y^2}+\frac{\partial^2 \varepsilon_{yy}}{\partial x^2}, \qquad 2\frac{\partial^2 \varepsilon_{yz}}{\partial y\, \partial z} = \frac{\partial^2 \varepsilon_{yy}}{\partial z^2}+\frac{\partial^2 \varepsilon_{zz}}{\partial y^2} \qquad (1.26), (1.27)$$

$$2\frac{\partial^2 \varepsilon_{zx}}{\partial z\, \partial x} = \frac{\partial^2 \varepsilon_{zz}}{\partial x^2}+\frac{\partial^2 \varepsilon_{xx}}{\partial z^2} \qquad (1.28)$$

In compact Einsteinian notation, the compatibility equations are written as follows:

$$\varepsilon_{ij,kl} + \varepsilon_{kl,ij} - \varepsilon_{ik,jl} - \varepsilon_{jl,ik} = 0 \quad (i, j, k, l = x, y, z). \qquad (1.29)$$

However, in all of what follows herein, namely treating plates and beams, invariably the governing differential equations are placed in terms of displacements, and if the solutions are functions which are single-valued and continuous, it is not necessary to utilize the compatibility equations.

1.9 Summary

It can be shown that both the stress and strain tensor quantities are symmetric, i.e.,

$$\sigma_{ij} = \sigma_{ji} \text{ and } \varepsilon_{ij} = \varepsilon_{ji} \quad (i, j = x, y, z). \tag{1.30}$$

Therefore, for the elastic solid there are fifteen independent variables; six stress components, six strain components and three displacements. In the case where compatibility is satisfied, there are fifteen equations: three equilibrium equations, six constitutive relations and six strain-displacement equations.

For a rather complete discussion [1.7] of the equations of elasticity for anisotropic materials, see Chapter 10 of this text.

1.10 References

1.1. Sokolnikoff, I.S. (1956) *Mathematical Theory of Elasticity*, McGraw-Hill Book Company, 2nd Edition, New York.
1.2. Timoshenko, S. and Goodier, J.N. (1970) *Theory of Elasticity*, McGraw-Hill Book Company, New York.
1.3. Green, A.E. and Zerna, W. (1954) *Theoretical Elasticity*, Oxford University Press.
1.4. Love, A.E.H. (1934) *Mathematical Theory of Elasticity*, Cambridge University Press.
1.5. Muskhelishvili, N.I. (1953) *Some Basic Problems in the Mathematical Theory of Elasticity*, Noordhoff Publishing Company.
1.6. Love, A.E.H. (1944) *A Treatise on the Mathematical Theory of Elasticity*, Dover Publications, Fourth Edition, New York.
1.7. Vinson, J.R. and Sierakowski, R.L. (2002) *The Behavior of Structures Composed of Composite Materials*, Second Edition, Kluwer Academic Publishers, Dordrecht, The Netherlands.

1.11 Problems

1.1. Prove that the stresses are symmetric, i.e., $\sigma_{ij} = \sigma_{ji}$.
(Suggestion: take moments about the x, y and z axes.)

1.2. When $\nu = 0.5$ a material is called 'incompressible'. Prove that for $\nu = 0.5$, under any set of stresses, the control volume of Figure 1.1 will not change volume when subjected to applied stresses.

1.3. An elastic body has the following strain field:

$$\begin{aligned} \varepsilon_{xx} &= 2x^2 + 3xy + 4y^2 & \varepsilon_{xy} &= 0 \\ \varepsilon_{yy} &= x^2 - 2y^2 + z^2 & \varepsilon_{yz} &= 2y^2 - 3z^2 \\ \varepsilon_{zz} &= 2y^2 - z^2 & \varepsilon_{xz} &= 3z^2 - 2y^2 \end{aligned}$$

Does this strain field satisfy compatibility? Note: compatibility is not satisfied if any one or more of the compatibility equations is violated.

CHAPTER 2

DERIVATION OF THE GOVERNING EQUATIONS FOR ISOTROPIC RECTANGULAR PLATES

This approach in this chapter is to systematically derive the governing equations for an isotropic classical, thin elastic isotropic rectangular plate. Analogous derivations are given in [2.1 - 2.8].

2.1 Assumptions of Plate Theory

In classical, linear thin plate theory, there are a number of assumptions that are necessary in order to reduce the three dimensional equations of elasticity to a two dimensional set that can be solved. Consider an elastic body shown in Figure 2.1, comprising the region $0 \leq x \leq a$, $0 \leq y \leq b$ and $-h/2 \leq z \leq h/2$, such that $h << a$ and $h << b$. This is called a plate.

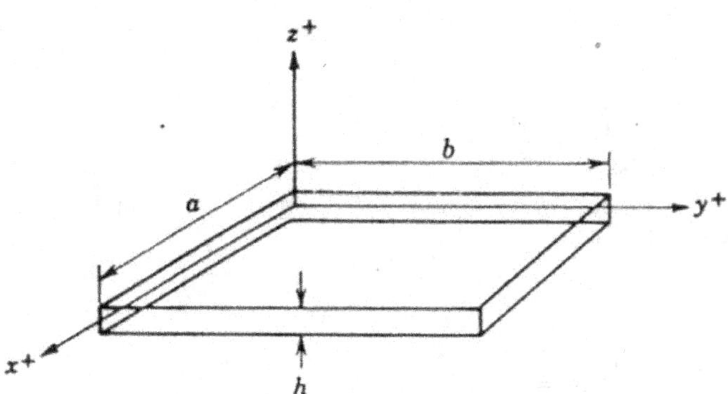

Figure 2.1. Rectangular plate.

The following assumptions are made.

1. A lineal element of the plate extending through the plate thickness, normal to the mid surface, x-y plane, in the unstressed state, upon the application of load:
    a. undergoes at most a translation and a rotation with respect to the original coordinate system;
    b. remains normal to the deformed middle surface.
2. A plate resists lateral and in-plane loads by bending, transverse shear stresses, and in-plane action, not through block like compression or tension in the plate in the

thickness direction. This assumption results from the fact that $h/a \ll 1$ and $h/b \ll 1$.

From 1a the following is implied:

3. A lineal element through the thickness does not elongate or contract.
4. The lineal element remains straight upon load application.

In addition,

5. St. Venant's Principle applies.

It is seen from 1a that the most general form for the two in-plane displacements is:

$$u(x, y, z) = u_0(x, y) + z\overline{\alpha}(x, y) \tag{2.1}$$

$$v(x, y, z) = v_0(x, y) + z\overline{\beta}(x, y) \tag{2.2}$$

where u_0 and v_0 are the in-plane middle surface displacements ($z = 0$), and α and β are rotations as yet undefined. Assumption 3 requires that $\varepsilon_z = 0$, which in turn means that the lateral deflection w is at most (from Equation 1.18)

$$w = w(x, y). \tag{2.3}$$

Also, Equations (1.11) is ignored.

Assumption 4 requires that for any z, both $\boldsymbol{\varepsilon}_{xz} = \text{constant}$ and $\varepsilon_{yz} = \text{constant}$ at any specific location (x, y) on the plate middle surface for all z. Assumption 1b requires that the constant is zero, hence

$$\varepsilon_{xz} = \varepsilon_{yz} = 0.$$

Assumption 2 means that $\boldsymbol{\sigma}_z = 0$ in the stress strain relations.

Incidentally, the assumptions above are identical to those of thin classical beam, ring and shell theory.

2.2 Derivation of the Equilibrium Equations for a Rectangular Plate

Figure 2.2 shows the positive directions of stress quantities to be defined when the plate is subjected to lateral and in-plane loads.

The stress couples are defined as follows:

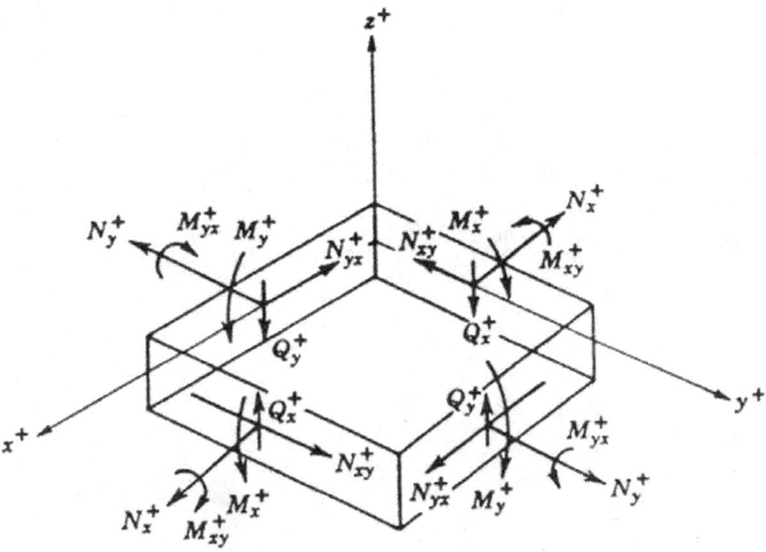

Figure 2.2. Positive directions of stress resultants and couples.

$$M_x = \int_{-h/2}^{+h/2} \sigma_x z \, dx \tag{2.4}$$

$$M_y = \int_{-h/2}^{+h/2} \sigma_y z \, dz \tag{2.5}$$

$$M_{xy} = \int_{-h/2}^{+h/2} \sigma_{xy} z \, dz \tag{2.6}$$

$$M_{yx} = \int_{-h/2}^{+h/2} \sigma_{yx} z \, dz = M_{xy} \, . \tag{2.7}$$

Physically, it is seen that the stress couple is the summation of the moment about the middle surface of all the stresses shown acting on all of the infinitesimal control elements through the plate thickness at a location (x, y). In the limit the summation is replaced by the integration.

Similarly, the shear resultants are defined as,

$$Q_x = \int_{-h/2}^{+h/2} \sigma_{xz} \, dz \tag{2.8}$$

$$Q_y = \int_{-h/2}^{+h/2} \sigma_{yz}\, \mathrm{dz}\,. \tag{2.9}$$

Again the shear resultant is physically the summation of all the shear stresses in the thickness direction acting on all of the infinitesimal control elements across the thickness of the plate at the location (x, y).

Finally, the stress resultants are defined to be:

$$N_x = \int_{-h/2}^{+h/2} \sigma_x\, \mathrm{dx} \tag{2.10}$$

$$N_y = \int_{-h/2}^{+h/2} \sigma_y\, \mathrm{dz} \tag{2.11}$$

$$N_{xy} = \int_{-h/2}^{+h/2} \sigma_{xy}\, \mathrm{dz} \tag{2.12}$$

$$N_{yx} = \int_{-h/2}^{+h/2} \sigma_{yx}\, \mathrm{dz} = N_{xy} \tag{2.13}$$

These then are the sum of all the in-plane stresses acting on all of the infinitesimal control elements across the thickness of the plate at x, y.

Thus, in plate theory, the details of each control element under consideration are disregarded when one integrates the stress quantities across the thickness h. Instead of considering stresses at each material point one really deals with the integrated stress quantities defined above. The procedure to obtain the governing equations for plates from the equations of elasticity is to perform certain integrations on them.

Proceeding, multiply Equation (1.5) by z dz and integrate between $-h/2$ and $+h/2$, as follows:

$$\int_{-h/2}^{+h/2} \left(z\frac{\partial \sigma_x}{\partial x} + z\frac{\partial \sigma_{xy}}{\partial y} + z\frac{\partial \sigma_{xz}}{\partial z} \right) \mathrm{dz} = 0$$

$$\frac{\partial}{\partial x}\int_{-h/2}^{+h/2} \sigma_x z\, \mathrm{dz} + \frac{\partial}{\partial y}\int_{-h/2}^{+h/2} \sigma_{xy} z\, \mathrm{dz} + \int_{-h/2}^{+h/2} z\frac{\partial \sigma_{xz}}{\partial z}\, \mathrm{dz} = 0$$

$$\left[\frac{\partial M_x}{\partial x} + \frac{\partial M_{xy}}{\partial y} + z\sigma_{xz} \right]_{-h/2}^{+h/2} - \int_{-h/2}^{+h/2} \sigma_{xz}\, \mathrm{dz} = 0.$$

In the above, the order of differentiation and integration can be reversed because x and z are orthogonal one to the other. Looking at the third term, $\sigma_{xz} = \sigma_{zx} = 0$ when there are no shear loads on the upper or lower plate surface. If there are surface shear stresses then defining $\tau_{1x} = \sigma_{xz}(+h/2)$ and $\tau_{2x} = \sigma_{xz}(-h/2)$, the results are shown below in Equation (2.14). It should also be noted that for plates supported on an edge, σ_{xz} may not go to zero at $\pm h/2$, and so the theory is not accurate at that edge, but due to St. Venant's Principle, the solutions are satisfactory away from the edge supports.

$$\frac{\partial M_x}{\partial x} + \frac{\partial M_{xy}}{\partial y} + \frac{h}{2}(\tau_{1x} + \tau_{2x}) - Q_x = 0. \tag{2.14}$$

Likewise Equation (1.6) becomes

$$\frac{\partial M_{xy}}{\partial x} + \frac{\partial M_y}{\partial y} + \frac{h}{2}(\tau_{1y} + \tau_{2y}) - Q_y = 0 \tag{2.15}$$

where

$$\tau_{1y} = \sigma_{yz}(+h/2) \text{ and } \tau_{2y} = \sigma_{yz}(-h/2).$$

These two equations describe the moment equilibrium of a plate element. Looking now at Equations (1.7), multiplying it by dz, and integrating between $-h/2$ and $+h/2$, results in

$$\int_{-h/2}^{+h/2} \left(\frac{\partial \sigma_{zx}}{\partial x} + \frac{\partial \sigma_{zy}}{\partial y} + \frac{\partial \sigma_z}{\partial z} \right) dz = 0$$

$$\frac{\partial Q_x}{\partial x} + \frac{\partial Q_y}{\partial y} + \sigma_z \int_{-h/2}^{+h/2} = 0$$

$$\frac{\partial Q_x}{\partial x} + \frac{\partial Q_y}{\partial y} + p_1(x, y) - p_2(x, y) = 0 \tag{2.16}$$

where $p_1(x, y) = \sigma_z(+h/2)$, $p_2(x, y) = \sigma_z(-h/2)$.

One could also derive (2.16) by considering vertical equilibrium of a plate element shown in Figure 2.3.

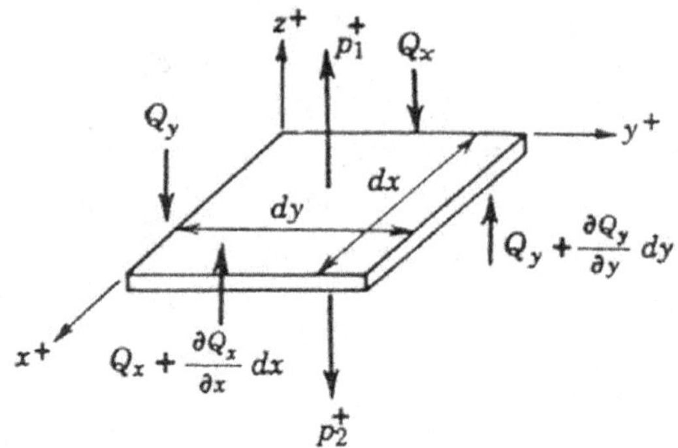

Figure 2.3. Vertical forces on a plate element.

One may ask why use is made of σ_z in this equation and not in the stress-strain relation? The foregoing is not really inconsistent, since σ_z does not appear explicitly in Equation (2.16) and once away from the surface the normal surface traction is absorbed by shear and in-plane stresses rather than by σ_z in the plate interior, as stated previously in Assumption 2.

Similarly, multiplying Equations (1.5) and (1.6) by dz and integrating across the plate thickness results in the plate equilibrium equations in the x and y directions respectively, in terms of the in-plane stress resultants and the surface shear stresses.

$$\frac{\partial N_x}{\partial x} + \frac{\partial N_{xy}}{\partial y} + (\tau_{1x} - \tau_{2x}) = 0 \tag{2.17}$$

$$\frac{\partial N_{xy}}{\partial x} + \frac{\partial N_y}{\partial y} + (\tau_{1y} - \tau_{2y}) = 0. \tag{2.18}$$

2.3 Derivation of Plate Moment-Curvature Relations and Integrated Stress Resultant-Displacement Relations

Now, the plate equations must be derived corresponding to the elastic stress strain relations. The strains ε_x, ε_y and ε_{xy} will not be used explicitly since the stresses have been averaged by integrating through the thickness. Hence, displacements are utilized. Thus, combining (1.9) through (1.21) gives the following, remembering that σ_z has been assumed zero in the interior of the plate and excluding Equation (1.11) for reasons given previously.

$$\frac{\partial u}{\partial x} = \frac{1}{E}[\sigma_x - \nu\sigma_y] \tag{2.19}$$

$$\frac{\partial v}{\partial y} = \frac{1}{E}[\sigma_y - \nu\sigma_x] \tag{2.20}$$

$$\frac{1}{2}\left(\frac{\partial u}{\partial y} + \frac{\partial v}{\partial x}\right) = \frac{1}{2G}\sigma_{xy} \tag{2.21}$$

$$\frac{1}{2}\left(\frac{\partial v}{\partial z} + \frac{\partial w}{\partial y}\right) = \frac{1}{2G}\sigma_{yz} \tag{2.22}$$

$$\frac{1}{2}\left(\frac{\partial w}{\partial x} + \frac{\partial u}{\partial z}\right) = \frac{1}{2G}\sigma_{xz}. \tag{2.23}$$

Next, recall the form of the admissible displacements resulting from the plate theory assumptions, given in (2.1) through (2.3):

$$u = u_0(x, y) + z\overline{\alpha}(x, y) \tag{2.24}$$

$$v = v_0(x, y) + z\overline{\beta}(x, y) \tag{2.25}$$

$$w = w(x, y) \text{ only.} \tag{2.26}$$

In plate theory it is remembered that a lineal element through the plate will experience translations, rotations, but no extensions or contractions. For these assumptions to be valid, the lateral deflections are restricted to being small compared to the plate thickness. It is noted that if a plate is very thin, lateral loads can cause lateral deflections many times the thickness and the plate then behaves largely as a membrane because it has little or no bending resistance, i.e., $D \to 0$.

The assumptions of classical plate theory require that transverse shear deformation be zero. If $\varepsilon_{xz} = \varepsilon_{yz} = 0$ then from Equations (1.20) and (1.21)

$$\frac{1}{2}\left(\frac{\partial u}{\partial z} + \frac{\partial w}{\partial x}\right) = 0 \quad \text{or} \quad \frac{\partial u}{\partial z} = -\frac{\partial w}{\partial x}, \quad \text{likewise}$$

$$\frac{\partial v}{\partial z} = -\frac{\partial w}{\partial y}.$$

Hence, from Equations (2.24) through (2.26) and the above, it is seen that the rotations are

$$\overline{\alpha} = -\frac{\partial w}{\partial x} \tag{2.27}$$

$$\overline{\beta} = -\frac{\partial w}{\partial y} \tag{2.28}$$

Using (2.24) and (2.19), multiplying (2.19) through by z dz and integrating from –h/2 to +h/2, one obtains

$$\int_{-h/2}^{+h/2} \frac{\partial u_0}{\partial x} z\,\mathrm{d}z + \int_{-h/2}^{+h/2} z^2 \frac{\partial \overline{\alpha}}{\partial x}\,\mathrm{d}z = \int_{-h/2}^{+h/2} \frac{1}{E}[\sigma_x - \nu\sigma_y] z\,\mathrm{d}z. \tag{2.29}$$

Likewise (2.25) and (2.20) result in

$$\int_{-h/2}^{+h/2} \frac{\partial v_0}{\partial y} z\,\mathrm{d}z + \int_{-h/2}^{+h/2} z^2 \frac{\partial \overline{\beta}}{\partial y}\,\mathrm{d}z = \int_{-h/2}^{+h/2} \frac{1}{E}[\sigma_y - \nu\sigma_x] z\,\mathrm{d}z \tag{2.30}$$

and Equations (2.24), (2.25) and (2.21) give

$$\int_{-h/2}^{+h/2} \left(\frac{\partial u_0}{\partial y} + \frac{\partial v_0}{\partial x} \right) z\,\mathrm{d}z + \int_{-h/2}^{+h/2} \left(z^2 \frac{\partial \overline{\alpha}}{\partial y} + z^2 \frac{\partial \overline{\beta}}{\partial x} \right) \mathrm{d}z = \int_{-h/2}^{+h/2} \frac{1}{G} \sigma_{xy} z\,\mathrm{d}z. \tag{2.31}$$

Integrating (2.29), (2.30) and (2.31), and using (2.27) and (2.28)

$$\frac{h^3}{12}\frac{\partial \overline{\alpha}}{\partial x} = \frac{1}{E}\left[\boldsymbol{M}_x - \nu \boldsymbol{M}_y\right] = -\frac{h^3}{12}\frac{\partial^2 w}{\partial x^2} \tag{2.32}$$

$$\frac{h^3}{12}\frac{\partial \overline{\beta}}{\partial y} = \frac{1}{E}\left[\boldsymbol{M}_y - \nu \boldsymbol{M}_x\right] = -\frac{h^3}{12}\frac{\partial^2 w}{\partial y^2} \tag{2.33}$$

$$\frac{h^3}{12}\left(\frac{\partial \overline{\alpha}}{\partial y} + \frac{\partial \overline{\beta}}{\partial x} \right) = \frac{1}{G} M_{xy} = -\frac{h^3}{6}\frac{\partial^2 w}{\partial x\,\partial y}. \tag{2.34}$$

Since $G = E/2(1+\nu)$ (2.35)

$$M_{xy} = -(1-\nu) D \frac{\partial^2 w}{\partial x\,\partial y} \quad \text{where} \quad D = \frac{Eh^3}{12(1-\nu^2)}. \tag{2.36}$$

Solving (2.32) and (2.33) for M_x and M_y results in,

$$M_x = -D\left[\frac{\partial^2 w}{\partial x^2} + \nu \frac{\partial^2 w}{\partial y^2} \right] \tag{2.37}$$

$$M_y = -D\left[\frac{\partial^2 w}{\partial y^2} + \nu \frac{\partial^2 w}{\partial x^2}\right]. \tag{2.38}$$

Equations (2.36) through (2.38) are known as the moment-curvature relations, and D is seen to be the flexural stiffness of the plate per unit width. It is seen also that the curvatures in these moment-curvature relations for classical theory are:

$$\kappa_x = \frac{\partial \bar{\alpha}}{\partial x} = -\frac{\partial^2 w}{\partial x^2}, \qquad \kappa_y = \frac{\partial \bar{\beta}}{\partial y} = -\frac{\partial^2 w}{\partial y^2}$$

$$\kappa_{xy} = \left(\frac{\partial \bar{\alpha}}{\partial y} + \frac{\partial \bar{\beta}}{\partial x}\right) = -2\frac{\partial^2 w}{\partial x \partial y} \tag{2.39}$$

Likewise, substituting (2.37) and (2.38) into Equations (2.14) and (2.15) results in

$$Q_x = -D\frac{\partial}{\partial x}(\nabla^2 w) + \frac{h}{2}(\tau_{1x} + \tau_{2x})$$

$$Q_y = -D\frac{\partial}{\partial y}(\nabla^2 w) + \frac{h}{2}(\tau_{1y} + \tau_{2y}). \tag{2.40}$$

In the above the two dimensional Laplacian operator ∇^2 is defined as follows:

$$\nabla^2 w = \frac{\partial^2 w}{\partial x^2} + \frac{\partial^2 w}{\partial y^2}.$$

Also using Equations (2.24) and (2.25) substituting them into Equations (2.19) through (2.21), then multiplying the latter three equations by dz, integrating across the thickness, results in the following integrated stress-strain relationships:

$$N_x = K\left[\frac{\partial u_0}{\partial x} + \nu \frac{\partial v_0}{\partial y}\right] \tag{2.41}$$

$$N_y = K\left[\frac{\partial v_0}{\partial y} + \nu \frac{\partial u_0}{\partial x}\right] \tag{2.42}$$

$$N_{xy} = N_{yx} = Gh\left[\frac{\partial u_0}{\partial y} + \frac{\partial v_0}{\partial x}\right], \tag{2.43}$$

where $Eh/(1-\nu^2) = K$, the plate extensional stiffness. Equations (2.41) through (2.43) describe the in-plane force and deformation behavior.

2.4 Derivation of the Governing Differential Equations for a Plate

The equations governing the lateral deflections, and the bending and shearing action of a plate can be summarized as follows:

$$\frac{\partial M_x}{\partial x}+\frac{\partial M_{xy}}{\partial y}-Q_x+\frac{h}{2}(\tau_{1x}+\tau_{2x})=0 \tag{2.44}$$

$$\frac{\partial M_{xy}}{\partial x}+\frac{\partial M_y}{\partial y}-Q_y+\frac{h}{2}(\tau_{1y}+\tau_{2y})=0 \tag{2.45}$$

$$\frac{\partial Q_x}{\partial x}+\frac{\partial Q_y}{\partial y}+p_1-p_2=0 \tag{2.46}$$

$$M_x=-D\left[\frac{\partial^2 w}{\partial x^2}+\nu\frac{\partial^2 w}{\partial y^2}\right] \tag{2.47}$$

$$M_y=-D\left[\frac{\partial^2 w}{\partial y^2}+\nu\frac{\partial^2 w}{\partial x^2}\right] \tag{2.48}$$

$$M_{xy}=-D(1-\nu)\frac{\partial^2 w}{\partial x\,\partial y}. \tag{2.49}$$

The equations governing the in-plane stress resultants and in-plane midsurface displacements are:

$$\frac{\partial N_x}{\partial x}+\frac{\partial N_{xy}}{\partial y}+(\tau_{1x}-\tau_{2x})=0 \tag{2.50}$$

$$\frac{\partial N_{xy}}{\partial x}+\frac{\partial N_y}{\partial y}+(\tau_{1y}-\tau_{2y})=0 \tag{2.51}$$

$$N_x=K\left[\frac{\partial u_0}{\partial x}+\nu\frac{\partial v_0}{\partial y}\right] \tag{2.52}$$

$$N_y=K\left[\frac{\partial v_0}{\partial y}+\nu\frac{\partial u_0}{\partial x}\right] \tag{2.53}$$

$$N_{xy} = Gh\left[\frac{\partial u_0}{\partial y} + \frac{\partial v_0}{\partial x}\right]. \tag{2.54}$$

It should be noted that in classical, thin plate theory the equations related to bending and shear, Equations (2.44) through (2.49), are completely uncoupled from the equations dealing with in-plane loads and displacements, Equations (2.50) through (2.54). [*Note*: in Chapter 6, we shall see that when in-plane loads are sufficiently large, the in-plane loads do indeed cause lateral displacements (buckling), but a more sophisticated theory will be evolved at that time].

It should also be noted that the flexural stiffness D of the plate corresponds closely to the EI in beam theory, but is in terms of a unit width, and incorporates the Poisson's ratio effect. Likewise a similar correspondence exists between the extensional stiffness K and the EA in beam theory.

Equations (2.44) through (2.54) are the eleven governing plate equations. First note that the plate can only tell the difference between normal tractions on the upper and lower surface. Hence, one can define $p(x, y)$ as

$$p_1(x,y) - p_2(x,y) = p(x,y). \tag{2.55}$$

Substituting (2.44) and (2.45) into (2.46) results in the following for the case of no shear stresses on the plate upper and lower surfaces:

$$\frac{\partial^2 M_x}{\partial x^2} + 2\frac{\partial^2 M_{xy}}{\partial x\,\partial y} + \frac{\partial^2 M_y}{\partial y^2} + p(x,y) = 0. \tag{2.56}$$

Substituting (2.47) through (2.49) into (2.56) results in:

$$D\left[\frac{\partial^4 w}{\partial x^4} + 2\frac{\partial^4 w}{\partial x^2\,\partial y^2} + \frac{\partial^4 w}{\partial y^4}\right] = p(x,y)$$

or

$$D\nabla^4 w = p(x,y), \tag{2.57}$$

where

$$\nabla^2(\;) = \frac{\partial^2(\;)}{\partial x^2} + \frac{\partial^2(\;)}{\partial y^2} \quad \text{and} \quad \nabla^4(\;) = \nabla^2(\nabla^2(\;)). \tag{2.58}$$

∇^2, the Laplacian operator, is really the sum of the curvatures in two orthogonal directions at the location x, y in the plate. ∇^4, the biharmonic operator, is the Laplacian of the Laplacian, and is physically, then, the sum of the curvatures of the sum of the curvatures in orthogonal directions. One might say that it is a measure of 'bulginess'.

Next, treating Equations (2.50) through (2.54) by substituting Equations (2.52) through (2.54) into the two equilibrium equations, becomes, after considerable manipulation, and for the case of no surface shear stresses,

$$\nabla^4 u_0 = 0 \tag{2.59}$$

$$\nabla^4 v_0 = 0. \tag{2.60}$$

One never needs to use (2.59) and (2.60) but it shows a certain correspondence to (2.57) above.

For the bending vibrations of a plate, an inertial load per unit platform area is added as an equivalent force per unit area, i.e., d'Alembert's Principle, resulting in Equations (2.57) being modified, as seen below:

$$D\nabla^4 w = p - \rho_m h \frac{\partial^2 w}{\partial t^2} \tag{2.61}$$

where ρ_m is the *mass* density of the plate material, and t is the coordinate of time. Here, $w = w(x, y, t)$ and $p = p(x, y, t)$. This modification can be made because the theory is linear and superposition is possible.

In a plate of varying thickness, $h = h(x, y)$, the following equation is derived rather than (2.57):

$$\nabla^2 (D\nabla^2 w) - (1-\nu)\Diamond^4 (D, w) = p(x, y) \tag{2.62}$$

where $\Diamond^4$ is the die operator defined as

$$\Diamond^4 (D, w) = \frac{\partial^2 D}{\partial x^2}\frac{\partial^2 w}{\partial y^2} - 2\frac{\partial^2 D}{\partial x \partial y}\frac{\partial^2 w}{\partial x \partial y} + \frac{\partial^2 D}{\partial y^2}\frac{\partial^2 w}{\partial x^2}. \tag{2.63}$$

If a plate is on an elastic foundation, in which a linear foundation modulus k, in units of lbs/in/in^2 can be defined, then Equation (2.57) is altered by adding in the additional lateral force per unit platform area:

$$D\nabla^4 w = p(x, y) - kw. \tag{2.64}$$

Since classical linear elasticity is involved herein, superposition permits the writing of a vibrating plate on an elastic foundation as follows:

$$D\nabla^4 w = p(x, y, t) - \rho_m h \frac{\partial^2 w}{\partial t^2} - kw. \tag{2.65}$$

2.5 Boundary Conditions for a Rectangular Plate

First, the boundary conditions for the bending of a plate subjected to lateral loads, Equation (2.57), will be discussed. Additional boundary conditions for Equations (2.59) and (2.60) for a plate subjected also to in-plane loads will be discussed later.

Since (2.57) is a fourth order partial differential equation in x and y describing the bending of a plate, four boundary conditions are needed on the x edges and four are needed on the y edges, i.e., two on each edge. For the clamped and simply supported edges, knowledge of beam theory dictates the following:

For a clamped edge | *For a simply supported edge*

$$w = 0 \qquad\qquad w = 0$$
$$\frac{\partial w}{\partial n} = 0 \qquad\qquad M_n = 0 \tag{2.66}$$

where n is the direction normal to the edge.

For a Free Edge

Consider an x = constant free edge. Since by definition a free edge has no loads applied to it, Figure 2.2 shows that $\boldsymbol{M_x}$, M_{xy} and Q_x all are zero on that edge. Hence, six boundary conditions must be satisfied on the two x = constant plate edges. However, the plate equation is only fourth order in x, hence one cannot specify more than two boundary conditions on each edge. [*Note*: In a more advanced plate theory that includes the effects of transverse shear deformation, $\boldsymbol{\varepsilon_{xz}} \neq 0$ and $\varepsilon_{yz} \neq 0$, the governing equations are sixth order in both x and y and the problem discussed here does not occur.]

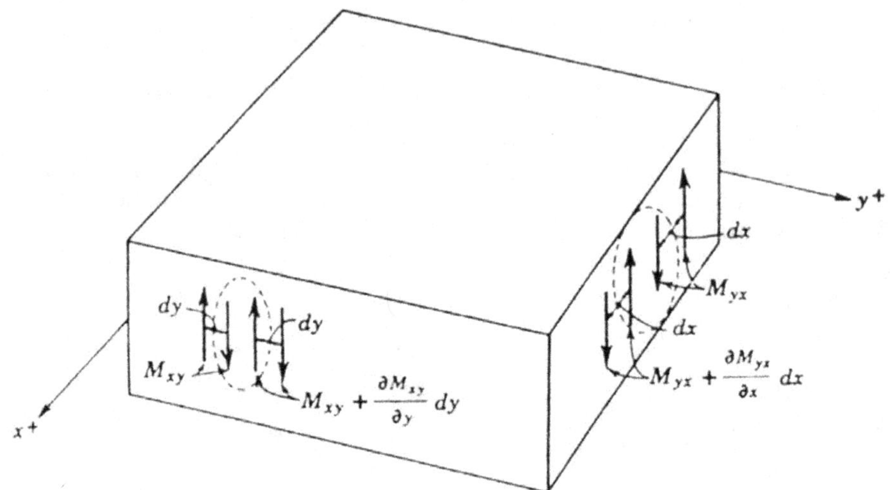

Figure 2.4. Development of the Kirchhoff boundary conditions for a free edge.

To eliminate the problem, Kirchhoff proceeded as follows: On the free x = constant edge, M_x is set equal to zero. M_{xy}, the twisting stress couple is considered to be a couple consisting of two forces of magnitude M_{xy} separated by a small distance dy, as shown in Figure 2.4. Since the stress couple M_{xy} is not constant in general along an edge, nearby is another couple, $M_{xy} + (\partial M_{xy} / \partial y)\,\mathrm{d}y$. It too can be regarded as two forces of magnitude $M_{xy} + (\partial M_{xy} / \partial y)\,\mathrm{d}y$, separated by a distance dy. Therefore, considering an infinitesimal region of the edge shown within the dotted line, it is seen that there is a force M_{xy} positive downward, a force $M_{xy} + (\partial M_{xy} / \partial y)\,\mathrm{d}y$ positive upwards as well as the force due to the transverse shear resultant, $\boldsymbol{Q_x}$ **dy**, acting positive upwards. These must equal zero, hence,

$$-M_{xy} + M_{xy} + \frac{\partial M_{xy}}{\partial y}\mathrm{d}y + Q_x \mathrm{d}y = 0$$

or

$$V_x = Q_x + \frac{\partial M_{xy}}{\partial y} = 0 \tag{2.67}$$

where V_x is called the effective shear resultant on a free edge.

Physically it is seen that on the free edge neither Q_x nor M_{xy} are zero, only the relationship given by (2.67) is zero. However, this approximation was found to have sufficient accuracy that it has been widely used in plate analysis and is known as the Kirchhoff Free Edge Boundary Condition.

Likewise on a y = constant free edge

$$V_y = Q_y + \frac{\partial M_{xy}}{\partial x} = 0 \tag{2.68}$$

and of course on either edge the other boundary condition is

$$M_n = 0 \tag{2.69}$$

where n refers to the directional normal to the edge.

Edge Elastically Supported Against Deflection

Suppose there exists a linear spring support at an edge of magnitude c lbs/in^2. Then the boundary conditions become:

$$M_n = 0 \tag{2.70}$$

$$V_n + cw = 0 \tag{2.71}$$

or using (2.67) and (2.36)

$$Q_n + \frac{\partial M_{ns}}{\partial s} + cw = \frac{\partial^3 w}{\partial n^3} + (2-\nu)\frac{\partial^3 w}{\partial n\,\partial s^2} - \frac{cw}{D} = 0 \tag{2.72}$$

where s refers to the direction parallel to the edge.

Edge Elastically Restrained Against Rotation

Suppose there exists a torsional spring support at an edge of r in lbs/in. Then the boundary conditions would be:

$$V_n = 0 \tag{2.73}$$

$$M_n + r\frac{\partial w}{\partial n} = 0$$

or using (2.37)

$$\frac{\partial^2 w}{\partial n^2} + \nu\frac{\partial^2 w}{\partial s^2} - \frac{r}{D}\frac{\partial w}{\partial n} = 0. \tag{2.74}$$

In-Plane Boundary Conditions

In Section 2.4, it was seen that the governing equations involving the in-plane forces and midsurface displacements are completely uncoupled from the equations, involving bending, shear, lateral forces and lateral displacements, the boundary conditions for which have been discussed above.

In the case of a plate not subjected to any prescribed in-plane loads or prescribed midsurface displacements at the boundaries, the solutions to Equations (2.59) and (2.60) are simply as follows for all values of x and y.

$$u_0 = v_0 = 0.$$

For other cases, the details of the in-plane boundary conditions of the plate structure being analyzed must be studied in detail, to specify which boundary conditions should be prescribed. However through the use of variational procedures, which will be discussed in Chapter 9, it can be shown that the boundary conditions to use in solving Equations (2.59) and (2.60) are:

For an x = constant edge:

$$\begin{aligned}&\text{Either } u_0 \text{ is prescribed or } \boldsymbol{N_x} = 0\\&\text{and}\\&\text{Either } v_0 \text{ is prescribed or } N_{xy} = 0\end{aligned} \tag{2.75}$$

For a y = constant edge:

$$\begin{aligned}&\text{Either } v_0 \text{ is prescribed or } N_y = 0\\&\text{and}\\&\text{Either } u_0 \text{ is prescribed or } N_{yx} = 0.\end{aligned} \tag{2.76}$$

2.6 Stress Distribution within a Plate

In plate theory because all equations are integrated across the thickness only integrated stress quantities are obtained. For stresses on a control element or material point within a plate, one must *assume* a stress distribution. This is done by means of an analogy to beam theory. Thus,

$$\sigma_x = \frac{M_x z}{h^3/12} + \frac{N_x}{h} \tag{2.77}$$

$$\sigma_y = \frac{M_y z}{h^3/12} + \frac{N_y}{h} \tag{2.78}$$

$$\sigma_{xy} = \frac{M_{xy} z}{h^3/12} + \frac{N_{xy}}{h} \tag{2.79}$$

$$\sigma_{xz} = \frac{3Q_x}{2h}\left[1 - \left(\frac{z}{h/2}\right)^2\right] - \frac{S_x}{4} \tag{2.80}$$

$$\sigma_{yz} = \frac{3Q_y}{2h}\left[1 - \left(\frac{z}{h/2}\right)^2\right] - \frac{S_y}{4} \tag{2.81}$$

where

$$S_x = \tau_{1x}\left[1 - 2\left(\frac{z}{h/2}\right) - 3\left(\frac{z}{h/2}\right)^2\right] + \tau_{2x}\left[1 + 2\left(\frac{z}{h/2}\right) - 3\left(\frac{z}{h/2}\right)^2\right] \tag{2.82}$$

$$S_y = \tau_{1y}\left[1 - 2\left(\frac{z}{h/2}\right) - 3\left(\frac{z}{h/2}\right)^2\right] + \tau_{2y}\left[1 + 2\left(\frac{z}{h/2}\right) - 3\left(\frac{z}{h/2}\right)^2\right] \tag{2.83}$$

It can easily be shown that these distributions satisfy the definitions of Equations (2.4) through (2.13). Equally important they satisfy the equilibrium equations of elasticity (1.5) and (1.6) exactly, and Equation (1.7) on the average. Thus the stresses obtained through the use of plate theory (or beam, shell and ring theory) are not exact, in the sense of being three dimensional elasticity theory solutions, but they are very close to the exact solution.

2.7 References

2.1. Timoshenko, S. and Woinowsky-Krieger, A. (1959) *Theory of Plates and Shells*, McGraw-Hill Book Company, Inc., 2nd Edition, New York.

2.2. Marguerre, K. and Woernle, H.T. (1970) *Elastic Plates*, Blaisdell Publishing Company.

2.3. Mansfield, E.H. (1964) *The Bending and Stretching of Plates*, Pergamon Press.

2.4. Jaeger, L.G. (1964) *Elementary Theory of Elastic Plates*, Pergamon Press.

2.5. Morley, L.S.D. (1963) *Skew Plates and Structures*, Pergamon Press.

2.6. Vlasov, V.Z. and Leont'ev, N.N. (1966) *Beams, Plates and Shells on Elastic Foundations*, published for NASA and NSF by the Israel Program for Scientific Translations.

2.7. Vinson, J.R. (1974) *Structural Mechanics: The Behavior of Plates and Shells*, John Wiley and Sons, New York.

2.8. Vinson, J.R. and Chou, T.W. (1975) *Composite Materials and Their Use in Structures*, Applied Science Publishers, London.

2.8 Problems

2.1. The governing equation for a rectangular plate subjected to a lateral distributed load $p(x, y)$ are given by Equations (2.57). However, when the plate is subjected to surface shear stresses τ_{1x}, τ_{2x}, τ_{1y} and τ_{2y}, additional terms are added which are functions of those surface shear stresses, such that the equations can be written as:

$$D\nabla^4 w = p(x,y) + e(\tau_{1x}, \tau_{2x}, \tau_{1y}, \tau_{2y})$$
$$K\nabla^4 u_0 = f(\tau_{1x}, \tau_{2x}, \tau_{1y}, \tau_{2y})$$
$$K\nabla^4 v_0 = g(\tau_{1x}, \tau_{2x}, \tau_{1y}, \tau_{2y}).$$

Starting with (2.44) on, and retaining the surface shear stress terms, find the functions e, f and g.

2.2. Derive Equations (2.59) and (2.60), starting with Equations (2.50) through (2.54).

2.3. Show that the stress distributions of Sections 2.6 do in fact satisfy the definitions of Equations (2.4) through (2.13).

2.4. Show that the stress distributions of Section 2.6 satisfy Equations (1.5) and (1.6) where the body forces $\boldsymbol{F_i} = 0$. Do they satisfy Equations (1.7) with $F_z = 0$? Do they satisfy (1.7) on the average, i.e.,

$$\frac{1}{h}\int_{-h/2}^{+h/2}\left[\frac{\partial\sigma_{xz}}{\partial x} + \frac{\partial\sigma_{yz}}{\partial y} + \frac{\partial\sigma_{z}}{\partial z}\right]dz = 0?$$

2.5. Starting with the pertinent elasticity equations, derive Equations (2.50) and (2.52).

2.6. Consider the plate shown in Figure 2.1. The plate is subjected to a constant in-plane load in the y-direction, $N_y = N_0$, only.

a. What are the stresses $\boldsymbol{\sigma_x}$, σ_y and σ_{xy} in the plate?

b. What are the displacements u, v and w in the plate? Assume $\boldsymbol{v_0} = 0$ along the $y = 0$ edge and $\boldsymbol{u_0} = 0$ along the $x = 0$ edge.

CHAPTER 3

SOLUTIONS TO PROBLEMS OF ISOTROPIC RECTANGULAR PLATES

3.1 Some General Solutions of the Biharmonic Equation

The governing equation for the bending of an isotropic, constant thickness, rectangular plate subjected to lateral distributed loads is given by (2.57) and repeated below.

$$\nabla^4 w = \frac{\partial^4 w}{\partial x^4} + \frac{2\partial^4 w}{\partial x^2 \partial y^2} + \frac{\partial^4 w}{\partial y^4} = \frac{p(x,y)}{D}. \tag{3.1}$$

First, the homogeneous equation, $\nabla^4 w = 0$ is investigated. It is interesting to do this in order to identify the functions that are characteristic of the two dimensional biharmonic equation in a Cartesian coordinate system.

One of the most common methods used to solve this homogeneous equation is by separation of variables. This process can be attempted when the boundary conditions are homogeneous. We cannot count upon the separation of variables to yield all of the complete exact solutions, but it will give all the separable solutions. There *may* be others. Let

$$w(x,y) = X(x)Y(y). \tag{3.2}$$

From (3.1) and (3.2),

$$X^{IV}Y + 2X''Y'' + XY^{IV} = 0.$$

Dividing by *XY* gives

$$\frac{X^{IV}}{X} + 2\frac{X''}{X}\frac{Y''}{Y} + \frac{Y^{IV}}{Y} = 0. \tag{3.3}$$

The variables are still not separated, hence, let

$$\frac{X^{IV}}{X} = f(x), \quad \frac{X''}{X} = g(x), \quad \frac{Y''}{Y} = k(y), \quad \frac{Y^{IV}}{Y} = p(y).$$

Equation (3.3) becomes

$$f(x) + 2g(x)k(y) + p(y) = 0. \tag{3.4}$$

Differentiating with respect to x gives,

$$f'(x) + 2g'(x)k(y) = 0$$

or,

$$\frac{f'(x)}{g'(x)} + 2k(y) = 0.$$

For this to be true, then $f'(x)/g'(x) =$ constant and $k(y) =$ constant. Thus,

$$-\lambda^2 = k(y) = \text{constant}. \tag{3.5}$$

Similarly differentiating (3.4) with respect to y gives

$$2g(x)k'(y) + p'(y) = 0$$

$$\frac{p'(y)}{k'(y)} + 2g(x) = 0.$$

Hence, the following must be true.

$$-\gamma^2 = g(x) = \text{constant}. \tag{3.6}$$

Case 1 $k(y) = -\lambda^2$

Case 1a $\lambda^2 > 0$

$$k(y) = \frac{Y''}{Y} = -\lambda^2$$

$$Y'' + \lambda^2 Y = 0 \text{ or } Y = \begin{Bmatrix} \cos \lambda y \\ \sin \lambda y \end{Bmatrix}.$$

Substituting (3.5) into (3.3) gives,

$$\frac{X^{IV}}{X} - 2\lambda^2 \frac{X''}{X} + \lambda^4 = 0$$

$$X^{IV} - 2\lambda^2 X'' + \lambda^4 x = 0 \qquad (3.7)$$

Let $X = e^{\alpha x}$

$$\alpha^4 - 2\lambda^2\alpha^2 + \lambda^4 = 0 = (\alpha^2 - \lambda^2)^2$$

$$\text{or } \alpha = \pm\lambda, \pm\lambda \text{ or } X = \begin{Bmatrix} \cosh \lambda x \\ x \cosh \lambda x \\ \sinh \lambda x \\ x \sinh \lambda x \end{Bmatrix}.$$

So, there are eight such products as solutions of where

$$w(x, y) = X(x)Y(y) \text{ and } k(y) = -\lambda^2.$$

Case 1b $\lambda^2 = 0$

$$k(y) = \frac{Y''}{Y} = 0 \quad Y = \begin{Bmatrix} 1 \\ y \end{Bmatrix}.$$

Substituting (3.5) into (3.3) gives,

$$\frac{X^{IV}}{X} - 2\lambda^2 \frac{X''}{X} + \lambda^4 = 0$$

If $\qquad (3.8)$

$$\lambda^2 = 0$$

$$X^{IV} = 0, \; X = \begin{Bmatrix} 1 \\ x \\ x^2 \\ x^3 \end{Bmatrix}.$$

Hence, another eight products are found to be solutions to $\nabla^4 w = 0$ where $w = X(x)Y(y)$ and $k(y) = 0$.

Case Ic $\lambda^2 < 0$ (λ is imaginary). Hence, let $\lambda = i\bar{\lambda}$

$$Y'' + \lambda^2 Y = 0 \quad Y'' - \bar{\lambda}^2 Y = 0 \quad Y = \begin{Bmatrix} \sinh \bar{\lambda} y \\ \cosh \bar{\lambda} y \end{Bmatrix},$$

as before $\alpha = \pm\lambda, \pm\lambda$ where λ is imaginary, let

$$X = e^{\alpha x} \quad \text{so}$$

$$\frac{X^{IV}}{X} - 2\lambda^2 \frac{X^4}{X} + \lambda^4 = 0$$

$$\alpha^4 + 2\bar{\lambda}^2 \alpha^2 + \bar{\lambda}^4 = 0 \quad X = \begin{Bmatrix} \cos \bar{\lambda} x \\ x \cos \bar{\lambda} x \\ \sin \bar{\lambda} x \\ x \sin \bar{\lambda} x \end{Bmatrix}$$

$$\alpha = \pm i\bar{\lambda}, \pm i\bar{\lambda}. \tag{3.9}$$

Case II

$$g(x) = -\gamma^2 \tag{3.10}$$

Case IIa $\gamma^2 > 0$

$$g(x) = \frac{X''}{X} = -\gamma^2$$

$$X'' + \gamma^2 X = 0 \quad X = \begin{Bmatrix} \cos \gamma x \\ \sin \gamma x \end{Bmatrix}$$

and as before

$$Y = \begin{Bmatrix} \cosh \gamma y \\ y \cosh \gamma y \\ \sinh \gamma y \\ y \sinh \gamma y \end{Bmatrix}. \tag{3.11}$$

Case IIb $\gamma^2 = 0$

then

$$\left.\begin{aligned} X &= \begin{Bmatrix} 1 \\ X \end{Bmatrix} \\ Y &= \begin{Bmatrix} 1 \\ y \\ y^2 \\ y^3 \end{Bmatrix}. \end{aligned}\right\} \quad (3.12)$$

Case IIc $\gamma^2 < 0$

let $\gamma = i\bar{\gamma}$

$$\left.\begin{aligned} X &= \begin{Bmatrix} \cosh \bar{\gamma} x \\ \sinh \bar{\gamma} x \end{Bmatrix} \\ Y &= \begin{Bmatrix} \sin \bar{\gamma} y \\ y \sin \bar{\gamma} y \\ \cos \bar{\gamma} y \\ y \cos \bar{\gamma} y \end{Bmatrix}. \end{aligned}\right\} \quad (3.13)$$

So in each case there are eight possible products to satisfy any particular case. These solutions comprise all the possible separable solutions of the homogeneous two-dimensional biharmonic equation in a Cartesian coordinate system. In the solution of any particular problem one can attempt to find the solution through exploiting the particular boundary conditions and loading, intuition and experience. However, if that fails then one can resort to trying each of the above solutions for the homogeneous solution.

3.2 Double Series Solution (Navier Solution)

In plate problems one can usually obtain solutions using a doubly infinite series, such as

$$w(x,y) = \sum_{m=1}^{\infty}\sum_{n=1}^{\infty} A_{mn} f_m(x) g_n(y).$$

Such solutions are often inefficient to compute with due to the very slow convergence of the series. As an alternative one may obtain a solution where the function of only one spatial variable is summed such that, in this case:

$$w(x,y) = \sum_{n=1}^{\infty} \phi_n(y) f_n(x).$$

This approach is particularly useful when two opposite edges are simply supported, because then the function $f_n(x)$ above can be a half range sine series. This is discussed in the next section.

In assuming the functions $f_m(x)$ and $g_n(y)$ for the double series solution (Navier Solution), or assuming the functions $f_n(x)$ for the single series solution (the M. Levy Solution), the functions must be complete in order that the lateral deflection can be adequately represented. Furthermore, it is most convenient from a computational point of view that the functions be orthogonal. Also of course they must satisfy the boundary conditions for the problem. One straightforward approach to selecting such functions is to use the vibration modes or buckling modes for a beam of constant cross section with the same boundary conditions as those on opposite edges of the plate, because all such modes comprise a complete, orthogonal set. The beam vibration modes for all boundary conditions and their properties have been conveniently catalogued by Young and Felgar [3.1] and Felgar [3.2].

The doubly infinite series approach will be treated first. Consider a rectangular plate simply supported on all four edges in the region $0 \le x \le a$, $0 \le y \le b$, $-h/2 \le z \le h/2$.

The governing equation is:

$$\nabla^4 w = p(x,y)/D$$

The solution for the lateral displacement can be written as

$$\text{Let } w(x,y) = \sum_{m=1}^{\infty}\sum_{n=1}^{\infty} A_{mn} \sin\frac{m\pi x}{a} \sin\frac{n\pi y}{b}, \tag{3.14}$$

because these functions are complete, orthogonal and they satisfy the boundary conditions of the problem. The lateral load must be expanded in the same series solution:

$$p(x, y) = \sum_{m=1}^{\infty} \sum_{n=1}^{\infty} B_{mn} \sin \frac{m\pi x}{a} \sin \frac{n\pi y}{b} \tag{3.15}$$

where, following the usual Fourier series procedures,

$$B_{mn} = \frac{4}{ab} \int_0^a \int_0^b p(x, y) \sin \frac{m\pi x}{a} \sin \frac{n\pi y}{b}\, dy\, dx. \tag{3.16}$$

Substituting these series representations of the load and lateral deflection into the governing differential equation results in the following:

$$\sum \sum A_{mn} \pi^4 \left\{ \frac{m^4}{a^4} + 2\frac{m^2}{a^2}\frac{n^2}{b^2} + \frac{n^4}{b^4} \right\} \sin \frac{m\pi x}{a} \sin \frac{n\pi y}{b}$$

$$= \frac{1}{D} \sum \sum B_{mn} \sin \frac{m\pi x}{a} \sin \frac{n\pi y}{b}.$$

For the left hand side to equal the right hand side for the above doubly infinite series requires that an equality exists for each *m* and *n* combination in the series. Looking at the *m*th and *n*th term, A_{mn} is easily found to be

$$A_{mn} = \frac{B_{mn}}{D\pi^4 \left\{ \frac{m^2}{a^2} + \frac{n^2}{b^2} \right\}^2}. \tag{3.17}$$

Thus, the solution is easily found for this case, because B_{mn} is determined from (3.16), and A_{mn} if then found from the equation above, hence $w(x, y)$ is then known everywhere from (3.14). From this, all slopes, stress couples and shear resultants can be calculated at any location x, y. As mentioned previously, the doubly infinite series solution usually converges slowly. Moreover, the derivatives of $w(x, y)$ needed to obtain stress couples and shear resultants always converge still slower than the deflection function itself.

An example for obtaining B_{mn} can be briefly given. Consider a plate simply supported on all four edges subjected to a uniform constant lateral loading p_0.

$$\begin{aligned}
B_{mn} &= \frac{4p_0}{ab}\int_0^a\int_0^b \sin\frac{m\pi x}{a}\sin\frac{n\pi y}{b}\,dy\,dx \\
&= \frac{4p_0}{ab}\left\{\frac{a}{m\pi}\left[-\cos\frac{m\pi x}{a}\right]_0^a\right\}\left\{\frac{b}{n\pi}\left[-\cos\frac{n\pi y}{b}\right]_0^b\right\} \\
&= \frac{4p_0}{mn\pi^2}(1-\cos m\pi)(1-\cos n\pi) \qquad (3.18)\\
&= \frac{4p_0}{mn\pi^2}[1-(-1)^m][1-(-1)^n] \\
&= \frac{16p_0}{mn\pi^2} \quad \text{(if, } m, n \text{ odd only)}
\end{aligned}$$

A similar procedure is followed for any other lateral load over all or part of the plate surface.

3.3 Single Series Solution (Method of M. Levy)

Consider a plate with opposite edges simply supported, as shown in Figure 3.1.

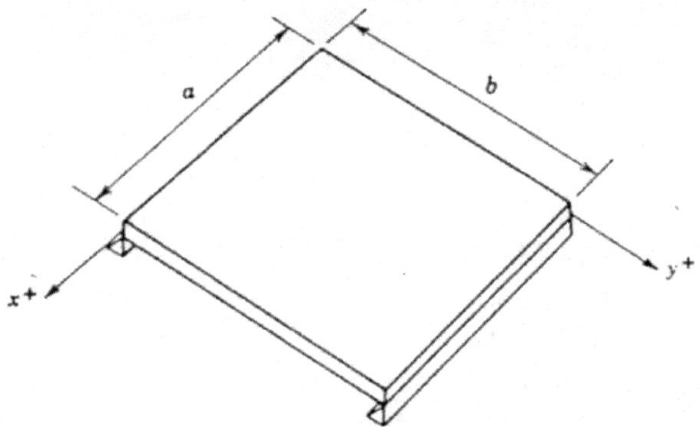

Figure 3.1. Plate simply supported on opposite edges.

Again, the governing differential equation is:

$$\nabla^4 w = \frac{p(x,y)}{D}.$$

The boundary conditions on the y edges are:

$$\begin{aligned}
w(x,0) &= w(x,b) = 0 \\
M_y(x,0) &= M_x(x,b) = 0.
\end{aligned} \qquad (3.19)$$

From (2.38), the stress couple is given by

$$M_y = -D\left[\frac{\partial^2 w}{\partial y^2} + \nu \frac{\partial^2 w}{\partial x^2}\right].$$

Hence on the $y = 0$ and $y = b$ edges,

$$\frac{\partial^2 w}{\partial y^2}\left(x, \begin{matrix} 0 \\ b \end{matrix}\right) + \nu \frac{\partial^2 w}{\partial x^2}\left(x, \begin{matrix} 0 \\ b \end{matrix}\right) = 0.$$

However, on each of those edges,

$$\frac{\partial^2 w}{\partial x^2}\left(x, \begin{matrix} 0 \\ b \end{matrix}\right) = 0$$

because the curvature is zero parallel to the simply supported edge. Therefore, for the $y =$ 0 and b simply supported edges,

$$\frac{\partial^2 w}{\partial y^2}\left(x, \begin{matrix} 0 \\ b \end{matrix}\right) = 0. \tag{3.20}$$

Assume a form of the solution to be as follows, which satisfies the boundary condition on the y edges given by (3.19) and (3.20):

$$w(x, y) = \sum_{n=1}^{\infty} \phi_n(x) \sin \frac{n\pi y}{b}. \tag{3.21}$$

For this example, the lateral distributed load is taken to be the following:

$$p(x, y) = g(x) h(y) \tag{3.22}$$

where $g(x)$ and $h(y)$ are given. It is necessary to expand $h(y)$ in a series solution that corresponds to (3.21), hence,

$$h(y) = \sum_{n=1}^{\infty} A_n \sin \frac{n\pi y}{b} \tag{3.23}$$

where

$$A_n = \frac{2}{b} \int_0^b h(y) \sin \frac{n\pi y}{b} \, dy.$$

Substituting (3.21) through (3.23) into (3.1) gives:

$$\sum_{n=1}^{\infty}\left\{\phi_n^{IV} - 2\lambda_n^2\phi_n'' + \lambda_n^4\phi_n\right\}\sin\frac{n\pi y}{b} = \frac{1}{D}\sum_{n=1}^{\infty}A_n g_n(x)\sin\frac{n\pi y}{b} \tag{3.24}$$

where $\lambda_n = n\pi / b$.

As before, for this to be true, the series must be equated term by term.

$$\phi_n^{IV}(x) - 2\lambda_n^2\phi_n''(x) + \lambda_n^4\phi_n(x) = \frac{1}{D}A_n g_n(x). \tag{3.25}$$

Note, at this point the boundary conditions on the other two edges have not been specified. Thus, any time a problem has two opposite edges simply supported, one can arrive at (3.25) without other information regarding the x = constant edges.

Proceeding to solve (3.25) in the customary way, let $\phi_n = e^{sx}$ such that the homogeneous solution becomes:

$$s^4 - 2\lambda_n^2 s^2 + \lambda_n^4 = 0$$

$$(s^2 - \lambda_n^2)(s^2 - \lambda_n^2) = 0 \quad \text{where } \lambda_n^2 > 0$$

$$s = \pm\lambda_n, \pm\lambda_n$$

So, the complementary solution is:

$$\phi_n(x) = (C_1 + C_2 x)\cosh\lambda_n x + (C_3 + C_4 x)\sinh\lambda_n x. \tag{3.26}$$

Equation (3.26) is the form of the homogeneous solution for $\boldsymbol{\lambda_n(x)}$ for any set of boundary conditions on the x-edges. The boundary conditions on the x = constant edges are used to determine the constants C_1 through C_4 above.

3.3.1 Example: Plate Simply Supported on All Four Edges and p = p(y) Only

On the x = constant edges, the boundary conditions are:

$$\begin{aligned} w(0,y) &= w(a,y) = 0 \\ M_x(0,y) &= M_x(a,y) = 0. \end{aligned} \tag{3.27}$$

Because of no curvature along the x = constant edges, i.e.,

$$\frac{\partial^2 w}{\partial y^2}\left(\frac{0}{a}, y\right) = 0$$

the bending moment boundary conditions can be written as:

$$\frac{\partial^2 w}{\partial x^2}(0, y) = \frac{\partial^2 w}{\partial x^2}(a, y) = 0. \qquad (3.28)$$

Also since $p = p(y)$ in this example, simply let $g(x) = 1$ in (3.22), and from (3.25) the particular solution can be written as:

$$\phi_{n_p}(x) = \frac{A_n}{D\lambda_n^4}. \qquad (3.29)$$

Therefore for this example, the complete solution for $\boldsymbol{\phi_n(x)}$ is:

$$\phi_n(x) = (C_1 + C_2 x)\cosh\lambda_n x + (C_3 + C_4 x)\sinh\lambda_n x + \frac{A_n}{D\lambda_n^4}. \qquad (3.30)$$

Substituting (3.30), the complete solution for $\boldsymbol{\phi_n(x)}$, and its derivatives, into (3.27) and (3.28), the boundary conditions on the x edges, provides the values of the undetermined constants C_1 through C_4, for this problem. The results are:

$$\left.\begin{aligned} C_1 &= -\frac{A_n}{D\lambda_n^4} \\ C_2 &= -\frac{A_n}{2D\lambda_n^3}\,\frac{1-\cosh\lambda_n a}{\sinh\lambda_n a} \\ C_3 &= \frac{A_n a}{2D\lambda_n^3(1+\cosh\lambda_n a)}\left[\frac{2}{\lambda_n a}\sinh\lambda_n a - 1\right] \\ C_4 &= \frac{A_n}{2D\lambda_n^3} \end{aligned}\right\} \qquad (3.31)$$

Thus, the complete solution for the lateral deflection is:

$$w(x, y) = \sum_{n=1}^{\infty}\left[(C_1 + C_2 x)\cosh\lambda_n x + (C_3 + C_4 x)\sinh\lambda_n x + \frac{A_n}{D\lambda_n^4}\right]\sin\lambda_n y \qquad (3.32)$$

where C_1 through C_4 are given by (3.31).

It should be noted that we could have solved this problem by assuming the deflection any one of these following ways, because the plate is simply supported on all four edges:

$$w(x,y) = \sum_{m=1}^{\infty}\sum_{n=1}^{\infty} A_{mn} \sin\frac{m\pi x}{a} \sin\frac{n\pi y}{b}$$

$$w(x,y) = \sum_{n=1}^{\infty} \phi_n(x) \sin\frac{n\pi y}{b}$$

$$w(x,y) = \sum_{m=1}^{\infty} \psi_m(y) \sin\frac{m\pi x}{a}.$$

Using the first of these equations the convergence is slower than using the form given by the second and third equations to describe the lateral deflection.

For the case of the x edges being clamped or free, and with the same loading, $p = p(y)$ only, Equation (3.30) with the appropriate boundary conditions to obtain the solution may be used.

3.4 Example of a Plate with Edges Supported by Beams

The use of beams to support plate elements is very commonplace. Innovative and efficient design for that case often results in complex analytical procedures, so complicated in fact that doctoral dissertations have been written in this regard. For instance, complications can arise when (1) the plate mid-surface differs from the mid-surface of the support beams, (2) beam sections involve centers of twist in difficult locations, (3) discontinuous joining of the beams and the plate, etc.

Presented here is the simplest of beam-plate combinations, merely to introduce the concepts involved.

Consider a rectangular plate with the following boundary conditions: y = 0, b simply supported; $x = 0, a$ supported by beams; and a lateral load varying only in the y direction, given by $p(y) = \sum_{n=1}^{\infty} A_n \sin(n\pi y/b)$, as was done in (3.22) and (3.23) where $g(x) = 1$.

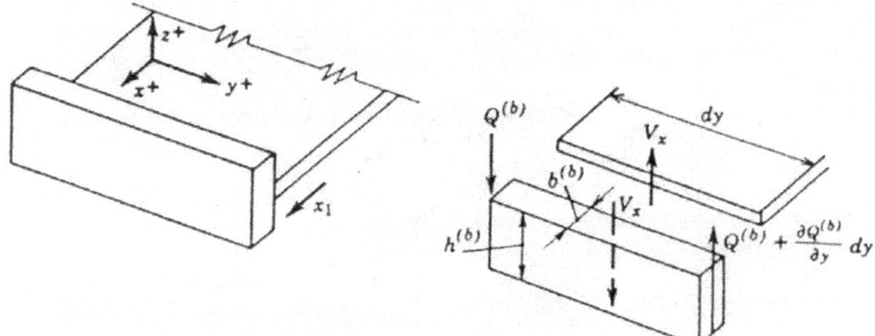

Figure 3.2. Vertical forces at beam supported edge.

For one boundary condition on the x edge; consider an element of beam as a free body, as shown in Figure 3.2. A force balance in the z direction provides one plate boundary condition.

Looking at the details of Figure 3.2, the superscripted b quantities refer to a beam, whose flexural stiffness is $(EI)^b$, which is mechanically joined to the edge of the plate denoted by $x = x_1$ such that the middle surface of both the plate and beam are identical, to retain simplicity in this example. Hence, the lateral deflection of the beam and plate are identical at their common boundary, $x = x_1$. Therefore, the force balance is given by the following, where the shear resultant of the beam is

$$Q^{(b)}(y) = b^{(b)} \int_{-h^{(b)}/2}^{+h^{(b)}/2} \sigma_{yz}\, \mathrm{d}z$$

and the Kirchoff 'effective' shear resultant is used for the plate.

$$\sum F_z^{(b)} = 0 = -Q^{(b)} - V_x\, \mathrm{d}y + Q_b^{(b)} + \frac{\mathrm{d}Q^{(b)}}{\mathrm{d}y}\mathrm{d}y = 0$$

$$(V_x)_{x=x_1} = \frac{\mathrm{d}Q^{(b)}}{\mathrm{d}y}(x_1, y) = -(EI)^{(b)}\left(\frac{\partial^4 w^{(b)}}{\partial y^4}\right)_{x=x_1} = -(EI)^b\left(\frac{\partial^4 w}{\partial y^4}\right)_{x=x_1}$$

since $w_b = w$ at $x = x_1$ and since $\boldsymbol{Q^{(b)} = \mathrm{d}M^{(b)}/\mathrm{d}y}$, and $\boldsymbol{M^{(b)} = -(EI)^{(b)}\partial^2 w^{(b)}/\partial y^2}$.

For the plate

$$(V_x)_{x=x_1} = -D\left[\frac{\partial^3 w}{\partial x^3} + (2-\nu)\frac{\partial^3 w}{\partial x\, \partial y^2}\right]_{x=x_1}.$$

The second boundary condition, the balancing of twisting moments, provides the requirement. The beam has a torsional stiffness $(GJ)^b$ (Figure 3.3).

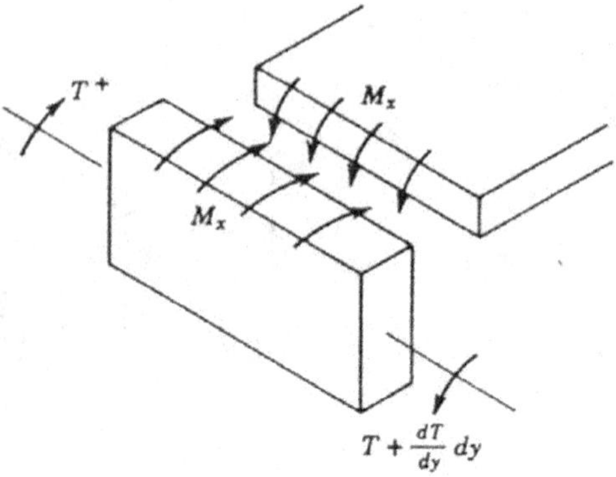

Figure 3.3. Stress couples at beam supported edges.

For beam, at $x = x_1$, moment equilibrium requires that

$$-T - M_x \,\mathrm{d}y + T + \frac{\mathrm{d}T}{\mathrm{d}y}\mathrm{d}y = 0$$

or,

$$M_x = \frac{\mathrm{d}T}{\mathrm{d}y} = -(GJ)^{(b)}\frac{\mathrm{d}^2\theta^{(b)}}{\mathrm{d}y^2} = -(GJ)^{(b)}\left(\frac{\partial^3 w}{\partial x\,\partial y^2}\right)$$

$$= -D\left[\frac{\partial^2 w}{\partial x^2} + \nu\frac{\partial^2 w}{\partial y^2}\right]$$

since

$$T = -(GJ)^{(b)}\frac{\mathrm{d}\theta^{(b)}}{\mathrm{d}y} \quad \text{and} \quad \theta^{(b)} = \frac{\partial w^{(b)}}{\partial x} \quad \text{and} \quad \frac{\mathrm{d}w^{(b)}}{\mathrm{d}x} \approx \frac{\mathrm{d}w}{\mathrm{d}x} \quad \text{at } x = x_1.$$

So the two plate boundary conditions at the junction between the plate and the beam support are:

$$D\left[\frac{\partial^3 w}{\partial x^3} + (2-\nu)\frac{\partial^3 w}{\partial x\,\partial y^2}\right] = (EI)^{(b)}\frac{\partial^4 w}{\partial y^4} \quad \text{at } x = x_1 \tag{3.33}$$

$$D\left[\frac{\partial^2 w}{\partial x^2} + \nu\frac{\partial^2 w}{\partial y^2}\right] = (GJ)^{(b)}\frac{\partial^3 w}{\partial x\,\partial y^2} \quad \text{at } x = x_1. \tag{3.34}$$

Once the solution for the plate deflection is found which satisfies these boundary conditions at $x = x_1$, the other plate boundary conditions for a given lateral load, then one also knows the deflection of the beam, and so the complete solution of the beam is determined also.

For all practical cases one can assume that the beam end conditions are the same as the plate end conditions, and assume that the ends of the beam are completely restrained against rotation.

It is easy to see how the boundary conditions can become more complex with more complicated beam – plate joints. However, the same philosophy as used above can be used to solve those problems.

3.5 Isotropic Plates Subjected to a Uniform Lateral Load

For isotropic plates several textbooks such as Timoshenko and Woinowsky-Krieger [2.1] and Vinson [3.3] have provided expressions for the maximum deflection, w_{max}, and the maximum stress couple, M, a plate attains when subjected to a constant laterally distributed load p_0, such as,

$$w_{max} = \frac{C_0 p_0 a^4}{Eh^3} \tag{3.35}$$

$$M_{max} = C_1 p_0 a^2 \tag{3.36}$$

where a and b are plate side dimensions; E is the modulus of elasticity of the plate; h is the plate thickness; and $\nu = 0.3$.

The dimensionless constants C_0 and C_1 are given in tabular form for various boundary conditions, and these are repeated herein for completeness in Tables 3.1 through 3.4. Table 3.5 also provides information for the case wherein the plate is subjected to an hydraulic head. These tables and procedures are well known and well used.

Table 3.1. Coefficients for determining Maximum Deflections and Maximum Stresses for a Rectangular Plate, with $b > a$, Simply Supported at the Edges, under Uniform Pressure Loading p_0 with Sufficient Corner Forces to Hold it Down on the Foundation ($\nu = 0.3$).

b/a	1	1.2	1.4	1.6	1.8	2	3	4	5	∞
C_0	0.044	0.062	0.077	0.091	0.102	0.111	0.134	0.140	0.142	0.142
C_1	0.048	0.063	0.075	0.086	0.095	0.102	0.119	0.124	0.125	0.125

Table 3.2. Coefficients for determining Maximum Deflections and Maximum Stresses for Rectangular Plate, under Uniform Load p_0, with the a Edges Clamped and b Edges Simply Supported **($\nu = 0.3$)**.

b/a	∞	3	2	1.6	1.3	1	0.75	0.50	0.25
C_0	0.142	0.128	0.099	0.066	0.042	0.021	0.0081	0.00177	0.00011
C_1	0.125	0.125	0.119	0.109	0.094	0.070	0.045	0.021	0.0052

Table 3.3. Coefficients for Determining Maximum Deflections and Maximum Stresses for a Rectangular Plate, under Uniform Loading p_0, Clamped along the a Edge, and Simply Supported along the Three Remaining Edges $(\nu = 0.3)$.

b/a	∞	2	1.5	1.2	1	0.75	0.50	0.25
C_0	0.142	0.101	0.070	0.047	0.030	0.0133	0.0033	0.0002
C_1	0.125	0.122	0.112	0.098	0.084	0.058	0.031	0.0077

Table 3.4. Coefficients for determining Maximum Deflections and Maximum Stresses for a Rectangular Plate, under Uniform Loading p_0, Clamped on All Four Edges $(\nu = 0.3)$.

b/a	1	1.2	1.4	1.6	1.8	2.0	2.2
C_0	0.0138	0.0188	0.0226	0.0251	0.0267	0.0277	0.0285
C_1	0.0513	0.0639	0.0726	0.0780	0.0812	0.0829	0.0833

Table 3.5. Coefficients for determining Maximum Deflections and Maximum Stresses for a Rectangular Plate, Simply Supported on All Four Sides, Subjected to a Linearly Increasing Hydraulic Pressure along the a Edges, one b side having Zero Pressure, the Opposite b side having p_0. [The Maximum Deflections occurs just off the middle of the plate toward the p_0 side (at about $0.55a$), the Maximum Stress somewhat farther off to the side $(\nu = 0.3)$.]

b/a	∞	4	3	2	1.5	1	0.75	0.50	0.25
C_0	0.071	0.070	0.067	0.055	0.042	0.022	0.012	0.0037	0.0004
C_1	0.064	0.063	0.061	0.053	0.043	0.026	0.021	0.0139	0.0051

Of course, for the isotropic plate, the flexural stiffness is given by

$$D = \frac{Eh^3}{12(1-\nu^2)}. \tag{3.37}$$

and the maximum bending stress, which occurs on the top and bottom surfaces of the plate, is

$$\sigma_{\max}(\pm h/2) = \pm\frac{6M_{\max}}{h^2} \tag{3.38}$$

Also for Tables 3.1 through 3.5, the numerical coefficients correspond to a Poisson's ratio of $\nu = 0.3$ wherein $1-\nu^2 = 0.91$. Therefore, for materials with other Poisson ratios, ν, Equation (3.35) must be changed to

$$w_{max} = \frac{C_0 p_0 a^4}{Eh^3}\left(\frac{1-\nu^2}{0.91}\right). \tag{3.39}$$

It is seen that for an isotropic plate design,

(1) The plate must not be overstressed, i.e., the maximum stress is determined from the use of Equations (3.36) and (3.38) to determine the maximum stress couple, M, and the maximum stress. The determined maximum stress cannot exceed some allowable stress, σ_{all}, defined by the material's ultimate stress or yield stress divided by a factor of safety on ultimate stress or yield stress, whichever is smaller. This requires a certain value of plate thickness, h, which in analysis is specified from which one determines if the plate is overstressed. In design, using the allowable stress, the thickness, h, is found.
(2) The monocoque plate must not be over deflected determined by Equation (3.35). This is sometimes specified, but in other cases the plate deflection cannot exceed the plate thickness or some fraction thereof. If the maximum plate deflections reaches a value of the plate thickness, h, the equations discussed herein become inapplicable because the plate behavior becomes increasingly nonlinear which requires that other equations be used. Again, to prevent over-deflection, a plate thickness, h, is determined by Equation (3.35).

Therefore, in plate design, the plate thickness, h, is determined either from a strength or stiffness requirement, whichever requires the larger thickness.

3.6 Summary

In this chapter the two basic approaches to solving problems of isotropic rectangular plates subjected to lateral loads have been treated. Also, a more complicated boundary condition example was investigated than the classical boundary conditions of Section 2.5. In rectangular plates with more difficult boundary conditions than simply supported edges References 3.1 and 3.2 provide functions suitable for either the Navier or the Levy Method.

Many solutions to plate problems are known, and are catalogued in numerous references such as Timoshenko and Woinowsky-Krieger [2.1], Marguerre and Woernle [2.2] and Mansfield [2.3].

3.7 References

3.1. Young, D. and Felgar, R.P. Jr. (1949) *Tables of Characteristic Functions Representing Normal Modes of Vibration of a Beam*, The University of Texas Engineering Research Series Report, No. 44, July 1.

3.2. Felgar, R.P. Jr. (1950) *Formulas for Integrals Containing Characteristic Functions of a Vibrating Beam*, The University of Texas Bureau of Engineering Research Circular, No. 14.

3.3. Vinson, J.R. (1989) *The Behavior of Thin Walled Structures: Beams, Plates and Shells*, Kluwer Academic Publishers, Dordrecht, The Netherlands.

3.8 Problems

3.1. Consider a rectangular isotropic plate occupying the region $0 \le x \le a$, $0 \le y \le b$, and $-h/2 \le z \le h/2$. The plate is simply supported on the edges $y = 0$ and b. The plate is subjected to a laterally distributed load given by Equations (3.22) and (3.23). If $g(x) = 1$, the solution is given by Equation (3.32). In the plate clamped along the edges $x = 0$ and a, determine the constants C_1 through C_4.

3.2. In problem 3.1 above, if the plate is free along the edges $x = 0$ and a, determine the constants C_1 through C_4.

3.3. In problem 3.1 above, if the plate is simply supported at $x = 0$ and clamped at $x = a$, determine the constants C_1 through C_4.

3.4. In problem 3.1 above, if the plate is simply supported at $x = 0$ and free along $x = a$, determine the constants C_1 through C_4.

3.5. In problem 3.1 above, for the plate clamped along $x = 0$ and free along $x = a$, determine the constants C_1 through C_4.

3.6. Consider a floor slab whose geometry is described in problem 3.1. The slab is square, simply supported on all edges, and is loaded with sand in such a way that the load can be approximated by

$$p(x, y) = p_0 \sin\frac{\pi x}{a} \sin\frac{\pi y}{b}.$$

Determine the location and magnitude of the maximum deflection, the maximum bending stresses in both directions and the maximum shear stresses in each direction.

3.7. A certain window in an aircraft is approximated by a square plate of dimensions a on each side, simply supported on all four edges and subjected to a uniform cabin pressure p_0. Using the Navier solution for a square plate of length and width a, the solution is given by Equation (3.14). The maximum value of the lateral deflection can be written as

$$w_{\max} = C_1 p_0 a^4 / D$$

for the plate subjected to a constant lateral loading, p_0. Determine the numerical coefficient C_1 to three significant figures.

The maximum bending moment $M_{x_{max}} = M_{y_{max}}$ can be written as

$$M_{max} = C_2 p_0 a^2.$$

Find C_2 to three significant figures, if the Poisson's ratio of the window material is $\nu = 0.3$.

3.8. A certain hull plate on the flat bottom of a ship may be considered to be a rectangular plate under uniform loading, p_0, from the water pressure, and clamped along all edges. A $1/2''$ steel plate four feet in width is to be used for the bottom plate in the ship draws 13 ½ feet of water maximum. If the maximum allowable stress in the steel is 20,000 psi, what is the maximum plate length, i.e., bulkhead spacing that can be used in the ship design, and what is the corresponding maximum deflection of the hull plate. Salt water weighs 64 lbs/ft^3, $E_{steel} = 30\times10^6$ psi, and $\nu_{steel} = 0.3$.

For a plate clamped on all four edges, subjected to a lateral load p_0, the maximum deflection and maximum stress couple can be found using Table 3.4. Linear interpolation is permitted.

3.9. A rectangular wing panel component, $8''\times5''$ is made of aluminum, and under the most severe maneuver conditions can be subjected to a uniform lateral load of 20 psi. This wing panel can be approximated by a flat plate simply supported on all four edges. What thickness must the panel be, and what is the resulting maximum deflection under this maneuver condition? Use a Table from Section 3.5.

The aluminum used has an allowable stress of 20,000 psi, and $E = 10\times10^6$ psi and $\nu = 0.3$.

3.10. A rectangular steel plate is used as part of a flood control structure, and is mounted vertically under water such that it is subjected to a hydraulic loading

$$p(x,y) = p_0 + p_1\frac{y}{b}$$

where p_0 and p_1 are constants associated with the pressure heads. Find the Euler coefficient B_{mn} for this loading in Equation (3.16).

3.11. A glass manufacturer has been asked to construct plate glass windows for a new modern office building. The windows must be 10 ft. wide and 20 ft. high. Design the windows so that they can withstand wind forces due to air velocities of 150 miles/hour. State all assumptions and physical constants clearly.

3.12. A flat portion of a wind tunnel measuring $30''\times54''$ will be subjected to a maximum uniform wind load of 10 psi. If the steel to be used has an allowable stress of 40,000 psi, and a Poisson's ratio of $\nu = 0.3$, what plate thickness is required if the plate is

(a) Simply supported on all four edges.
(b) Clamped on all four edges.
Use the Tables from Section 3.5.

3.13. A portion of the cover on a hover craft is to be rectangular measuring $8' \times 4'$ in planform, and is to be simply supported on all four edges. It is calculated that the maximum air pressure the panel will be subjected to is 20 psi.

(1) How thick must the panel be if it is constructed of aluminum ($E = 10 \times 10^6$ psi, $\nu = 0.3$) if the allowable stress is limited to 30,000 psi?
(2) How thick must the plate be if it is constructed of steel ($E = 30 \times 10^6$ psi, $\nu = 0.3$) if the allowable stress is limited to 60,000 psi?
(3) If the weight density of steel is 0.283 lbs/in^3 and that of aluminum is 0.1 lbs/in^3, which material should be selected to minimize weight?
(4) Suppose the aluminum plate of (1) above were clamped on all four edges, what thickness is required?

Use the Tables from Section 3.5.

3.14. A rectangular steel plate, used as a footing, rests on the ground is subjected to a uniform lateral pressure, $p(x, y) = -p_0$ (psi). The ground deflects linearly below the footing with a spring constant k (lbs/in^2/in) under this loading and deflection.

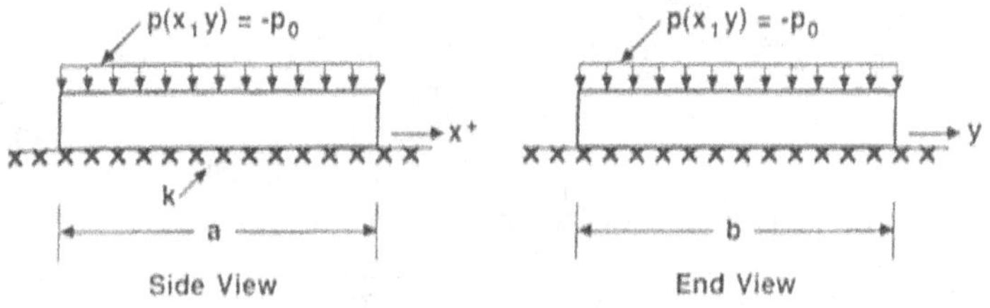

Figure 3.4. Elasticically supported footing.

(a) What is the governing differential for the bending of this plate on an elastic foundation?
Hint: One can consider the effect of the elastic foundation to be analogous to an infinite set of springs such that it acts like a lateral load analogous to $p(x,y)$.
(b) What are the boundary conditions on the $x = 0$ and $x = a$ edges?
(c) What are the boundary conditions on the $y = 0$ and $y = b$ edges?

3.15. A designer is faced with the problem of designing a rectangular plate cover over an opening that is 9 feet by 3 feet. The design load is a lateral pressure of 10 psi. If steel is used ($E = 30 \times 10^6$ psi, $\nu = 0.3$, $\sigma_{all} = 35{,}000$ psi, $\rho = 0.283$ lbs/in^3):

(a) If the plate is clamped on all four edges, what will it weigh?
(b) If the plate is simply supported on all four edges, what will it weigh?

3.16. A rectangular aluminum plate, measuring $40'' \times 20''$, is subjected to a uniform lateral pressure of 10 psi. Using the maximum stress theory, if the allowable stress is 30,000 psi, what is the plate thickness required if:

(a) All edges of the plate are simply supported?
(b) All edges of the plate are clamped?

(c) If the plate were made of steel with the same allowable stress as the aluminum above, would the required thickness differ from that of the aluminum plate?

(d) If the plate were made of steel with the same allowable stress as the aluminum above, would the maximum deflection differ from that of the aluminum plate?

3.17. Consider a plate clamped on all four edges made of the same steel as in Problem 3.12. The plate is subjected to a uniform later load of $p = 10$ psi. If the plate is 10″ wide and 16″ long, what thickness h is required to prevent overstressing or a maximum deflection of 0.1″?

CHAPTER 4

THERMAL STRESSES IN PLATES

4.1 General Considerations

Consider any elastic body with a constant coefficient of thermal expansion, α, in the units of **in/in/°F**, or equivalent units, at a uniform temperature wherein the body is assumed to be free of any thermal stresses and strains. If the body is free to deform, and the temperature is raised slowly to a temperature of ΔT degrees from the stress free temperature, the thermal strains produced at any material point can be written as

$$\varepsilon_{ij_{\text{th}}} = \alpha\,\Delta T(x_i)\delta_{ij} \tag{4.1}$$

where x_i are the coordinate direction, and δ_{ij} is the Kronecker delta ($\delta_{ij} = 1$ for $i = j$, $\delta_{ij} = 0$ for $i \neq j$). It should be noted that thermal strains are purely dilatational ($i = j$); thermal shear strains do not exist.

In Equation (4.1), ΔT is positive when the temperature of the material point is above the stress free temperature. The coefficient of thermal expansion α is positive for almost all isotropic engineering materials, i.e., the body expands when it is heated. However, there are some graphite materials which have negative coefficients of thermal expansion.

In many thermoelastic bodies, the changes in temperature within the body tend to result in strains which do not satisfy the compatibility equations. In that case isothermal strains, $\varepsilon_{ij_{\text{iso}}}$, the strains discussed in Chapter 1, are induced such that the total strain, $\varepsilon_{ij_{\text{tot}}}$, satisfies compatibility.

$$\varepsilon_{ij_{\text{tot}}} = \varepsilon_{ij_{\text{iso}}} + \varepsilon_{ij_{\text{th}}}. \tag{4.2}$$

In that case the 'thermal stresses' are induced due to the isothermal strains induced to insure compatibility. This can occur, for example, in 'thermal shock' from very rapid or localized heating.

A second way that thermal stresses occur is through displacement restrictions on the elastic body. One simple example of this occurs when a bar is placed between immovable end grips and subsequently heated. There, compressive thermal stresses result.

Hence, thermal stresses are caused by two mechanisms: one by displacement restrictions, the other through induced isothermal strains to maintain compatibility.

Next, consider an unrestricted thin rod at a uniform temperature. If the rod is slowly heated uniformly, such that the thermal strains satisfy the compatibility equations, the heated rod has thermal strains but no thermal stresses. Now if the unheated thin rod is placed in immovable end grips such that the rod cannot increase in length, slowly heating the rod uniformly will result in thermal stresses and no thermal strains.

In the latter case, if the compressive axial thermal stresses reach a value equal to the Euler buckling load (discussed later in Chapter 6), the rod will buckle. This is called thermal buckling.

4.2 Derivation of the Governing Equations for a Thermoelastic Plate

In deriving the governing equations for a thermoelastic plate, the equilibrium equations and the strain-displacement equations are not altered from those of the isothermal plate of Chapter 1, because in the former the equations involve force balances, and the latter are purely kinematic relationships involving total strains.

However, the stress strain relations, Equations (1.9) and (1.10), are modified in accordance with Equations (4.2) and (4.1):

$$\varepsilon_{x_{\text{iso}}} = \varepsilon_{x_{\text{tot}}} - \alpha\Delta T = \frac{1}{E}[\sigma_x - \nu\sigma_y]$$

$$\varepsilon_{y_{\text{iso}}} = \varepsilon_{y_{\text{tot}}} - \alpha\Delta T = \frac{1}{E}[\sigma_y - \nu\sigma_x]$$

or

$$\varepsilon_x = \frac{1}{E}[\sigma_x - \nu\sigma_y] + \alpha\Delta T \tag{4.3}$$

$$\varepsilon_y = \frac{1}{E}[\sigma_y - \nu\sigma_x] + \alpha\Delta T. \tag{4.4}$$

In Equations (4.3) and (4.4) and in all that follows the subscript for total strains is dropped, and all strains noted explicitly are those which satisfy compatibility, i.e., the total strains, and which appear in the strain-displacement relations. Hence, in Equations (4.3) and (4.4) the first terms on the right-hand side are really isothermal strains, and the second terms on the right-hand side are thermal strains for this isotropic material.

Proceeding as in Chapter 1, employing the strain displacement relations, Equations (1.16) and (1.17) and Equations (2.24) through (2.28), Equations (4.3) and (4.4) after multiplying by E become

$$\sigma_x - \nu\sigma_y + E\alpha\Delta T = E\frac{\partial u_0}{\partial x} - Ez\frac{\partial^2 w}{\partial x^2} \tag{4.5}$$

$$\sigma_y - \nu\sigma_x + E\alpha\Delta T = E\frac{\partial v_0}{\partial y} - Ez\frac{\partial^2 w}{\partial y^2} \tag{4.6}$$

Now, two quantities N^* and M^*, known as the thermal stress resultant and the thermal stress couple, respectively, are defined as

$$N^T = \int_{-h/2}^{+h/2} E\alpha\Delta T \,\mathrm{d}z, \quad M^T = \int_{-h/2}^{+h/2} E\alpha\Delta Tz \,\mathrm{d}z. \tag{4.7}$$

Multiplying Equations (4.5) and (4.6) by dz and integrating across the thickness of the plate, then multiplying them by z dz and also integrating them, provides the integrated stress strain relations for a thermoelastic isotropic plate. It should be remembered from the discussion of Section 4.1 that the shear stress-strain relations are not altered by the inclusion of thermoelastic effects.

$$N_x - \nu N_y + N^T = Eh\frac{\partial u_0}{\partial x}$$

$$N_y - \nu N_x + N^T = Eh\frac{\partial v_0}{\partial y}$$

$$M_x - \nu M_y + M^T = \frac{-Eh^3}{12}\frac{\partial^2 w}{\partial x^2}$$

$$M_y - \nu M_x + M^T = \frac{-Eh^3}{12}\frac{\partial^2 w}{\partial y^2}$$

$$N_{xy} = \frac{K(1-\nu)}{2}\left[\frac{\partial u_0}{\partial y} + \frac{\partial v_0}{\partial x}\right] \tag{4.8}$$

$$M_{xy} = -D(1-\nu)\frac{\partial^2 w}{\partial x\,\partial y}. \tag{4.9}$$

Rearranging the first four of the above results in

$$N_x = K\left[\frac{\partial u_0}{\partial x} + \nu\frac{\partial v_0}{\partial y}\right] - \frac{N^T}{(1-\nu)} \tag{4.10}$$

$$N_y = K\left[\frac{\partial v_0}{\partial y} + \nu\frac{\partial u_0}{\partial x}\right] - \frac{N^T}{(1-\nu)} \tag{4.11}$$

$$M_x = -D\left[\frac{\partial^2 w}{\partial x^2} + \nu \frac{\partial^2 w}{\partial y^2}\right] - \frac{M^T}{(1-\nu)} \tag{4.12}$$

$$M_y = -D\left[\frac{\partial^2 w}{\partial y^2} + \nu \frac{\partial^2 w}{\partial x^2}\right] - \frac{M^T}{(1-\nu)}. \tag{4.13}$$

Introducing these thermoelastic stress-strain relations into the equilibrium Equations (2.44) through (2.46) and (2.50) and (2.51), the governing differential equations for a thermoelastic isotropic plate are determined. For the case of no surface shear stresses these become:

$$D\nabla^4 w = p(x,y) - \frac{1}{(1-\nu)}\nabla^2 M^T \tag{4.14}$$

$$K\nabla^4 u_0 = \frac{1}{(1-\nu)}\frac{\partial}{\partial x}(\nabla^2 N^T) \tag{4.15}$$

$$K\nabla^4 v_0 = \frac{1}{(1-\nu)}\frac{\partial}{\partial y}(\nabla^2 N^T). \tag{4.16}$$

Also for completeness, other useful relationships are catalogued below:

$$\begin{aligned} Q_x &= -D\frac{\partial}{\partial x}(\nabla^2 w) - \frac{1}{(1-\nu)}\frac{\partial M^T}{\partial x} \\ Q_y &= -D\frac{\partial}{\partial y}(\nabla^2 w) - \frac{1}{(1-\nu)}\frac{\partial M^T}{\partial y} \end{aligned} \tag{4.17}$$

$$\begin{aligned} V_x &= -D\left[\frac{\partial^3 w}{\partial x^3} + (2-\nu)\frac{\partial^3 w}{\partial x\,\partial y^2}\right] - \frac{1}{(1-\nu)}\frac{\partial M^T}{\partial x} \\ V_y &= -D\left[\frac{\partial^3 w}{\partial y^3} + (2-\nu)\frac{\partial^3 w}{\partial y\,\partial x^2}\right] - \frac{1}{(1-\nu)}\frac{\partial M^T}{\partial y} \end{aligned} \tag{4.18}$$

Due to the inclusion of thermal quantities, the expressions for various normal stresses in the plate become [4.1]:

$$\sigma_x = \frac{1}{h}\left[N_x + \frac{N^T}{(1-\nu)}\right] + \frac{z}{h^3/12}\left[M_x + \frac{M^T}{(1-\nu)}\right] - \frac{E\alpha\Delta T}{(1-\nu)} \tag{4.19}$$

$$\sigma_y = \frac{1}{h}\left[N_y + \frac{N^T}{(1-\nu)}\right] + \frac{z}{h^3/12}\left[M_y + \frac{M^T}{(1-\nu)}\right] - \frac{E\alpha\Delta T}{(1-\nu)}.$$

The inclusion of the N^T and M^T terms in (4.19) are easy to visualize since they are thermal stress resultants and couples analogous to N_x, N_y, M_x and M_y, which are caused by lateral and in-plane 'mechanical' loads. The last terms in (4.19) and (4.20) can be visualized by the following in which it is assumed that the first two terms do not contribute. Suppose at some value of (x, y) in a plate, the upper surface is heated while the lower surface is cooled. Thus the value of ΔT in the upper portion of the plate is positive and is negative in the lower plate portion, as shown in the sketch below.

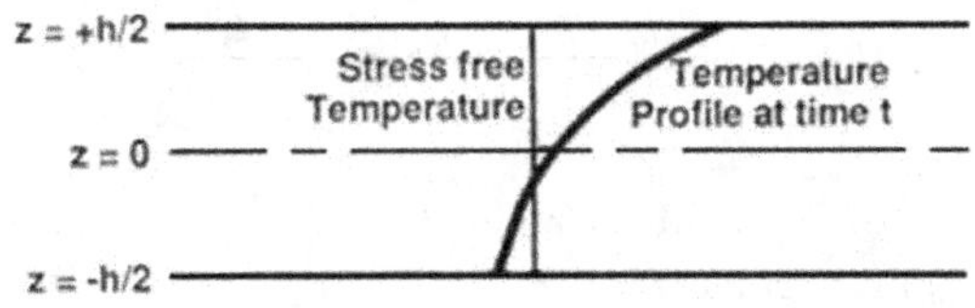

The last term of (4.19) shows that in the upper portion of the plate compressive stresses exist while in the lower portion there are tensile stresses. Physically, the material points in the upper portion of the plate want to expand considerably but are being restrained by those in the cooler areas of the plate, hence, tending to cause high compressive stresses there. Likewise, in the cooler portion of the plate the material points wish to contract, but are being extended by the hotter portions of the plate, hence, thrown into tension. Such thermal stresses can result in material failure just as stresses caused by mechanical loads.

As discussed before, shear stresses and strains are not affected by thermal effects, hence remain the same as in Chapter 2:

$$\sigma_{xy} = \frac{N_{xy}}{h} + \frac{M_{xy}z}{h^3/12}$$

$$\sigma_{xz} = \frac{3Q_x}{2h}\left[1-\left(\frac{z}{h/2}\right)^2\right], \qquad \sigma_{yz} = \frac{3Q_y}{2h}\left[1-\left(\frac{z}{h/2}\right)^2\right]. \tag{4.20}$$

Of course if there exists shear stresses applied to the upper or lower surfaces of the plate, the latter two expressions must be modified as in Section 2.6.

To proceed with solutions of thermoelastic plates using Equations (4.14) through (4.16), one now proceeds using the same solution techniques that were introduced in Chapter 3. However, the additions of thermal effects do introduce certain difficulties with boundary conditions that cause some analytical difficulties. These are discussed in the next section. Also, because of thermal expansions and contractions, solutions usually involve solving for the in-plane displacements u_0 and v_0 in addition to solving for the lateral deflections, $w(x, y)$. Excellent texts include [4.1, 4.2 and 4.3].

4.3 Boundary Conditions

Looking now at the boundary conditions associated with a thermoelastic plate, comparisons are made with an isothermal plate, where again n denotes normal to the edge and s denotes along the edge:

Simply Supported Edge

$$w = 0, \qquad \boldsymbol{M_n = 0}. \tag{4.21}$$

From Equations (4.12) and (4.13), the latter equation above is in fact

$$M_n = 0 = -D\left[\frac{\partial^2 w}{\partial n^2} + \nu \frac{\partial^2 w}{\partial s^2}\right] - \frac{M^T}{(1-\nu)}.$$

Since there is no curvature parallel to the simply supported edge (i.e., $\partial^2 w / \partial s^2 = 0$), this equation becomes

$$\frac{\partial^2 w}{\partial n^2} = -\frac{M^T}{D(1-\nu)}. \tag{4.22}$$

Hence, the boundary conditions for a simply supported thermoelastic plate are nonhomogeneous.

Clamped Edge

$$w = 0, \qquad \frac{\partial w}{\partial n} = 0. \tag{4.23), (4.24}$$

These remain the same as those for the isothermal plate.

Free Edge

The boundary conditions are

$$M_n = 0 \text{ and } V_n = 0.$$

Hence, the first condition is given by Equation (4.22) and the latter is seen to be from Equation (4.18).

$$\frac{\partial^3 w}{\partial n^3} + (2-\nu)\frac{\partial^3 w}{\partial n \, \partial s^2} = -\frac{1}{D(1-\nu)} \frac{\partial M^T}{\partial n}. \tag{4.25}$$

Here the boundary conditions are seen to be nonhomogeneous.

General

In many problems involving thermoelastic plates, it is seen that the boundary conditions are nonhomogeneous. Why is this important? In solving linear partial differential equations, separation of variables cannot be used with nonhomogeneous boundary conditions. Fortunately, methods are available to transform either homogeneous or nonhomogeneous partial differential equations with nonhomogeneous boundary conditions to nonhomogeneous partial differential equations with homogeneous boundary conditions, so that separation of variables may be used. A generalized method is presented in the next section.

4.4 General Treatment of Plate Nonhomogeneous Boundary Conditions

Consider a plate with the $y = 0$, b edges simply supported. The governing equation for the lateral deflection is given by Equation (4.14). From Equations (4.21) and (4.22), the boundary conditions are:

$$w(x,0) = w(x,b) = 0 \tag{4.26}$$

$$\frac{\partial^2 w}{\partial y^2}(x,0) = -\frac{M^T(x,0)}{D(1-\nu)} = -\frac{M_1^T(x)}{D(1-\nu)} \tag{4.27}$$

$$\frac{\partial^2 w}{\partial y^2}(x,b) = -\frac{M^T(x,b)}{D(1-\nu)} = -\frac{M_2^T(x)}{D(1-\nu)} \tag{4.28}$$

where $\boldsymbol{M_1^T(x) = M^T(x,0)}$ and $\boldsymbol{M_2^T(x) = M^T(x,b)}$.

We now introduce a function $\psi(x,y)$, which satisfies homogeneous boundary conditions on the $y = 0$ and $y = b$ edges. Let

$$\mathbf{w(x,y) = \psi(x,y) + f_1(y)M_1^T(x) + f_2(y)M_2^T(x)}. \tag{4.29}$$

where for this problem we take $\psi(x,y)$ to be of the Levy form:

$$\psi(x,y) = \sum_{n=1}^{\infty} \phi_n(x) \sin\frac{n\pi y}{b}. \tag{4.30}$$

Also in Equation (4.29) $f_1(y)$ and $f_2(y)$ are to be determined to satisfy the boundary conditions (4.26) through (4.28).

Substituting Equation (4.29) into Equations (4.26) through (4.28) results in

$$w(x,0) = \psi(x,0) + f_1(0)M_1^T(x) + f_2(0)M_2^T(x) = 0$$

$$w(x,b) = \psi(x,b) + f_1(b)M_1^T(x) + f_2(b)M_2^T(x) = 0$$

$$\frac{\partial^2 w(x,0)}{\partial y^2} = \frac{\partial^2 \psi(x,0)}{\partial y^2} + f_1''(0)M_1^T(x) + f_2''(0)M_2^T(x) = -\frac{M_1^T(x)}{D(1-\nu)}$$

$$\frac{\partial^2 w(x,b)}{\partial y^2} = \frac{\partial^2 \psi(x,b)}{\partial y^2} + f_1''(b)M_1^T(x) + f_2''(b)M_2^T(x) = -\frac{M_1^T(x)}{D(1-\nu)}.$$

Since it is required that $\psi(x,y)$ satisfy homogeneous boundary conditions at $y = 0$ and $y = b$, then

$$\psi(x,0) = \psi(x,b) = \frac{\partial^2 \psi(x,0)}{\partial y^2} = \frac{\partial^2 \psi(x,b)}{\partial y^2} = 0. \tag{4.31}$$

Hence, from the above, the following is required:

$$\begin{array}{ll} f_1(0) = 0 & f_2(0) = 0 \\ f_1(b) = 0 & f_2(b) = 0 \\ f_1''(0) = -1/D(1-\nu) & f_2''(0) = 0 \\ f_1''(b) = 0 & f_2''(b) = -1/D(1-\nu). \end{array} \tag{4.32}$$

These are the only requirements on $f_1(y)$ and $f_2(y)$. Since there are four conditions on each function, each can be assumed to be a third order polynomial.

Let

$$f_1(y) = C_0 + C_1 y + C_2 y^2 + C_3 y^3 \tag{4.33}$$

and

$$f_2(y) = k_0 + k_1 y + k_2 y^2 + k_3 y^3. \tag{4.34}$$

Substituting Equations (4.33) and (4.34) into (4.32), the result is

$$f_1(y) = \frac{1}{6bD(1-\nu)}(2b^2 y - 3by^2 + y^3) \tag{4.35}$$

$$f_2(y) = \frac{1}{6bD(1-\nu)}(b^2 y - y^3). \tag{4.36}$$

Using Equations (4.35) and (4.36), the substitution of Equation (4.29) into (4.14) results in the following:

$$\begin{aligned} D\nabla^4\psi &= p(x,y) - \frac{1}{(1-\nu)}\nabla^2 M^T \\ &- \nabla^4\left\{\frac{1}{6(1-\nu)b}(2b^2 y - 3by^2 + y^3)M_1^T(x)\right\} \\ &- \nabla^4\left\{\frac{b^2 y - y^3}{6(1-\nu)b}M_2^T(x)\right\}. \end{aligned} \tag{4.37}$$

Looking at Equation (4.37) it is seen that the original problem, which was Equation (4.14), with nonhomogeneous boundary conditions, given by Equations (4.26) through (4.28), has been transformed into a problem involving a 'lateral deflection' ψ, with homogeneous boundary conditions (4.31) and an 'altered loading', given by the right-hand side of Equation (4.37), which shall now simply be written as $H(x, y)$. Hence,

$$D\nabla^4\psi = H(x,y). \tag{4.38}$$

Here, $\psi(x, y)$ is given by Equation (4.30) and $H(x, y)$ must be expanded correspondingly into a Fourier series as

$$H(x,y) = \sum_{n=1}^{\infty} h_n(x)\sin\lambda_n y \quad \text{where} \quad \lambda_n = n\pi/b. \tag{4.39}$$

Substituting (4.30) and (4.39) into Equation (4.38) gives

$$D\sum_{n=1}^{\infty}\left\{\phi_n^{IV} - 2\lambda_n^2\phi_n'' + \lambda_n^4\phi_n\right\}\sin\lambda_n y = \sum_{n=1}^{\infty} h_n(x)\sin\lambda_n y.$$

Hence,

$$\phi_n^{IV} - 2\lambda_n^2\phi_n'' + \phi_n\lambda_n^4 = \frac{h_n(x)}{D}. \tag{4.40}$$

It is seen that this has the same form of the ordinary differential equation in the Section 3.3 discussion of the Levy method. Now the boundary conditions at $x = 0$ and $x = a$ can be considered. For the sake of a specific example, consider them to be simply supported also. Then,

$$w(0,y) = w(a,y) = 0, \qquad \boldsymbol{M_x(0,y) = M_x(a,y) = 0}$$

or

$$\frac{\partial^2 w(0,y)}{\partial x^2} = -\frac{M^T(0,y)}{D(1-\nu)}, \qquad \frac{\partial^2 w(a,y)}{\partial x^2} = -\frac{M^T(a,y)}{D(1-\nu)}. \tag{4.41}$$

Substituting Equation (4.29) into the above results in

$$w(0,y) = \psi(0,y) + f_1(y)M_1^T(0) + f_2(y)M_2^T(0) = 0$$

$$w(a,y) = \psi(a,y) + f_1(y)M_1^T(a) + f_2(y)M_2^T(a) = 0$$

$$\frac{\partial^2 w(0,y)}{\partial x^2} = \frac{\partial^2 \psi(0,y)}{\partial x^2} + f_1(y)M_1^{T''}(0) + f_2(y)M_2^{T''}(0) = -\frac{M_1^T(0,y)}{D(1-\nu)}$$

$$\frac{\partial^2 w(a,y)}{\partial x^2} = \frac{\partial^2 \psi(a,y)}{\partial x^2} + f_1(y)M_1^{T''}(a) + f_2(y)M_2^{T''}(a) = -\frac{M_1^T(a,y)}{D(1-\nu)}.$$

Rearranging the above produces

$$\psi(0,y) = -[f_1(y)M_1^T(0) + f_2(y)M_2^T(0)]$$

$$\psi(a,y) = -[f_1(y)M_1^T(a) + f_2(y)M_2^T(a)]$$

$$\frac{\partial^2 \psi(0,y)}{\partial x^2} = -\left[\frac{M^T(0,y)}{D(1-\nu)} + f_1(y)M_1^{T''}(0) + f_2(y)M_2^{T''}(0)\right] \tag{4.42}$$

$$\frac{\partial^2 \psi(a,y)}{\partial x^2} = -\left[\frac{M^T(a,y)}{D(1-\nu)} + f_1(y)M_1^{T''}(a) + f_2(y)M_2^{T''}(a)\right].$$

Remembering that $\psi(x,y)$ is given by Equation (4.30), it is logical to make the following expansions:

$$f_1(y) = \sum_{n=1}^{\infty} A_n \sin\lambda_n y \quad M^T(0,y) = \sum_{n=1}^{\infty} C_n \sin\lambda_n y$$

$$f_2(y) = \sum_{n=1}^{\infty} B_n \sin\lambda_n y \quad M^T(a,y) = \sum_{n=1}^{\infty} E_n \sin\lambda_n y. \tag{4.43}$$

Here, A_n, B_n, C_n and E_n are easily found by the usual Fourier procedures. Substituting Equations (4.43) and (4.30) into (4.42) and equating all coefficients results in

$$\phi_n(0) = -[A_n M_1^T(0) + B_n M_2^T(0)]$$

$$\phi_n(a) = -[A_n M_1^T(a) + B_n M_2^T(a)]$$

$$\phi_n''(0) = -\left[\frac{C_n}{D(1-\nu)} + A_n M_1^{T''}(0) + B_n M_2^{T''}(0)\right] \tag{4.44}$$

$$\phi_n''(a) = -\left[\frac{E_n}{D(1-\nu)} + A_n M_1^{T''}(a) + B_n M_2^{T''}(a)\right].$$

Hence, these four boundary values provide the necessary information to determine the constants K_1 through K_4 in the solution of Equation (4.40), which is

$$\phi_n(x) = (K_1 + K_2 x)\cosh\lambda_n x + (K_3 + K_4 x)\sinh\lambda_n x + \eta_n(x). \tag{4.45}$$

Here, $\boldsymbol{\eta_n(x)}$ is the particular solution, i.e., the right hand side of Equations (4.40). Using (4.45), Equation (4.30) is completely solved, and in turn Equation (4.29) is solved. Subsequently, stress couples, shear resultants and stresses can be determined everywhere using Equations (4.19) and (4.20).

This general approach can be used to solve any plate problem that involves nonhomogeneous boundary conditions.

4.5 Thermoelastic Effects on Beams

For the thermoelastic beam, it can be easily shown from Chapters 2 and 3, and Section 4.2 that the governing differential equations are as follows: Assume the length of the beam is in the x direction, that nothing varies in the y direction, and hence, all equations can be multiplied by the beam width, b. Remembering that all derivatives with respect to y are zero then,

$$P + P^T = EA\frac{du_0}{dx} \tag{4.46}$$

$$M_b + M_b^T = -EI\frac{d^2 w}{dx^2} \tag{4.47}$$

$$V_b = -EI\frac{d^3w}{dx^3} - \frac{dM^T}{dx} \tag{4.48}$$

$$EI\frac{d^4w}{dx^4} = q(x) - \frac{d^2M^T}{dx^2} \tag{4.49}$$

$$\sigma_x = \frac{1}{A}[P + P^T] + \frac{z}{I}[M_b + M^T] - E\alpha\Delta T \tag{4.50}$$

$$P^T = \int_{-h/2}^{+h/2} Eb\alpha\Delta T \, dz, \quad M^T = \int_{-h/2}^{+h/2} Eb\alpha\Delta Tz \, dz. \tag{4.51}$$

where
P = in-plane load
A = beam cross sectional area = bh if rectangular
b = beam width
$M_b = bM_x$
I = moment of inertia = $bh^3/12$ if rectangular
$V_b = bQ_x$
$q(x) = bp(x)$

Using these equations, solutions are easily obtained. As in the case of plates, the expressions for boundary conditions of simply supported and free edges for a thermoelastic beam will be nonhomogeneous. However, for beams (unlike plates) this causes no particular problem, because ordinary differential equations are involved, not partial differential equations, hence, separation of variables is not needed.

4.6 Self-Equilibration of Thermal Stresses

In Section 4.1 the two mechanisms by which thermal stresses are introduced into a thermoelastic solid body are discussed, namely, by displacement restrictions caused by the boundary conditions, the other by the introduction of the isothermal strains in Equations (4.2) in order that $\varepsilon_{ij_{\text{total}}}$ satisfy compatibility when the thermal strain, $\varepsilon_{ij_{\text{th}}}$ do not satisfy compatibility.

One other physical phenomenon occurs that is very important in the structural mechanics of planar bodies such as beams and plates, namely, if the planar body is not restricted by the boundary conditions the thermal stresses in the body are self-equilibrating: i.e., the average stress across the thickness is zero,

$$\sigma_{i_{\text{avg}}} = \frac{1}{h}\int_{-h/2}^{+h/2} \sigma_i \, dz = 0 \quad (i = x, y) \tag{4.52}$$

This can be exemplified in the easiest and shortest way by considering a load free beam lying on a friction free flat surface and heated or cooled such that at any time t, the temperature change is given by:

$$\Delta T(z) = \sum_{n=0}^{\infty} a_n \left(\frac{z}{h}\right)^n = a_0 + a_1 \frac{z}{h} + a_2 \frac{z^2}{h^2} + \ldots \tag{4.53}$$

It is seen that the term a_0 is merely a uniform heating or cooling of the entire beam; $(a_0 + a_1 z)$ is a steady state heat transfer situation and in that case the temperature is linear in the z direction. The entire expression given by (4.53) represents a temperature situation that involves additional terms and occurs during transient heating or cooling and/or internal heat generation. In the following example, it is sufficient to consider only the first three terms of (4.53) to illustrate the point. From (4.53) and (4.51),

$$P^T = EA\alpha\left[a_0 + \frac{a_2}{12}\right] \quad M^T = \frac{E\alpha b h^2}{12} a_1. \tag{4.54}$$

For the illustrative problem, the lateral deflection is

$$w(x) = C_0 + C_1 x + C_2 x^2 + C_3 x^3 \tag{4.55}$$

where this form is the solution to Equation (4.49) for this case. The boundary conditions for this example are seen to be simply supported: i.e.,

$$w(0, L) = M_b(0, L) = 0. \tag{4.56}$$

With these boundary conditions, the constants of (4.55) are found to be

$$C_0 = 0 \quad C_1 = \frac{\alpha a_1 L}{2h} \quad C_2 = -\frac{\alpha a_1}{2h} \quad C_3 = 0 \tag{4.57}$$

and the lateral deflection is seen to be

$$w(x) = \frac{\alpha a_1}{2h} x(L - x). \tag{4.58}$$

It is seen that only if a_1 (or any $\boldsymbol{a_n}$, n odd) is non-zero will the beam deflect at all in the thickness direction.

From (4.50), it is seen that

$$\sigma_x = E\alpha \frac{a_2}{12}\left[1 - \frac{12z^2}{h^2}\right] \quad (i = x, y). \tag{4.59}$$

It is extremely important to see that thermal stresses will occur in the beam if and only if a_2 (or any other n even) is non-zero. Therefore, it is seen that for a steady state temperature distribution ($\boldsymbol{a_n}, n \geq 2 = 0$) the beam in this example is stress free, whether a deflection occurs or not.

Now if (4.59) is substituted into (4.52) it results in, for all cases

$$\sigma_{t_{avg}} = 0. \tag{4.60}$$

Therefore, in those cases where there are no boundary conditions constraining the thin beam of plate, so that it is free to expand or contract, the average stress across the thickness is zero, i.e., the stresses are self-equilibrating. This is an important concept to remember in the design of flat plates and beam structures of all kinds.

4.7 References

4.1. Boley, B.A. and Wiener, J.H. (1962) *Theory of Thermal Stresses*, John Wiley and Sons, Inc., New York.
4.2. Johns, D.J. (1965) *Thermal Stress Analyses*, Pergamon Press.
4.3. Nowinski, J.L. (1978) *Theory of Thermoelasticity with Applications*, Sijthoff and Noordhoff Publishers.

4.8 Problems

4.1. A flat structural panel on the wing of a supersonic fighter of thickness 0.2 inches is considered to be unstressed at $70\,^{\circ}\mathrm{F}$. After a considerable time at cruise speed such that a steady state temperature distribution is reached, the temperature on the heated side is measured to be $140\,^{\circ}\mathrm{F}$, the temperature on the cooler side is measured at $80\,^{\circ}\mathrm{F}$, and the temperature gradient through the plate is considered to be linear. Calculate the thermal stress resultant, $\boldsymbol{N^T}$, and the thermal stress couple, $\boldsymbol{M^T}$, where for aluminum $\boldsymbol{E = 10 \times 10^6}$ **psi**, $\boldsymbol{\alpha = 10 \times 10^{-6}}$ **in/in/°F, and** $\boldsymbol{\nu = 0.3}$.

4.2. The same aluminum panel as in Problem 4.1 is now heated symmetrically from both the top side and the bottom side. After 10 seconds thermocouples placed on both surfaces of the panel read $160\,^{\circ}\mathrm{F}$, and a thermocouple at the mid-surface reads $80\,^{\circ}\mathrm{F}$. Assuming the temperature profile in the panel to be parabolic (i.e., a second degree polynomial), what is the thermal stress resultant N^T and the thermal stress couple M^T at this time?

4.3. In Problem 4.1 at what location across the plate thickness are the absolute values of σ_x and σ_y a maximum and what is that stress; assuming $N_x = N_y = M_x = M_y = 0$?

4.4. In Problem 4.2 at what location across the plate thickness are the absolute vales of σ_x and σ_y a maximum value and what is the value; assuming $N_x = N_y = M_x = M_y = 0$?

4.5. An aluminum panel 0.4″ thick, and stress free at 70°F is subjected to transient heating on one of its surfaces such that $T = T(z)$ only as in the previous problems. Thermocouples record at a critical time that $\boldsymbol{T(h/2) = 170°F}$, $\boldsymbol{T(0) = 130°F}$, and $T(-h/2) = 100°F$. Assuming a polynomial temperature distribution calculate N^T, M^T and $\sigma_{x\max} = \sigma_{y\max}$ assuming N_x, N_y, M_x, and $M_y = 0$.

4.6. Thermocouples are used to measure the temperature profile through a two inch thick plate, through three measurements: one on the upper surface, one at the mid-surface, and one on the lower surface. At one specific time the measurements were:

Location	Actual Temperature Measured
$h = +1''$	200°F
$h = 0''$	110°F
$h = -1''$	80°F

If the stress free temperature is 60°F, calculate N^T and M^T.

4.7. A plate is heated from both the top and bottom such that at a certain time, three thermocouples read $\boldsymbol{T(h/2) = 300°F}$, $\boldsymbol{T(0) = 80°F}$, and $\boldsymbol{T(-h/2) = 300°F}$. If the stress free temperature is 70°F, calculate N^T and M^T for a plate that is 2″ in thickness. This aluminum plate has $\boldsymbol{E = 10 \times 10^6}$ psi and $\alpha = 10 \times 10^{-6}$ in/in/°F.

4.8. A thin walled structure 1/2″ thick, i.e., $-1/4'' \leq z \leq +1/4''$, is composed of an aluminum with properties $\boldsymbol{E = 10 \times 10^6}$ psi and $\alpha = 10 \times 10^{-6}$ in/in/°F. As a certain time *t*, thermocouples on the top, at mid-surface and at the bottom record 90°F, 100°F, and 150°F. If the stress free temperature is 70°F, determine the equation for ΔT to perform *w* subsequent thermoelastic analysis.

CHAPTER 5

CIRCULAR ISOTROPIC PLATES

5.1 Introduction

In previous chapters, attention has been focused on rectangular plates. However, circular plate structural elements are encountered in all phases of engineering. It is, therefore, necessary to develop an understanding of the behavior of circular plates.

Consider the following element from a circular plate (Figure 5.1), with positive directions of stresses and deflections as shown.

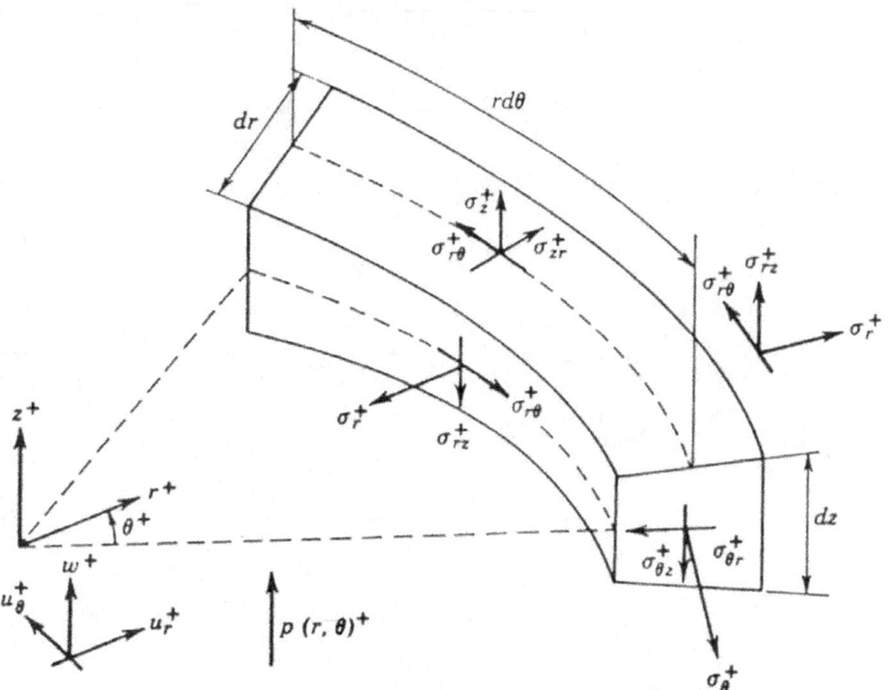

Figure 5.1. Circular plate element.

5.2 Derivation of the Governing Equations

The equations of elasticity can be derived in a circular cylindrical coordinate system, or could be obtained by transforming the elasticity equations given in Chapter 1 through the use of the relationships:

$$x = r\cos\theta, \quad y = r\sin\theta \quad \text{and} \quad x^2 + y^2 = r^2.$$

However, they are merely presented here in their final form.

Equilibrium Equations in Circular Cylindrical Coordinates

$$\frac{\partial \sigma_r}{\partial r} + \frac{1}{r}\frac{\partial \sigma_{r\theta}}{\partial \theta} + \frac{\partial \sigma_{rz}}{\partial z} + \frac{\sigma_r - \sigma_\theta}{r} = 0 \tag{5.1}$$

$$\frac{\partial \sigma_{r\theta}}{\partial r} + \frac{1}{r}\frac{\partial \sigma_\theta}{\partial \theta} + \frac{\partial \sigma_{\theta z}}{\partial z} + \frac{2}{r}\sigma_{r\theta} = 0 \tag{5.2}$$

$$\frac{\partial \sigma_{rz}}{\partial r} + \frac{1}{r}\frac{\partial \sigma_{\theta z}}{\partial \theta} + \frac{\partial \sigma_z}{\partial z} + \frac{1}{r}\sigma_{rz} = 0. \tag{5.3}$$

Stress-Strain Relations (after using classical plate assumptions)

$$\varepsilon_r = \frac{1}{E}[\sigma_r - \nu\sigma_\theta], \qquad \varepsilon_\theta = \frac{1}{E}[\sigma_\theta - \nu\sigma_r] \qquad (5.4), (5.5)$$

$$\varepsilon_{r\theta} = \frac{1}{2G}\sigma_{r\theta}, \qquad \sigma_z = \varepsilon_{rz} = \varepsilon_{\theta z} = 0. \qquad (5.6), (5.7)$$

Strain-Displacement Relations, General

$$\varepsilon_r = \frac{\partial u_r}{\partial r}; \quad \varepsilon_\theta = \frac{1}{r}\frac{\partial u_\theta}{\partial \theta} + \frac{u_r}{r}; \quad \varepsilon_z = \frac{\partial w}{\partial z} \tag{5.8}$$

$$\varepsilon_{r\theta} = \frac{1}{2}\left(\frac{1}{r}\frac{\partial u_r}{\partial \theta} + \frac{\partial u_\theta}{\partial r} - \frac{u_\theta}{r}\right) \tag{5.9}$$

$$\varepsilon_{\theta z} = \frac{1}{2}\left(\frac{\partial u_\theta}{\partial z} + \frac{1}{r}\frac{\partial w}{\partial \theta}\right), \qquad \varepsilon_{rz} = \frac{1}{2}\left(\frac{\partial w}{\partial r} + \frac{\partial u_r}{\partial z}\right). \qquad (5.10), (5.11)$$

Of course, for classical plate theory, $\boldsymbol{\varepsilon_z = \varepsilon_{rz} = \varepsilon_{\theta z}} = 0$ in (5.8), (5.10) and (5.11) above.

Similar to the case of rectangular plates, stress resultants, stress couples and shear resultants are defined as follows:

$$\begin{Bmatrix} N_r \\ N_\theta \\ N_{r\theta} \end{Bmatrix} = \int_{-h/2}^{+h/2} \begin{Bmatrix} \sigma_r \\ \sigma_\theta \\ \sigma_{r\theta} \end{Bmatrix} \mathrm{d}z, \qquad \begin{Bmatrix} M_r \\ M_\theta \\ M_{r\theta} \end{Bmatrix} = \int_{-h/2}^{+h/2} \begin{Bmatrix} \sigma_r \\ \sigma_\theta \\ \sigma_{r\theta} \end{Bmatrix} z\,\mathrm{d}z \tag{5.12), (5.13}$$

$$\begin{Bmatrix} Q_r \\ Q_\theta \end{Bmatrix} = \int_{-h/2}^{+h/2} \begin{Bmatrix} \sigma_{rz} \\ \sigma_{\theta z} \end{Bmatrix} \mathrm{d}z. \tag{5.14}$$

In developing the governing equations for a circular plate, one proceeds as in Chapter 2, multiplying Equations (5.1) through (5.3) by dz and integrating the equations across the thickness of the plate: then multiplying (5.1) and (5.2) by z dz and again integrating these across the plate thickness. For the isothermal circular plate, the results are:

$$\frac{\partial N_r}{\partial r} + \frac{1}{r}\frac{\partial N_{r\theta}}{\partial \theta} + \frac{N_r - N_\theta}{r} = 0 \tag{5.15}$$

$$\frac{\partial N_{r\theta}}{\partial r} + \frac{1}{r}\frac{\partial N_\theta}{\partial \theta} + \frac{2}{r} N_{r\theta} = 0 \tag{5.16}$$

$$\frac{\partial Q_r}{\partial r} + \frac{1}{r}\frac{\partial Q_\theta}{\partial \theta} + \frac{1}{r} Q_r + p(r,\theta) = 0 \tag{5.17}$$

$$\frac{\partial M_r}{\partial r} + \frac{1}{r}\frac{\partial M_{r\theta}}{\partial \theta} + \frac{M_r - M_\theta}{r} - Q_r = 0 \tag{5.18}$$

$$\frac{\partial M_{r\theta}}{\partial r} + \frac{1}{r}\frac{\partial M_\theta}{\partial \theta} + \frac{2}{r} M_{r\theta} - Q_\theta = 0. \tag{5.19}$$

If there are surface shear stresses applied $\tau_{1r}, \tau_{2r}, \tau_{1\theta}$ and $\tau_{2\theta}$ stresses will be appended to (5.15) through (5.19) identical to those for rectangular plates of Section 2.4, with appropriate subscripts.

Again, for the bending of a circular plate, displacements are taken in a form analogous to (2.1) through (2.3)

$$u_r = u_{0r} + \bar{\alpha} z, \quad u_\theta = u_{0\theta} + \bar{\beta} z, \quad \text{and} \quad w = w(r,\theta). \tag{5.20}$$

Since in a classical circular plate $\boldsymbol{\varepsilon_{rz} = \varepsilon_{\theta z} = 0}$, substituting (5.20) into Equations (5.10) and (5.11) results in

$$\bar{\alpha} = -\frac{\partial w}{\partial r} \quad \text{and} \quad \bar{\beta} = -\frac{1}{r}\frac{\partial w}{\partial \theta} \quad \text{or}$$

$$u_r = u_{0r} - z\frac{\partial w}{\partial r}, \quad u_\theta = u_{0\theta} - \frac{z}{r}\frac{\partial w}{\partial \theta}. \tag{5.21}$$

From Equations (5.21), (5.8) and (5.4),

$$\varepsilon_r = \frac{\partial u_r}{\partial r} = \frac{1}{E}[\sigma_r - \nu\sigma_\theta] = \frac{\partial u_{0r}}{\partial r} - z\frac{\partial^2 w}{\partial r^2}. \tag{5.22}$$

From Equations (5.21), (5.8) and (5.5),

$$\begin{aligned}\varepsilon_\theta &= \frac{1}{r}\frac{\partial u_\theta}{\partial \theta} + \frac{u_r}{r} = \frac{1}{E}[\sigma_\theta - \nu\sigma_r] \\ &= \frac{1}{r}\frac{\partial u_{0\theta}}{\partial \theta} + \frac{u_{0r}}{r} - \frac{z}{r^2}\frac{\partial^2 w}{\partial \theta^2} - \frac{z}{r}\frac{\partial w}{\partial r}.\end{aligned} \tag{5.23}$$

From Equations (5.21), (5.9) and (5.6),

$$\begin{aligned}\varepsilon_{r\theta} &= \frac{1}{2G}\sigma_{r\theta} = \frac{1+\nu}{E}\sigma_{r\theta} \\ &= \frac{1}{2}\left(\frac{1}{r}\frac{\partial u_{0r}}{\partial \theta} + \frac{\partial u_{0\theta}}{\partial r} - \frac{u_{0\theta}}{r}\right) - \frac{2z}{r}\frac{\partial^2 w}{\partial r\,\partial \theta} + \frac{2z}{r^2}\frac{\partial w}{\partial \theta}.\end{aligned} \tag{5.24}$$

Multiplying Equations (5.22) through (5.24) by dz, and integrating across the thickness of the plate, then multiplying them by z dz and again integrating them across the plate thickness and with some algebraic manipulation the stress resultant in-plane displacement relations and moment-curvature relations for a circular plate evolve (for the case of no surface shear stresses).

$$N_r = K\left[\frac{\partial u_{0r}}{\partial r} + \frac{\nu}{r}\frac{\partial u_{0\theta}}{\partial \theta} + \frac{\nu u_{0r}}{r}\right] \tag{5.25}$$

$$N_\theta = K\left[\frac{1}{r}\frac{\partial u_{0\theta}}{\partial \theta} + \frac{u_{0r}}{r} + \nu\frac{\partial u_{0r}}{\partial r}\right] \tag{5.26}$$

$$N_{r\theta} = K(1-\nu)\left[\frac{1}{r}\frac{\partial u_{0r}}{\partial \theta} + \frac{\partial u_{0\theta}}{\partial r} - \frac{u_{0\theta}}{r}\right] \tag{5.27}$$

$$M_r = -D\left[\frac{\partial^2 w}{\partial r^2} + \frac{\nu}{r}\frac{\partial w}{\partial r} + \frac{\nu}{r^2}\frac{\partial^2 w}{\partial \theta^2}\right] \tag{5.28}$$

$$M_\theta = -D\left[\frac{1}{r^2}\frac{\partial^2 w}{\partial\theta^2}+\frac{1}{r}\frac{\partial w}{\partial r}+\nu\frac{\partial^2 w}{\partial r^2}\right] \tag{5.29}$$

$$M_{r\theta} = -D(1-\nu)\left[\frac{1}{r}\frac{\partial^2 w}{\partial r\,\partial\theta}-\frac{1}{r^2}\frac{\partial w}{\partial\theta}\right] \tag{5.30}$$

where again $K = Eh/(1-\nu^2)$ and $D = Eh^3/12(1-\nu^2)$, the in-plane stiffness and flexural stiffness, respectively.

Solving (5.18) and (5.19) for Q_r and $\boldsymbol{Q_\theta}$, and substituting the result into Equation (5.17) provides an equations involving M_r, $\boldsymbol{M_\theta}$, $M_{r\theta}$ and $p(r,\theta)$. Substituting Equations (5.28) through (5.30) into that equation results in the final governing differential equations for the bending of a circular plate, again the biharmonic equation.

$$\boldsymbol{D\nabla^4 w = p(r,\theta)} \tag{5.31}$$

where

$$\nabla^2(\;) = \frac{\partial^2(\;)}{\partial r^2}+\frac{1}{r}\frac{\partial(\;)}{\partial r}+\frac{1}{r^2}\frac{\partial^2(\;)}{\partial\theta^2}. \tag{5.32}$$

Similarly, substituting (5.25) through (5.27) into (5.15) and (5.16) produces the equations for the stretching of a circular plate.

$$\nabla^4 u_{0r} = 0 \qquad \nabla^4 u_{0\theta} = 0. \tag{5.33}$$

Note that (5.31) and (5.33) are identical to (2.57), (2.59) and (2.60) for the rectangular plate. The biharmonic equations control plate behavior in both the Cartesian and the circular coordinate systems. Only the definition of the Laplacian operator changes with the coordinate system.

Of course once the plate solution is obtained, the stresses within the plate are given by:

$$\sigma_r = \frac{N_r}{h}+\frac{M_r z}{h^3/12}, \qquad \sigma_\theta = \frac{N_\theta}{h}+\frac{M_\theta z}{h^3/12}$$

$$\sigma_{r\theta} = \frac{N_{r\theta}}{h}+\frac{M_{r\theta} z}{h^3/12} \tag{5.34}$$

$$\sigma_{rz} = \frac{3Q_r}{2h}\left[1-\left(\frac{z}{h/2}\right)^2\right], \qquad \sigma_{\theta z} = \frac{3Q_\theta}{2h}\left[1-\left(\frac{z}{h/2}\right)^2\right].$$

For the case of surface shear stresses, the last two expressions above would be modified by the analogous expressions of (2.80) through (2.83), simply modified by changing x and y subscripts to r and θ.

Furthermore, to consider a thermoelastic circular plate, one merely adds appendages to Equations (5.25), (5.26), (5.28), (5.29), (5.31), (5.33) and the first two of (5.34), identical to the last terms of the N_x, N_y, M_x, M_y expression of (4.10) through (4.13), and the modifications for σ_x and σ_y in (4.19) with obvious subscript changes.

In the general case of no axial symmetry, the solution of Equation (5.31) results in Bessel functions and modified Bessel functions of the first and second kinds. Such problems will not be treated herein, but are treated in depth in various other texts dealing with circular plates. Because so often circular plates are subjected to axially symmetric loads, they are discussed below.

5.3 Axially Symmetric Circular Plates

When the plate is continuous in the θ direction, (i.e., is in the region $0 \leq \theta \leq 2\pi$), when the loading is not a function of θ, and when the boundary conditions do not vary around the circumference, the plate problem is said to be axially symmetric, and the following simplifications can be made:

$$\frac{\partial(\;)}{\partial\theta} = \frac{\partial^2(\;)}{\partial\theta^2} = M_{r\theta} = Q_\theta = 0. \tag{5.35}$$

The previous equations for the bending of a circular plate can therefore be simplified to the following, where primes denote differentiation with respect to r.

$$M_r = -D\left(w'' + \frac{\nu}{r}w'\right) \tag{5.36}$$

$$M_\theta = -D\left(\frac{1}{r}w' + \nu w''\right) \tag{5.37}$$

$$Q_r = -D(\nabla^2 w)' \tag{5.38}$$

$$D\nabla^4 w = p(r) \tag{5.39}$$

where

$$\nabla^2(\;) = (\;)'' + \frac{1}{r}(\;)' = \frac{1}{r}\frac{\mathrm{d}}{\mathrm{d}r}\left[r\frac{\mathrm{d}(\;)}{\mathrm{d}r}\right]. \tag{5.40}$$

Interestingly, Equation (5.39) can therefore be written as,

$$\nabla^4 w = \frac{1}{r}\frac{\mathrm{d}}{\mathrm{d}r}\left\{r\frac{\mathrm{d}}{\mathrm{d}r}\left[\frac{1}{r}\frac{\mathrm{d}}{\mathrm{d}r}\left(r\frac{\mathrm{d}w}{\mathrm{d}r}\right)\right]\right\} = \frac{p(r)}{D}. \tag{5.41}$$

5.4 Solutions for Axially Symmetric Circular Plates

Equation (5.41) can be made dimensionless by normalizing both the radial coordinate, r, and the later deflection, w, with respect to the radius of the circular plate, a, as follows:

$$\bar{r} = r/a, \quad \bar{w} = w/a. \tag{5.42}$$

Using (5.42) above, Equation (5.41) can be written as

$$\frac{1}{\bar{r}}\frac{\mathrm{d}}{\mathrm{d}\bar{r}}\left\{\bar{r}\frac{\mathrm{d}}{\mathrm{d}\bar{r}}\left[\frac{1}{\bar{r}}\frac{\mathrm{d}}{\mathrm{d}\bar{r}}\left(\bar{r}\frac{\mathrm{d}\bar{w}}{\mathrm{d}\bar{r}}\right)\right]\right\} = \frac{p(\bar{r})a^3}{D}. \tag{5.43}$$

One can proceed to obtain the homogeneous solution of Equation (5.43) above, by setting the right hand side equal to zero, and proceeding to integrate the left hand side, where below, C_0, C_1, C_2 and C_3 are the resulting constants of integration used to satisfy the boundary conditions.

Multiplying the homogeneous portion of Equation (5.43) by $\bar{r}$, then integrating once yields

$$\frac{\mathrm{d}}{\mathrm{d}\bar{r}}\left[\frac{1}{\bar{r}}\frac{\mathrm{d}}{\mathrm{d}\bar{r}}\left(\bar{r}\frac{\mathrm{d}\bar{w}}{\mathrm{d}\bar{r}}\right)\right] = \frac{C_0}{\bar{r}}.$$

Integrating once more and multiplying both sides by $\bar{r}$ provides

$$\frac{\mathrm{d}}{\mathrm{d}\bar{r}}\left(\bar{r}\frac{\mathrm{d}\bar{w}}{\mathrm{d}\bar{r}}\right) = C_0\bar{r}\ln\bar{r} + C_1\bar{r}.$$

To integrate the first term in the right hand side, let $\ln\bar{r} = y$, hence, $\bar{r} = e^y$, and $\mathrm{d}\bar{r} = e^y\,\mathrm{d}y$. Thus,

$$\int \bar{r}\ln\bar{r}\,\mathrm{d}\bar{r} = \int ye^{2y}\,\mathrm{d}y = \frac{ye^{2y}}{2} - \frac{e^{2y}}{4}.$$

Therefore, integrating the expression above, and dividing the results by $\bar{r}$ gives

$$\frac{d\bar{w}}{d\bar{r}} = C_0\left[\frac{\bar{r}\ln\bar{r}}{2} - \frac{\bar{r}}{4}\right] + \frac{C_1\bar{r}}{2} + \frac{C_2}{\bar{r}}$$

and finally, one more integration produces

$$\bar{w} = \frac{C_0}{2}\left[\frac{\bar{r}^2\ln\bar{r}}{2} - \frac{\bar{r}^2}{4}\right] - \frac{C_0\bar{r}^2}{8} + \frac{C_1\bar{r}^2}{4} + C_2\ln\bar{r} + C_3.$$

This final form of the homogeneous solution can be written more succinctly as

$$\bar{w} = A + B\ln\bar{r} + C\bar{r}^2 + E\bar{r}^2\ln\bar{r}. \tag{5.44}$$

Returning to Equation (5.43), the particular solution can be written as:

$$\bar{w}_p = \int\frac{1}{\bar{r}}\int\bar{r}\int\frac{1}{\bar{r}}\int\frac{p(\bar{r})a^3}{D}\bar{r}\,d\bar{r}\,d\bar{r}\,d\bar{r}\,d\bar{r}. \tag{5.45}$$

Thus, the complete solution for any circular plate under axially symmetric loading is given by Equations (5.44) and (5.45). It is easy to show that the particular solution for a plate with a uniform lateral load is:

$$\bar{w}_p = \frac{p_0 a^3\bar{r}^4}{64D} \quad \text{or} \quad w_p = \frac{p_0 r^4}{64D}. \tag{5.46}$$

For ease of calculation, the following quantities are given explicitly for the circular plate of radius a with uniform lateral loading $p(r) = p_0$.

$$\bar{w} = A + B\ln\bar{r} + C\bar{r}^2 + E\bar{r}^2\ln\bar{r} + \frac{p_0 a^3\bar{r}^4}{64D} \tag{5.47}$$

$$\frac{d\bar{w}}{d\bar{r}} = \frac{B}{\bar{r}} + 2C\bar{r} + E[2\bar{r}\ln\bar{r} + \bar{r}] + \frac{p_0 a^3\bar{r}^3}{16D} \tag{5.48}$$

$$M_r = -\frac{D}{a}\left[-\frac{B}{\bar{r}^2}(1-\nu) + 2C(1+\nu) + 2E(1+\nu)\ln\bar{r} + (3+\nu)E\right] - \frac{p_0 a^2\bar{r}^2(3+\nu)}{16} \tag{5.49}$$

$$Q_r = -\frac{D}{a^2}\left[\frac{4E}{\bar{r}}\right] - \frac{p_0 a \bar{r}}{2} \tag{5.50}$$

$$M_\theta = -\frac{D}{a}\left[\frac{B(1-\nu)}{\bar{r}^2} + 2C(1+\nu) + 2E\ln\bar{r}(1+\nu) + E(1+3\nu)\right] - \frac{p_0 a^2 \bar{r}^2 (1+3\nu)}{16}. \tag{5.51}$$

For other lateral loadings, the last terms only in each expression above would be changed, the homogeneous solution remains the same.

5.5 Circular Plate, Simply Supported at the Outer Edge, Subjected to a Uniform Lateral Loading, p_0

For the plate which is continuous from $0 \le r \le a$, and which contains no concentrated loading at $r = 0$, it is easy to see that $B = E = 0$; otherwise the lateral deflection and transverse shear resultant would be infinite at $r = 0$. At $\bar{r} = 1$ or $r = a$, the boundary conditions are

$$\bar{w}(1) = 0 \quad (\text{or } w(a) = 0), \qquad M_r(1) = 0$$

Hence,

$$A = \frac{(5+\nu)}{(1+\nu)}\frac{p_0 a^3}{64D}, \quad C = -\frac{p_0 a^3 (3+\nu)}{32D(1+\nu)}$$

and

$$w(\bar{r}) = \frac{p_0 a^4}{64D}\left[\frac{(5+\nu)}{(1+\nu)} - \frac{2(3+\nu)}{(1+\nu)}\bar{r}^2 + \bar{r}^4\right]. \tag{5.52}$$

5.6 Circular Plate, Clamped at the Outer Edge, Subjected to a Uniform Lateral Loading, p_0

Again $B = E = 0$. At the outer edge, $w(1) = 0$ and $\partial\bar{w}(1)/\partial\bar{r} = 0$. Hence,

$$A = \frac{p_0 a^3}{64D} \quad \text{and} \quad C = -\frac{p_0 a^3}{32D}.$$

Thus,

$$w(\bar{r}) = \frac{p_0 a^4}{64D}[1 - 2\bar{r}^2 + \bar{r}^4]. \tag{5.53}$$

5.7 Annular Plate, Simply Supported at the Outer Edge, Subjected to a Stress Couple, *M*, at the Inner Boundary

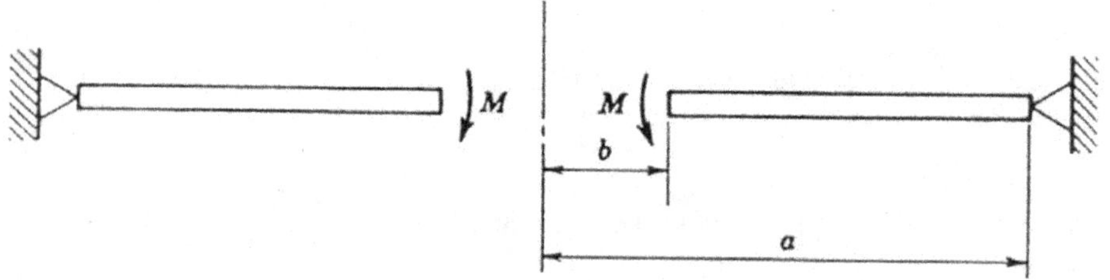

Figure 5.2. Annular circular plate.

Remembering that $\bar{r} = r/a$, and defining $s = b/a$, the governing differential equation in this case with no lateral load, $p(r)$, is

$$\nabla^4 w = 0$$

and the boundary conditions are:

$$\bar{w}(1) = 0 \quad M_r(s) = M$$

$$M_r(1) = 0 \quad Q_r(s) = 0.$$

The lateral deflection is found to be

$$w(\bar{r}) = \frac{Ma^2 s^2 \ln \bar{r}}{D(1-\nu)(1-s^2)} - \frac{Ma^2}{2D(1+\nu)}\left(\frac{s^2}{1-s^2}\right)(1-\bar{r}^2). \tag{5.54}$$

5.8 Annular Plate, Simply Supported at the Outer Edge, Subjected to a Shear Resultant, Q_0, at the Inner Boundary

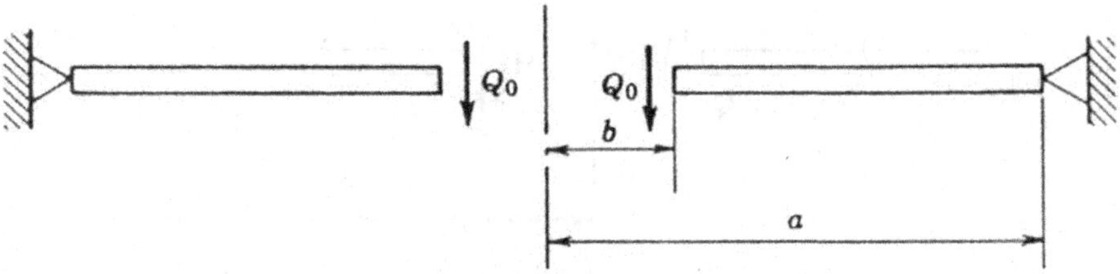

Figure 5.3. Annular circular plate.

Again, the governing differential equation is $\nabla^4 w = 0$, and the boundary conditions are,

$$\bar{w}(1) = 0 \quad M_r(s) = 0$$

$$M_r(1) = 0 \quad Q_r(s) = Q_0$$

The solution is:

$$\bar{w}(\bar{r}) = \frac{Q_0 s a^3}{4D}\left[-\frac{(3+\nu)(1-\bar{r}^2)}{2(1+\nu)} + \frac{s^2 \ln s}{(1-s^2)}(1-\bar{r}^2) - \frac{2(1+\nu)}{(1-\nu)}\left(\frac{s^2}{1-s^2}\right)\ln s \ln \bar{r} - \bar{r}^2 \ln \bar{r}\right]. \tag{5.55}$$

5.9 Some General Remarks

Of course the results given in Section 5.5 through 5.8 can be superimposed to form the solutions to other problems. Suppose the plate is subjected to a stress couple, M, on the inner boundary as well as a transverse shear resultant, Q_0, acting also at the inner edge, as shown in Figure 5.4.

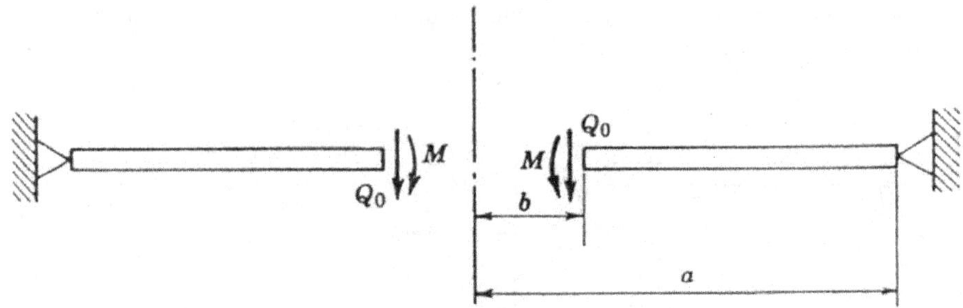

Figure 5.4. Annular circular plate.

The solution is the sum of (5.54) and (5.55). All other stress quantities are found by substituting this sum into (5.36) through (5.38) and (5.34).

Another example of using the previous examples as building blocks, consider the problem shown in Figure 5.5.

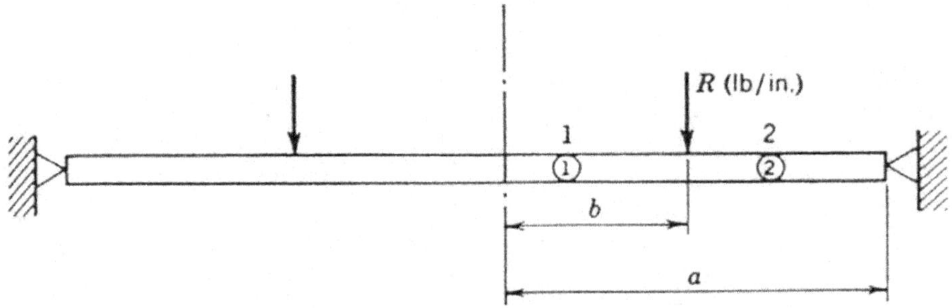

Figure 5.5. Circular plate with a ring load.

This simply supported circular plate is subjected to a ring load of R (lbs/in. of circumference). To solve this problem one first divides the plate problem into two parts: an inner solution 1 extending over the region $0 \leq r \leq b$, and an outer solution 2 over the region $b \leq r \leq a$. In each case the governing equation is

$$\nabla^4 w_1 = 0 \quad \text{and} \quad \nabla^4 w_2 = 0$$

and eight boundary conditions are needed. Since there is no lateral load $p(r)$, and the solution to each equation with suitable subscript, is (5.47) with $p_0 = 0$ (i.e., the homogeneous solution). From the reasoning of Sections 5.5 and 5.6, it is seen that $B_1 = E_1 = 0$. Likewise from the reasoning of Sections 5.5, 5.7 and 5.8, at $r = a$ or $r = 1$, $\overline{w}(1) = M_r(1) = 0$.

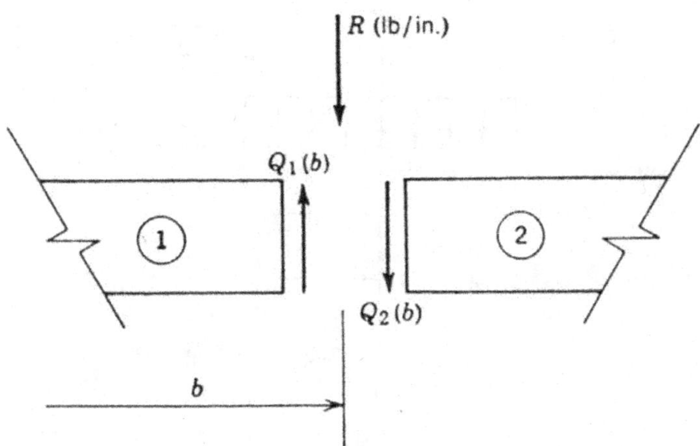

Figure 5.6. Equilibrium of plate with ring load.

At the junction of the two plate segments, it is obvious that the lateral deflection, the slopes and the stress couples must be equal for both plate segments; hence

$$w_1(b) = w_2(b) \quad \text{or} \quad \bar{w}_1(s) = \bar{w}_2(s)$$

$$\frac{dw_1(b)}{dr} = \frac{dw_2(b)}{dr} \quad \text{or} \quad \frac{d\bar{w}_1(s)}{d\bar{r}} = \frac{d\bar{w}_2(s)}{d\bar{r}}$$

$$M_{r1}(b) = M_{r2}(b) \quad \text{or} \quad M_{r1}(s) = M_{r2}(s).$$

For the eighth and last boundary condition is obtained by looking closely at the shear condition at $r = b$, as seen in Figure 5.6.

Hence, $\boldsymbol{R = Q_2(b) - Q_1(b)}$ is the eighth boundary condition. If one has either a discontinuity in load or a discontinuity in plate thickness, one must divide the plate into two segments. Examples of such problems are shown in Figure 5.7.

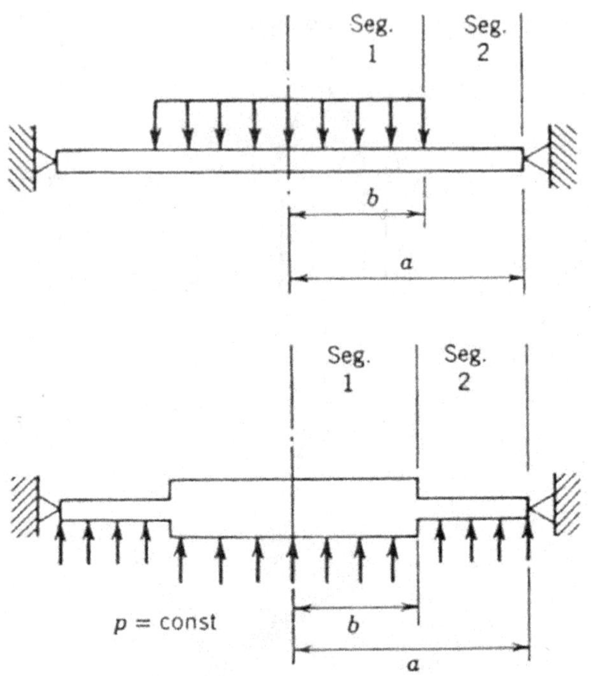

Figure 5.7. Circular plate with discontinuity of load or stiffness.

It should be noted that in the first of Figure 5.7, the lateral load over Segment 1 is a negative number, and the loading in the second example of Figure 5.7 is a positive number. Further, it should be noted that in each of these examples, because there is no concentrated load, R, as in the previous example, the eighth boundary condition here is $Q_1(b) = Q_2(b)$.

Of course if one had n structural and/or loading discontinuities, one must use $(n + 1)$ segments, and $4n$ boundary conditions.

Use of Equations (5.47) through (5.51), with the proper last terms (the particular solution) obtained through solving (5.45) reduces the problems to a straightforward procedure. Subsequently, stresses are found through (5.34).

Den Hartog [5.1] provides solutions for seventeen different isotropic circular plates on pages 128-132 of his text. One paper treating the vibrations of circular plates is by Oyibo and Brunelle [5.2]. Although it treats orthotropic plates, by letting $D_r = D_\theta = D_{r\theta} = D$, the results apply to isotropic circular plates.

5.10 Laminated Circular Thermoelastic Plates

As a practical example of the effects of thermal stresses on plates that are laminated, thermal stresses in laminated circular plates will now be addressed.

Consider a plate a radius L, composed of two laminae of different materials continuously bonded at their common face, such that no slippage can occur. The laminae may have different, but constant thicknesses limited only by plate theory assumptions. Constant material properties are assumed for each lamina. The coordinate system used is given in Figure 5.8, from which it is seen that the origin of the z axis for each lamina is at the middle surface of that lamina.

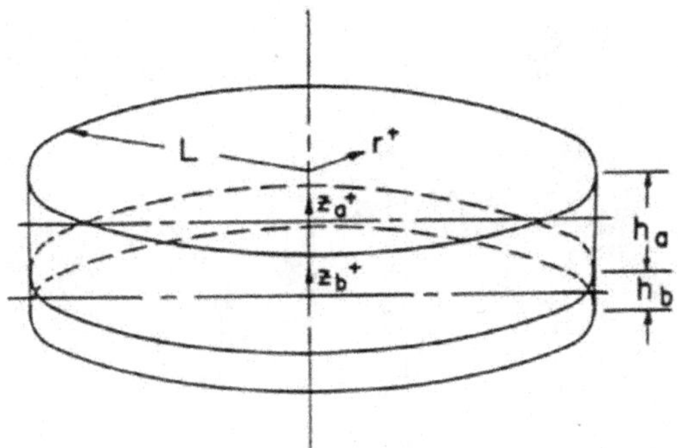

Figure 5.8. Plate Coordinate System.

The equations of equilibrium in cylindrical coordinates for the axi-symmetric case are:

$$\left.\begin{aligned} \frac{\partial \sigma_r}{\partial r} + \frac{\partial \sigma_{rz}}{\partial z} + \frac{\sigma_r - \sigma_\theta}{r} = 0 \\ \frac{\partial \sigma_{rz}}{\partial r} + \frac{\partial \sigma_z}{\partial z} + \frac{1}{r}\sigma_{rz} = 0 \end{aligned}\right\} \tag{5.56}$$

The strain-displacement relations are:

$$\varepsilon_r = \frac{\partial u_r}{\partial r}; \quad \varepsilon_\theta = \frac{u_r}{r}; \quad \varepsilon_{rz} = \frac{1}{2}\left(\frac{\partial w}{\partial r} + \frac{\partial u_r}{\partial z}\right) \tag{5.57}$$

The stress-strain relations for an isotropic thermoelastic solid are given by:

$$\left.\begin{aligned}\varepsilon_r &= \frac{1}{E}[\sigma_r - \nu\sigma_\theta] + \alpha\Delta T(r,z)\\ \varepsilon_\theta &= \frac{1}{E}[\sigma_\theta - \nu\sigma_r] + \alpha\Delta T(r,z)\\ \varepsilon_{rz} &= \frac{1}{2G}\sigma_{rz}\end{aligned}\right\} \tag{5.58}$$

where, σ_{ij} are the stresses, ε_{ij} the strains, u_r the displacement in the radial direction, w the lateral deflection, E the modulus of elasticity, ν Poisson's ratio, G the shear modulus, α the coefficient of thermal expansion and $\Delta T(r,t)$ the temperature increase or decrease from a reference temperature at which the plate is stress free thermally.

Stress resultants and couples are defined for each lamina as follows:

$$\begin{Bmatrix} N_{ri} \\ N_{\theta i} \\ Q_i \\ M_{ri} \\ M_{\theta i} \end{Bmatrix} = \int_{-h_i/2}^{+h_i/2} \begin{Bmatrix} \sigma_{ri} \\ \sigma_{\theta i} \\ \sigma_{rzi} \\ z_i\sigma_{ri} \\ z_i\sigma_{\theta i} \end{Bmatrix} dz_i \quad i = a, b. \tag{5.59}$$

The positive directions for these resultants and couples are shown in Figure 5.9.

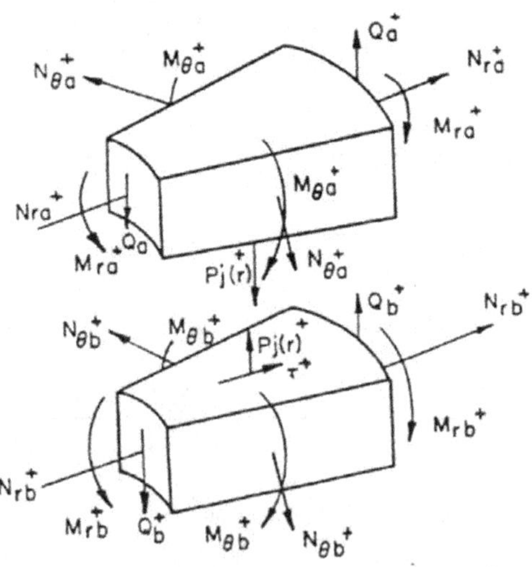

Figure 5.9. Stress Resultants and Stress Couples.

Equation (5.56) can now be multiplied by dz_i and by $z_i dz_i$ and be integrated across the thickness of each lamina to obtain the stress resultants and couples defined in (5.59). The non-vanishing boundary conditions are:

For lamina *a*:

$$\sigma_{rz_a}(-h_a/2) \quad \tau; \quad z_a\sigma_{rz_a}(-h_a/2) \quad -\frac{h_a\tau}{2}; \quad \sigma_{z_a}(-h_a/2) \quad p_j \tag{5.60}$$

For lamina *b*:

$$\sigma_{rz_b}(+h_b/2) \quad \tau; \quad z_b\sigma_{rz_b}(+h_b/2) \quad \frac{h_b\tau}{2}; \quad \sigma_{z_b}(+h_b/2) \quad p_j \tag{5.61}$$

where τ is the radial shear stress in the joint between the laminae, and p_j is the normal joint stress.

The resulting plate equilibrium equations are given below.

$$\frac{\mathrm{d}N_{ra}}{\mathrm{d}r} + \frac{N_{ra} - N_{\theta a}}{\tau} - \tau \quad 0 \tag{5.62}$$

$$\frac{\mathrm{d}M_{ra}}{\mathrm{d}r} + \frac{M_{ra} - M_{\theta a}}{r} - \frac{h_a\tau}{2} - Q_a \quad 0 \tag{5.63}$$

$$\frac{\mathrm{d}Q_a}{\mathrm{d}r} + \frac{Q_a}{r} - p_j \quad 0 \tag{5.64}$$

$$\frac{\mathrm{d}N_{rb}}{\mathrm{d}r} + \frac{N_{rb} - N_{\theta b}}{r} + \tau \quad 0 \tag{5.65}$$

$$\frac{\mathrm{d}M_{rb}}{\mathrm{d}r} + \frac{M_{rb} - M_{\theta b}}{r} - \frac{h_b\tau}{2} - Q_b \quad 0 \tag{5.66}$$

$$\frac{\mathrm{d}Q_b}{\mathrm{d}r} + \frac{Q_b}{r} + p_j \quad 0 \tag{5.67}$$

To determine stress-strain relations for the laminated plate, it is assumed that the displacements for each lamina will be of the form

$$\left.\begin{aligned} u_r(r,z) &\quad u_0(r) + z\bar{\alpha}(r) \\ w &\quad w(r) \end{aligned}\right\} \tag{5.68}$$

where u_0 is the in-plane displacement of the middle surface of the lamina, and $\bar{\alpha}$ is the rotation.

Neglecting transverse shear deformation, the last of (5.57) becomes

$$\frac{\partial w}{\partial r} = -\frac{\partial u_r}{\partial z}.$$

Substituting this into (5.68), it is seen that $\overline{\alpha} = -w'$, and the displacement relations become

$$\left.\begin{aligned} u_r &= u_0 - zw' \\ w &= w(r) \end{aligned}\right\} \tag{5.69}$$

where the prime denotes differentiation with respect to r. The strain-displacement relations for either lamina are then given by:

$$\left.\begin{aligned} \varepsilon_r &= u_0' - zw'' \\ \varepsilon_\theta &= \frac{u_0}{r} - \frac{z}{r}w' \end{aligned}\right\} \tag{5.70}$$

Upon substituting (5.70) into the first two equations of (5.58), integrating across the thickness of each lamina, and rearranging, the stress resultants and stress couples are found to be, for each lamina,

$$N_r = K\left(u_0' + \frac{\nu}{r}u_0\right) - \frac{1}{1-\nu}N^T \tag{5.71}$$

$$N_\theta = K\left(\nu u_0' + \frac{u_0}{r}\right) - \frac{1}{1-\nu}N^T \tag{5.72}$$

$$M_r = -D\left[w'' + \frac{\nu}{r}w'\right] - \frac{1}{1-\nu}M^T \tag{5.73}$$

$$M_\theta = -D\left[\frac{w'}{r} + \nu w''\right] - \frac{1}{1-\nu}M^T \tag{5.74}$$

where $N^T = \int_{-h/2}^{+h/2} E\alpha T\,dz$ is the "thermal stress resultant," $M^T = \int_{-h/2}^{+h/2} E\alpha Tz\,dz$ is the "thermal stress couple," $K = Eh/(1-\nu^2)$ and $D = Eh^3/12(1-\nu^2)$.

It is seen that we have six equilibrium equations, and eight equations derived from the strain-displacement and stress-strain relations. However, there are sixteen unknown variables; namely, N_{ra}, $N_{\theta a}$, M_{ra}, $M_{\theta a}$, Q_a, N_{rb}, $N_{\theta b}$, M_{rb}, $M_{\theta b}$, Q_b, w_a, w_b, u_{0a}, u_{0b}, τ and p_j. Hence, two more relationships are needed and are easily found.

First, because transverse normal deformation is neglected, and since $w_a(r,-h_a/2) = w_b(r,+h_b/2)$, because the laminae are bonded together,

$$w = w_a = w_b. \tag{5.75}$$

The second relationship expresses the fact that no slippage occurs in the joint between laminae, hence $\boldsymbol{u(r,-h_a/2) = u(r,+h_b/2)}$ and employing (5.69).

$$u_{0a} = u_{0b} - \frac{h_a + h_b}{2} w'. \tag{5.76}$$

The governing equations can now be combined into relations in terms of the lateral deflection, w, and the displacement, u_{0a}. In the following, it is assumed that Poisson's ratio for the material in each lamina is sufficiently similar to assume that $\nu_a = \nu_b = \nu$.

Solving for Q_a and Q_b in (5.63) and (5.66), and substituting them into (5.64) and (5.67) along with (5.73) and (5.74), two relations are produced which upon adding and subtracting result in the following:

$$(D_a + D_b)\nabla^4 w = -\frac{1}{1-\nu}\left[\boldsymbol{\nabla^2 M_a^T + \nabla^2 M_b^T}\right] + \frac{h_a + h_b}{2}\left(\tau' + \frac{\tau}{r}\right) \tag{5.77}$$

$$(D_a - D_b)\nabla^4 w = -\frac{1}{1-\nu}\left[\boldsymbol{\nabla^2 M_a^T - \nabla^2 M_b^T}\right] + \frac{h_a - h_b}{2}\left(\tau' + \frac{\tau}{r}\right) - 2p_j \tag{5.78}$$

where ∇^2 and ∇^4 are the Laplacian operators defined as

$$\nabla^2 \equiv \frac{1}{r}\frac{\mathrm{d}}{\mathrm{d}r}\left(r\frac{\mathrm{d}}{\mathrm{d}r}\right); \quad \nabla^4 \equiv \frac{1}{r}\frac{\mathrm{d}}{\mathrm{d}r}\left\{r\frac{\mathrm{d}}{\mathrm{d}r}\left[\frac{1}{r}\frac{\mathrm{d}}{\mathrm{d}r}\left(r\frac{\mathrm{d}}{\mathrm{d}r}\right)\right]\right\}$$

Adding and subtracting (5.62) and (5.65), and substituting (5.71), (5.72) and (5.76) into them results in

$$(K_a + K_b)\frac{\mathrm{d}}{\mathrm{d}r}\left[u_{oa}{}' + \frac{u_{oa}}{r}\right] + K_b\frac{h_a + h_b}{2}\frac{\mathrm{d}}{\mathrm{d}r}(\nabla^2 w) - \frac{1}{1-\nu}(T_a^{T'} \text{ to } T_b^{T'}) \square 0 \tag{5.79}$$

$$\tau = -\frac{K_a K_b (h_a + h_b)}{2(K_a + K_b)}\frac{\mathrm{d}}{\mathrm{d}r}(\nabla^2 w) - \frac{K_b T_a^{T'} - K_a T_b^{T'}}{(1-\nu)(K_a + K_b)}. \tag{5.80}$$

Substituting (5.80) and its first derivative into (5.77) yields

$$D_e\nabla^4 w = -\frac{1}{1-\nu}\left[\boldsymbol{C_0\nabla^2 T_a^T + C_1\nabla^2 T_b^T + \nabla^2 M_a^T + \nabla^2 M_b^T}\right] \tag{5.81}$$

where $C_0 = (h_a + h_b)K_b/2(K_a + K_b)$ and $C_1 = (h_a + h_b)K_a/2(K_a + K_b)$. The constant D_e is the effective flexural stiffness for a laminated plate and is given by

$$D_e = D_a + D_b + \frac{K_a K_b (h_a + h_b)^2}{4(K_a + K_b)}. \tag{5.82}$$

It should be noted that for the laminated plate, the flexural stiffness is greater than the sum of the rigidities of the individual laminae, as shown in (5.82). It is also interesting to note that due to the restraining action the thermal stress resultants enter into (5.81), while in the case of a plate of a homogeneous material only the thermal stress couple is involved. It should also be noted that (5.82) is also a valid expression for plates of any geometry, i.e., for rectangular plates as well.

Substituting (5.80) and its first derivative into (5.78), and rearranging, will directly produce the normal joint stress, p_j.

$$p_j = \frac{1}{2D_e(1-\nu)}\left[D_a - D_b + \frac{K_a K_b (h_a^2 - h_b^2)}{4(K_a + K_b)}\right] \times \left[C_0 \nabla^2 T_a^T + C_1 \nabla^2 T_b^T \right.$$
$$\left. + \nabla^2 M_a^T + \nabla^2 M_b^T\right] - \frac{1}{2(1-\nu)}\left[C_2 \nabla^2 T_a^T - C_3 \nabla^2 T_b^T + \nabla^2 M_a^T - \nabla^2 M_b^T\right]$$

(5.83)

where $C_2 = (h_a + h_b)K_b/2(K_a + K_b)$ and $C_3 = (h_a - h_b)K_a/2(K_a + K_b)$.

In (5.81), the lateral deflection w is the only unknown and can be easily solved for in terms of the temperature distribution in the laminae. In classical thermoelasticity, the stress and displacement fields are coupled to the temperature distribution but the temperature distribution is not affected by the stress and displacement fields in the solid. Hence, the temperature distribution in the body may be found independent of the stresses and displacements produced in the body as a result of the temperature distribution. It is assumed that for the cases discussed here the effects of the stress field on the temperature distribution in the body can be neglected.

Any axi-symmetric temperature distribution in the laminated plate may be represented by a Fourier series such that

$$N_a^T = \sum_{n=1}^{\infty} b_n \cos\lambda_n r \tag{5.84}$$

where

$$b_n = \frac{2}{L}\int_0^L N_a^T \cos\lambda_n r \, dr \quad \text{and} \quad \lambda_n = \frac{n\pi}{L}$$

and L is the radius of the circular plate in Figure 5.8.

Likewise,

$$N_b^T = \sum_{n=1}^{\infty} d_n \cos\lambda_n r \tag{5.85}$$

$$M_a^T = \sum_{n=1}^{\infty} e_n \cos\lambda_n r \tag{5.86}$$

$$M_b^T = \sum_{n=1}^{\infty} g_n \cos\lambda_n r \tag{5.87}$$

Integrating (5.81) and making use of (5.84) through (5.87) it is seen that

$$\nabla^2 w = -\frac{1}{D_e(1-\nu)} \sum_{n=1}^{\infty} \Phi_n \cos\lambda_n r + C_4 \ln r + C_5 \tag{5.88}$$

$$w' = \frac{1}{D_e(1-\nu)} \sum_{n=1}^{\infty} \frac{\Phi_n}{\lambda_n} \left(\sin\lambda_n r + \frac{1}{\lambda_n r} \cos\lambda_n r + \frac{C_4 r}{2} \left\{ \ln\frac{r}{a} - \frac{1}{2} \right\} + \frac{C_5 r}{2} + \frac{C_6}{r} \right) \tag{5.89}$$

where $\Phi_n = C_0 b_n + C_1 d_n + e_n + g_n$ and C_4, C_5 and C_6 are constants of integration.

For the case of a laminated plate with no hole at the center, the two boundary conditions at $r = 0$ are that the sum of the shear resultants, $Q_a + Q_b$, equal zero and that the slope equals zero.
Hence,

$$\frac{d}{dr}(\nabla^2 w) = 0 \quad \text{and} \quad w' = 0 \quad \text{at} \quad r = 0 \tag{5.90}$$

By proper substitution it is found that

$$C_4 = 0; \quad C_6 = \frac{1}{D_e(1-\nu)} \sum_{n=1}^{\infty} \frac{\Phi_n}{\lambda_n^2}. \tag{5.91}$$

The deflection is then given by

$$w = -\frac{1}{D_e(1-\nu)} \sum_{n=1}^{\infty} \frac{\Phi_n}{\lambda_n^2} \left[-\cos\lambda_n r + 2\ln\lambda_n r - \frac{(\lambda_n r)^2}{2 \cdot 2!} + \frac{(\lambda_n r)^4}{4 \cdot 4!} - \frac{(\lambda_n r)^6}{6 \cdot 6!} \cdots \right] + \frac{C_5 r^2}{4} + C_7 \tag{5.92}$$

So (5.92) is the deflection of a laminated plate, as shown in Figure 5.8, under any arbitrary axi-symmetric temperature distribution, when the constants C_5 and C_7 are determined by the particular method of support at $r > 0$.

Using (5.92), the displacement, u_{0a}, is determined by integrating (5.79), such that

$$
\begin{aligned}
u_{0a} = & -C_0 w' + \frac{1}{(K_a + K_b)(1-\nu)} \sum_{n=1}^{\infty} \frac{b_n + d_n}{\lambda_n} \times \left(\sin\lambda_n r + \frac{1}{\lambda_n r}\cos\lambda_n r \right) \\
& + \frac{C_8 r}{2} + \frac{C_9}{r}
\end{aligned} \tag{5.93}
$$

where C_8 and C_9 are constants of integration. Again, for the case of the plate with no hole at the center, it is seen that $u_{0a} = 0$ at $r = 0$, hence

$$
C_9 = -\frac{1}{(K_a + K_b)(1-\nu)} \sum_{n=1}^{\infty} \frac{b_n + d_n}{\lambda_n^2} \tag{5.94}
$$

Equation (5.93) then becomes

$$
\begin{aligned}
u_{0a} = & \frac{1}{(K_a + K_b)(1-\nu)} \sum_{n=1}^{\infty} \frac{b_n + d_n}{\lambda_n} \times \left(\sin\lambda_n r + \frac{1}{\lambda_n r}\cos\lambda_n r - \frac{1}{\lambda_n r} \right) \\
& - C_0 w' + \frac{C_8 r}{2}
\end{aligned} \tag{5.95}
$$

Since all the governing equations have been placed in terms of w and u_{0a}, substituting (5.92) and (5.95) and their derivatives into these equations will provide the solution for all stress resultants, stress couples and displacements.

Knowing the stress resultants and stress couples in each lamina, the stress distribution throughout each lamina may be determined. The radial stress and circumferential stresses in each lamina are given by

$$
\sigma_r = \frac{1}{h}\left[N_r + \frac{N^T}{(1-\nu)} \right] + \frac{z}{h^3/12}\left[M_r + \frac{M^T}{(1-\nu)} \right] - \frac{E\alpha\Delta T}{(1-\nu)} \tag{5.96}
$$

$$
\sigma_\theta = \frac{1}{h}\left[N_\theta + \frac{N^T}{(1-\nu)} \right] + \frac{z}{h^3/12}\left[M_\theta + \frac{M^T}{(1-\nu)} \right] - \frac{E\alpha\Delta T}{(1-\nu)} \tag{5.97}
$$

To find the shear stress σ_{rz} in each lamina, the distribution throughout each lamina must be such that the conditions for σ_{rza} and σ_{rzb} given in (5.60) and (5.61), and the relations for Q_a and Q_b given in (5.59) are each satisfied. The shear stress distributions are therefore:

$$\sigma_{rz\,a} = \frac{(3Q_a - \tau h_a)^2}{3h_a(2Q_a - \tau h_a)} - \frac{3(2Q_a - \tau h_a)}{h_a^3}\left[z + \frac{\tau h_a^2}{6(2Q_a - \tau h_a)}\right]^2 \tag{5.98}$$

$$\sigma_{rz\,b} = \frac{(3Q_b - \tau h_b)^2}{3h_b(2Q_b - \tau h_b)} - \frac{3(2Q_b - \tau h_b)}{h_b^3}\left[z + \frac{\tau h_b^2}{6(2Q_b - \tau h_b)}\right]^2 \tag{5.99}$$

The shear stresses and the normal stresses in the joint between the laminae are found to be:

$$\tau = \frac{1}{(K_a + K_b)(1-\nu)}\sum_{n=1}^{\infty}(K_b b_n - K_a d_n)\lambda_n \sin\lambda_n r - K_a C_0 \frac{\mathrm{d}}{\mathrm{d}r}(\nabla^2 w) \tag{5.100}$$

$$\begin{aligned} p_j = {} & \frac{1}{(1-\nu)}\sum_{n=1}^{\infty}\left[e_n + \frac{h_a}{2(K_a + K_b)}(K_b b_n - K_a d_n)\right] \times \lambda_n\left(\frac{1}{r}\sin\lambda_n r \right. \\ & \left. + \lambda_n \cos\lambda_n r\right) - \frac{K_a C_0 h_a}{2}(\nabla^4 w) - \frac{D_a(1-\nu)}{r^2}\frac{\mathrm{d}}{\mathrm{d}r}\left[r\left(w'' + \frac{\nu}{r}w'\right)\right] \end{aligned} \tag{5.101}$$

Thus, the general formulation of the stresses in and the displacements of a circular laminated plate subjected to an axi-symmetric temperature distribution has been found. To solve any particular case of interest, the constants, C_5, C_7 and C_8 must be solved for the boundary conditions at $r > 0$.

This analysis is useful in determining the thermal stresses produced in portions of multi-layer structures which can be approximated by a laminated plate. To obtain the complete stress condition in such a structure, the stresses due to a lateral pressure distribution may be calculated, and superposition of the pressure induced stresses and thermal stresses is possible as long as the plate is everywhere in the elastic range and the deflections are small.

In the case of a transient temperature distribution this analysis will describe the thermal stress condition at a specified time. Thus, the quasi-steady state analysis can be used at several particular times during a transient heat input to portray the stresses and deformations at those times.

This analysis is also useful in the solution of a circular plate made of one material subjected to a temperature distribution in which the transverse temperature gradient is so large that the mechanical properties of the material vary significantly across the thickness. The plate may be broken into "laminae," average material properties assigned to each, and this analysis employed.

5.11 References

5.1. Den Hartog, J.P. (1952) *Advanced Strength of Materials*, McGraw-Hill Publishers.
5.2. Oyibo, G.A. and Brunelle, E.J. (1985) Vibrations of Circular Orthotropic Plates in Affine Space, Vol. 23, No. 2, February, pp. 296-300.

5.12 Problems

5.1. The circular flooring in a silo of radius a is solidly supported at the walls such that the floor plate is considered to be clamped. If grain is poured onto the floor such that the floor loading is triangular in cross section as shown in Figure 5.10, what is the expression for the deflection at the center of the floor?

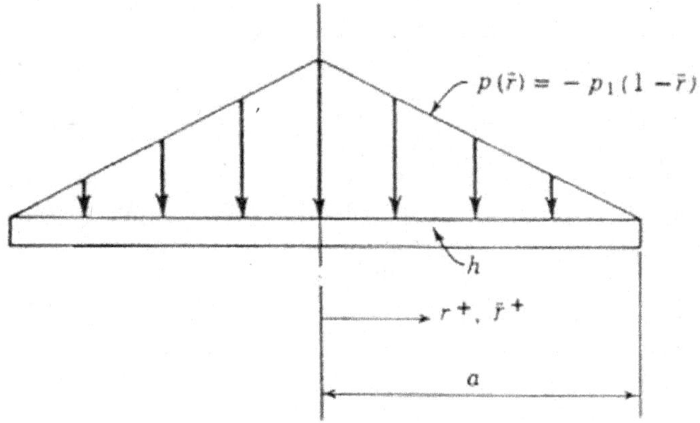

Figure 5.10. Circular plate with varying load.

5.2. Consider a circular plate of radius a, shown in Figure 5.11, loaded by an edge couple M_r M at the outer edge. Find the value of the stress couples M_r and M_θ throughout the plate.

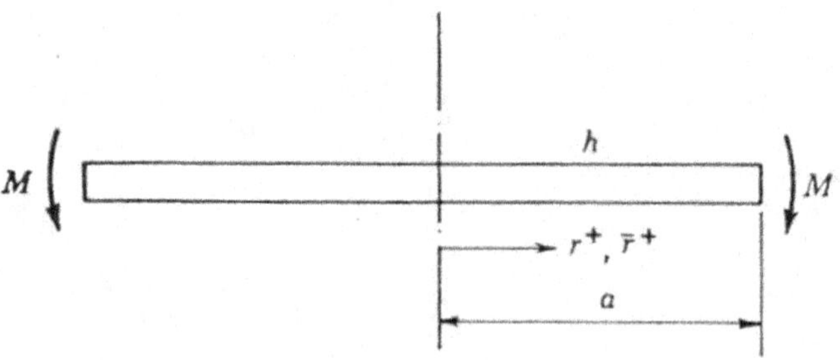

Figure 5.11. Circular plate with edge moment.

5.3. The flat head of a piston in an internal combustion engine is considered to be a plate of radius a, where the center support to the connecting rod is of radius b as shown in Figure 5.12. If the maximum down pressure is uniform of magnitude $p(r) = -p_0$, determine the location and magnitude of the maximum bending stress? Assume the head is clamped on both edges.

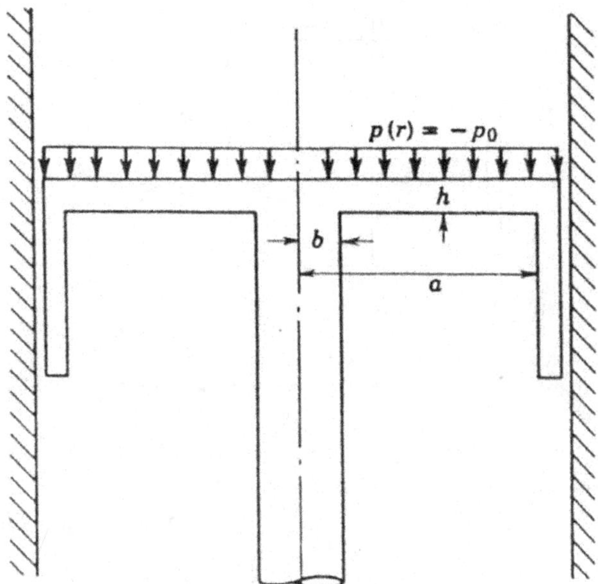

Figure 5.12. Piston in cylinder.

5.4. A certain pressure transducer is designed such that the pressure in the chamber deflects a thin circular plate, a rigid member joined to it at the center thus being deflected, and through a linkage mechanism shown in Figure 5.13, thus deflecting an arrow on a gage to the right. Determine the expression to relate the deflection of the gage to the pressure p_0 in the chamber. Assume that neither the rigid piece nor the linkages affect the deflection of the plate by their presence. Assume the circular plate is simply supported at the outer edge.

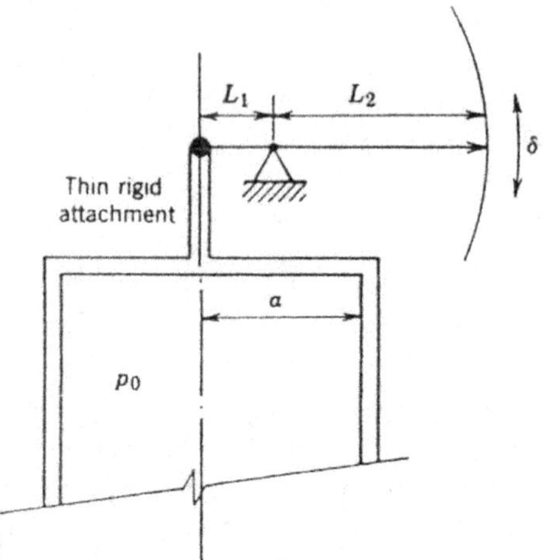

Figure 5.13. Pressure measurement device.

5.5. An underwater instrumentation canister, shown in Figure 5.14, is a cylinder with ends which are circular plates that can be considered clamped at the outer edge, $r = a$ or $\bar{r} = 1$. In order to design the ends, i.e., choose the thickness, h, for a given material system, one must determine the location and magnitude of the maximum stress when this canister is underwater with ambient pressure p_0. Assume that the cylindrical portion introduces no in-plane loads to the ends. Find the maximum radial stress. Also what is the maximum circumferential stress? What elastic properties of the material are involved in finding the maximum radial stress? The maximum circumferential stress?

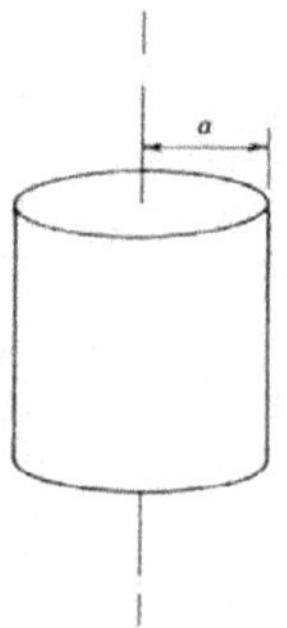

Figure 5.14. Underwater canister.

5.6. A flat circular plate roof is being designed to fit over an unused cave entrance. The outer radius is a, the thickness h, and the outer edge is considered clamped. If the weight density of the material used is ρ (lbs/in.^{3}), what is the maximum deflection, and the maximum stress in the plate due to gravity alone?

5.7. An air pump is constructed of a shaft, to which is clamped a disk of uniform thickness, at whose outer edge a soft gasket prevents air passage between the disk and the surrounding cylinder as shown in Figure 5.15.

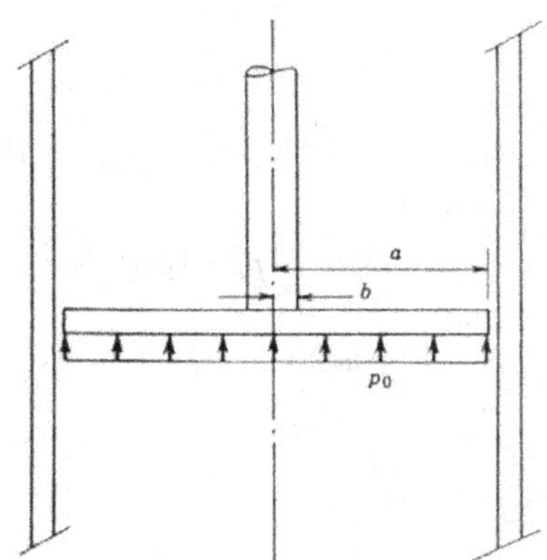

Figure 5.15. Air pump.

Assuming the disk is clamped on the shaft of radius b, and free at its outer edge of radius a, and a maximum differential pressure of p_0 (psi), what is the expression for the maximum radial stress? What is the value of the maximum shear resultant, Q_r (lbs/in.)? Could Q_r have been determined in another way?

5.8. In a chemical plant, a certain process involving high gaseous, pressures requires a blow-out diaphragm which will blow at 100 psi pressure in order that expensive equipment will not be damaged. The flat circular plate diaphragm is 10″ in radius and simply supported on its outer edge. Constructed of a brittle material with $\nu = 1/3$, and the ultimate tensile strength is 50,000 psi, what thickness is required to have the plate fracture when the lateral pressure reaches 100 psi?

5.9. A circular plate is used as a component in a pressure vessel. It has a 14″ diameter, is simply supported at the outer edge, and is composed of steel: $E = 30\times10^6$ psi, $\sigma_{\text{all}} = 40{,}000$ psi, $\nu = 0.3$. Using maximum principal stress theory, what thickness, h, is required for a pressure differential of 50 psi?

5.10. What is the maximum deflection for the solution of Problem 5.9 above?

5.11. What is the governing differential equation for the lateral deflection, w, for a circular plate subjected to a lateral pressure, $p(r)$, and a temperature distribution, $T(r, z)$?

5.12. A circular steel plate, used as a footing, rests on the ground and is subjected to a uniform lateral pressure, $p(r) = -p_0$ (psi). The ground deflects linearly with a spring constant of k (lbs/in./in.^{2}).

a. What is the governing differential equation for this problem?

b. What are the boundary conditions at the outer edge, $r = a$?

Hint: See Section 2.4.

5.13. In Problem 5.7, write explicitly the four boundary condition equations involving the constants *A*, *B*, *C* and *E*, etc. Do not bother to solve for the constants, however.

5.14. In Problem 5.8, the maximum radial and circumferential stresses are at the plate center and given by

$$\sigma_{r\max} = \sigma_{\theta\max} = \pm\frac{3}{8}(3+\nu)\frac{p_0 a^2}{h^2} \quad \text{at } z = \pm h/2.$$

If the plate is 10″ in radius, $\nu = 0.3$, $\boldsymbol{p_0 = 10}$ psi and the yield stress is 50,000 psi, what thickness is required if

a. the maximum principal stress failure theory is used?

b. the maximum distortion energy failure theory is used?

In each case assume the field is two dimensional, i.e., ignore σ_z, and refer to any text on structural mechanics to review the failure theories referred to.

CHAPTER 6

BUCKLING OF ISOTROPIC COLUMNS AND PLATES

6.1 Derivation of the Plate Governing Equations for Buckling

The governing equations for a thin plate subjected to both in-plane and lateral loads have been derived previously. In those equations, there was one governing equation describing the relationship between the lateral deflection and the laterally distributed loading,

$$D\nabla^4 w = p(x,y)$$

and other equations dealing with in-plane displacements, related to in-plane loads

$$\nabla^4 u_0 = \nabla^4 v_0 = 0.$$

As discussed previously, the equations involving lateral displacements and lateral loads is completely independent (uncoupled) from those involving the in-plane loadings and in-plane displacements.

However, it is true that when in-plane loads are compressive, upon attaining certain discrete values, these compressive loads do result in producing lateral displacements. Thus, there does occur a coupling between in-plane loads and lateral displacements, w. As a result, a more inclusive theory must be developed to account for this phenomenon, which is called *buckling* or *elastic instability*.

Unlike in developing the governing plate equations in Chapter 1, wherein the development began with the three dimensional equations of elasticity, the following shall begin with looking at the in-plane forces acting on a plate element, in which the forces are assumed to be functions of the midsurface coordinates x and y, as shown in Figure 6.1.

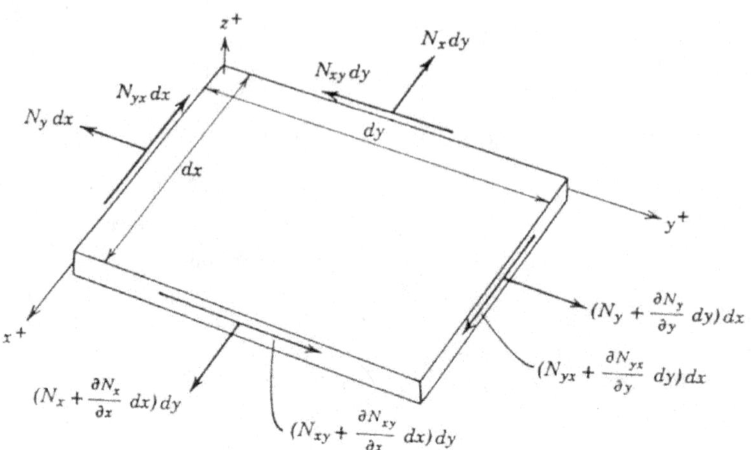

Figure 6.1. In-plane forces on a plate element.

Looking now at the plate element of Figure 6.2, viewed from the midsurface in the positive y direction, the relationship between forces and displacements is seen, when the plate is subjected to both lateral and in-plane forces, i.e., when there is a lateral deflection, w (note obviously that in the figure the deflection is exaggerated).

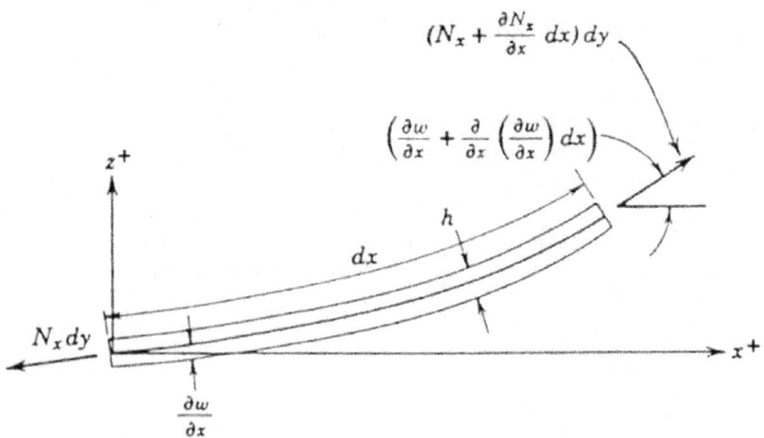

Figure 6.2. In-plane forces acting on a deflected plate element.

Hence, the z component of the N_x loading per unit area is, for small slopes (i.e., the sine of the angle equals the angle itself in radians):

$$\frac{1}{\mathrm{d}x\,\mathrm{d}y}\left[\left(N_x + \frac{\partial N_x}{\partial x}\mathrm{d}x\right)\mathrm{d}y\left(\frac{\partial w}{\partial x} + \frac{\partial^2 w}{\partial x^2}\mathrm{d}x\right) - N_x\,\mathrm{d}y\frac{\partial w}{\partial x}\right]$$

Neglecting terms of higher order, the force per unit planform area in the z direction is seen to be

$$N_x \frac{\partial^2 w}{\partial x^2} + \frac{\partial N_x}{\partial x}\frac{\partial w}{\partial x}. \tag{6.1}$$

Similarly, the z component of the N_y force per unit planform area is seen to be

$$N_y \frac{\partial^2 w}{\partial y^2} + \frac{\partial N_y}{\partial y}\frac{\partial w}{\partial y}. \tag{6.2}$$

Finally to investigate the z component of the in-plane resultants N_{xy} and N_{yx},

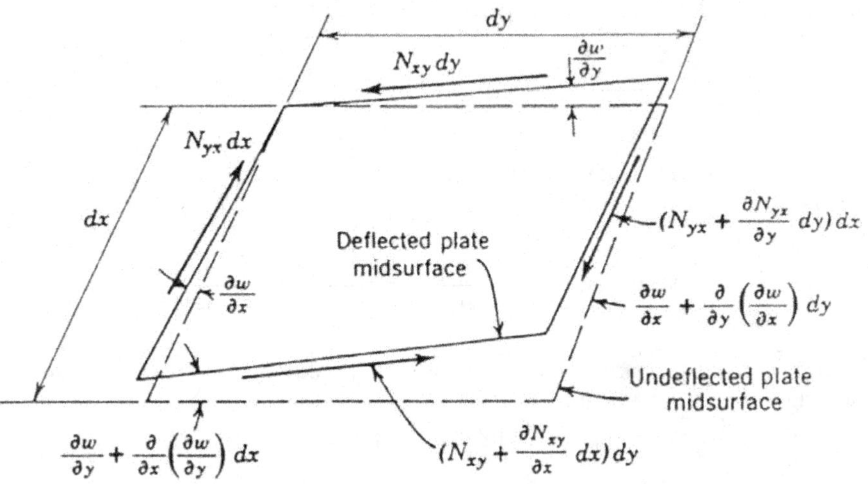

Figure 6.3. In-plane shear forces acting on a deflected plate element.

Hence, the z component per unit area of the in-plane shear resultant is:

$$\frac{1}{dx\,dy}\Bigg\{\left(N_{xy} + \frac{\partial N_{xy}}{\partial x}dx\right)\left(\frac{\partial w}{\partial y} + \frac{\partial^2 w}{\partial x\,\partial y}dx\right)dy$$
$$+\left(N_{yx} + \frac{\partial N_{yx}}{\partial y}dy\right)\left(\frac{\partial w}{\partial x} + \frac{\partial^2}{\partial x\,\partial y}dy\right)dx$$
$$-N_{xy}\frac{\partial w}{\partial y}dy - N_{yx}\frac{\partial w}{\partial x}dx\Bigg\}.$$

Neglecting higher order terms, this result in

$$N_{xy}\frac{\partial^2 w}{\partial x\,\partial y}+\frac{\partial N_{xy}}{\partial x}\frac{\partial w}{\partial y}+N_{yx}\frac{\partial^2 w}{\partial x\,\partial y}+\frac{\partial N_{yx}}{\partial y}\frac{\partial w}{\partial x}. \tag{6.3}$$

With all the above z components of forces per unit area, the governing plate equation can be modified to include the effect of these in-plane forces on the governing plate equations.

$$\begin{aligned} D\nabla^4 w = {} & p(x,y)+N_x\frac{\partial^2 w}{\partial x^2}+N_y\frac{\partial^2 w}{\partial y^2}+2N_{xy}\frac{\partial^2 w}{\partial x\,\partial y} \\ & +\frac{\partial N_x}{\partial x}\frac{\partial w}{\partial x}+\frac{\partial N_y}{\partial y}\frac{\partial w}{\partial y}+\frac{\partial N_{xy}}{\partial x}\frac{\partial w}{\partial y}+\frac{\partial N_{yx}}{\partial y}\frac{\partial w}{\partial x}. \end{aligned} \tag{6.4}$$

However, from in-plane force equilibrium, it is remembered from Equations (2.17) and (2.18), assuming no applied surface shear stresses, that

$$\frac{\partial N_x}{\partial x}+\frac{\partial N_{yx}}{\partial y}=0, \qquad \frac{\partial N_{xy}}{\partial x}+\frac{\partial N_y}{\partial y}=0 \qquad (6.5), (6.6)$$

Substituting these into the expression above, the final form of the equation is found to be:

$$D\nabla^4 w = p(x,y)+N_x\frac{\partial^2 w}{\partial x^2}+N_y\frac{\partial^2 w}{\partial y^2}+2N_{xy}\frac{\partial^2 w}{\partial x\,\partial y}. \tag{6.7}$$

Likewise, this governing plate equation can be reduced to the governing equation for a beam column by multiplying (6.7) by b (the width of the beam) and letting $\partial(\ )/\partial y = 0$, $v = 0$, $\overline{P} = -bN_x$ and $q(x) = bp(x)$, to provide

$$\frac{d^4 w}{dx^4}+k^2\frac{d^2 w}{dx^2}=\frac{q(x)}{EI} \quad \text{where } k^2 = \overline{P}/EI. \tag{6.8}$$

It should be noted that the load $\overline{P}$ defined above is an in-plane load which when positive produces compressive stresses, which differs from the convention used elsewhere throughout this text. However, it is commonly used in the literature on buckling, is convenient, so herein is described as a barred quantity.

6.2 Buckling of Columns Simply Supported at Each End

Solving Equation (6.8) by methods described previously, the solution can be written as:

$$w(x) = A\cos kx + B\sin kx + C + Ex + w_p(x) \tag{6.9}$$

where $w_p(x)$ is the particular solution for the loading $q(x)$. Consider, for example, the case wherein $q(x) = 0$, and the column is simply supported at each end. The boundary conditions, at $x = 0, L$, are then

$$w(0) = w(L) = 0$$
$$M_x\begin{pmatrix} L \\ 0 \end{pmatrix} = -\mathrm{EI}\frac{\mathrm{d}^2 w}{\mathrm{d}x^2} = 0 \text{ or } \frac{\mathrm{d}^2 w(0)}{\mathrm{d}x^2} = \frac{\mathrm{d}^2 w(L)}{\mathrm{d}x^2} = 0. \tag{6.10}$$

From the first boundary condition $A + C = 0$, from the third $A = 0$; hence, $C = 0$ also. From the second boundary condition $B\sin kL + EL = 0$, and from the fourth boundary condition

$$Bk^2 \sin kL = 0 = \frac{B\overline{P}}{\mathrm{EI}}\sin kL = 0. \tag{6.11}$$

Note that in Equation (6.11) when $kL \neq n\pi$, then $B = E = 0$; when $kL = n\pi$, then $E = 0$, $B \neq 0$ but is indeterminate and

$$\overline{P} = n^2\pi^2\frac{\mathrm{EI}}{L^2}. \tag{6.12}$$

It is thus seen that for most values of $\overline{P}$, the axial compressive loading, the lateral deflection w is zero ($A = B = C = E = 0$), and the in-plane and lateral forces and responses are uncoupled. However, for a countable infinity of discrete values of P, there is a lateral deflection, but it is of an indeterminate magnitude. Mathematically, this is referred to as an *eigenvalue problem* and the discrete values given in (6.12) are called *eigenvalues*. The resulting deflections, in this case, are

$$w(x) = B\sin kx$$

and are called *eigenfunctions*.

The natural vibration of elastic bodies are also eigenvalue problems, where in that case the natural frequencies are the eigenvalues and the vibration modes are the eigenfunctions. This is treated in the next chapter.

As to buckling, looking at Equation (6.12), as $\overline{P}$ increases, it is clear that the lowest buckling load occurs when $n = 1$, and at that particular load, the column will either inelastically deform and strain harden, or the column will fracture. Hence, $n > 1$ has no physical significance. The load

$$\overline{P} = \pi^2 \frac{\text{EI}}{L^2} \tag{6.13}$$

is therefore the critical buckling load for this column for these boundary conditions. In this particular case the buckling load is called the Euler buckling load, since the Swiss mathematician was the first to solve the problem successfully.

Another way to phrase the buckling problem is exemplified by solving Equation (6.8), letting $q(x) = q_0 =$ constant. The resulting particular solution, in this case, is $q_0 x^2/2P$. If the column is simply supported, solving the boundary value problem for the lateral deflection, results in

$$w(x) = \frac{q_0}{Pk^2 \sin kL}\left[\cos kx \sin kL - \cos kL - Lx \sin kL + k^2x^2 \sin kL\right]. \tag{6.14}$$

In Equation (6.14), the solution of a boundary value problem, when the axial load $\overline{P}$ has values given in (6.12) wherein $\sin kL = 0$, then $w(x)$ goes to infinity, or, more properly, since we have a small deflection linear mathematical model, $w(x)$ becomes indefinitely large.

Hence, whether we solve for the homogeneous solution of Equation (6.8), resulting in an eigenvalue problem, or we solve the nonhomogeneous Equation (6.8), resulting in a boundary value problem, the results are identical, when $\overline{P}$ has values given by (6.12), or physically where $\overline{P}$ attains the value given by (6.13), the column 'buckles'.

Note also that the buckling load, Equation (6.13), is not affected by any lateral load $q(x)$. The physical significance of a lateral load $q(x)$, however, is that the beam-column may deflect sufficiently, due to both the lateral and in-plane compressive loads, that the resulting curvature would cause bending stresses which in addition to the compressive stresses may fracture or yield the column at a load less than or prior to attaining the buckling load.

These elastic stability considerations are very important in analyzing or designing any structure in which compressive stresses result from the loading, because in addition to insuring that the structure is not merely overstressed or overdeflected, in this case a new failure mode has been added, i.e., buckling.

6.3 Column Buckling with Other Boundary Conditions

From the previous section, the critical compressive buckling load $\overline{P}_{cr}$ is given as

$$\overline{P}_{cr} = \pi^2 \frac{\mathrm{EI}}{L^2} \tag{6.15}$$

Numerous other texts derive critical buckling loads for columns with other boundary conditions, [6.1] through [6.4], and [12.2].

For ease of use in analysis and design, but without derivations, the following column buckling equations are listed for the other classical boundary conditions.

Column with both ends clamped

$$\overline{P}_{cr} = 4\pi^2 \frac{\mathrm{EI}}{L^2}. \tag{6.16}$$

Column with one end clamped and the other simply supported

$$\overline{P}_{cr} = \frac{\pi^2 \mathrm{EI}}{(0.669L)^2}. \tag{6.17}$$

Column with one end clamped and the other end free

$$\overline{P}_{cr} = \frac{\pi^2 \mathrm{EI}}{4L^2}. \tag{6.18}$$

6.4 Buckling of Isotropic Rectangular Plates Simply Supported on All Four Edges

Plate buckling qualitatively is analogous to column buckling, except that the mathematics is more complicated, and the conditions that result in the lowest eigenvalue (the actual buckling load) are not so lucid in many cases.

Whenever the in-plane forces are compressive, and are more than a few percent of the plate buckling loads (to be defined later), Equation (6.7) must be used rather than Equation (3.1) in the analysis of plates.

For the plate, just as the case of the beam-column, since the in-plane load that causes an elastic stability is not dependent upon a lateral load, to investigate the elastic stability we shall assume $p(x, y) = 0$ in Equation (6.7).

Consider, as an example, a simply supported plate subjected to constant in-plane loads N_x and N_y (let $N_{xy} = 0$), as shown in Figure 6.4.

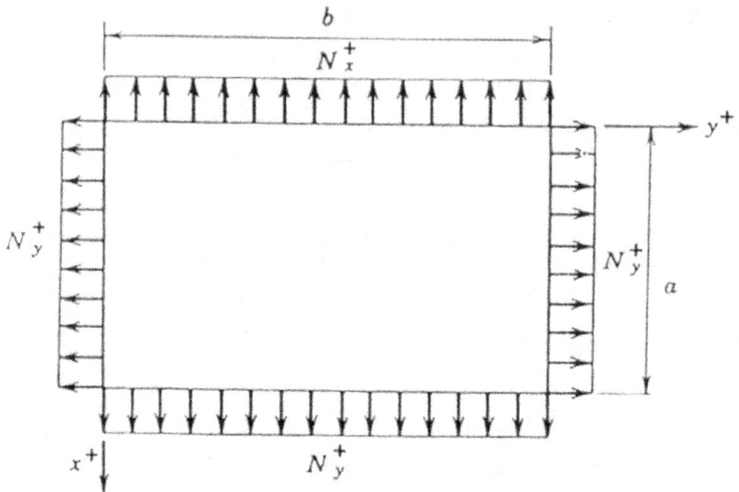

Figure 6.4. Rectangular plate subjected to in-plane loads.

Assume the solution of Equation (6.7) to be of the Navier form

$$w(x, y) = \sum_{m=1}^{\infty} \sum_{n=1}^{\infty} A_{mn} \sin\frac{m\pi x}{a} \sin\frac{n\pi y}{b}. \tag{6.19}$$

Substituting (6.19) into (6.7), it is convenient to define α here to be

$$\alpha = N_y / N_x . \tag{6.20}$$

The solution to the eigenvalue problem is found to be

$$N_{x\,\mathrm{cr}} = -D\pi^2 \frac{\left[\left(\frac{m}{a}\right)^2 + \left(\frac{n}{b}\right)^2\right]^2}{\left[\left(\frac{m}{a}\right)^2 + \left(\frac{n}{b}\right)^2 \alpha\right]}. \tag{6.21}$$

Here the subscript cr denotes that this is a critical load situation – the plate buckles. Also note that in (6.21) N_x is negative, i.e., a load that causes compressive stresses.

Equation (6.21) is the complete set of eigenvalues for the simply supported plate, analogous to Equation (6.12) for the column. In other words for these discrete values of N_x and N_y, Equation (6.7) has nontrivial solutions wherein the lateral deflection is given by (6.19); for other values $w(x, y) = 0$.

Since we know that as the load increases, the plate will buckle at the lowest buckling load (or eigenvalue) and all the rest of the eigenvalues have no physical meaning. So it is necessary to determine what values of the integers m and n (the number of half sine waves) make N_x a minimum.

Defining the length to width ratio of the plate to be $r = a/b$ Equation (6.21) can be rewritten as

$$N_x = -\frac{D\pi^2}{a^2}\frac{[m^2 + n^2r^2]^2}{[m^2 + n^2r^2\alpha]}. \tag{6.22}$$

Note if in Equation (6.22) $\alpha = 0$, $r = 1$ and $m = n = 1$, then

$$N_x = -\frac{4\pi^2 D}{a^2}. \tag{6.23}$$

Note the similarities between Equations (6.23) and (6.13).

The question remains; given a combination of N_x and N_y loadings, and a given

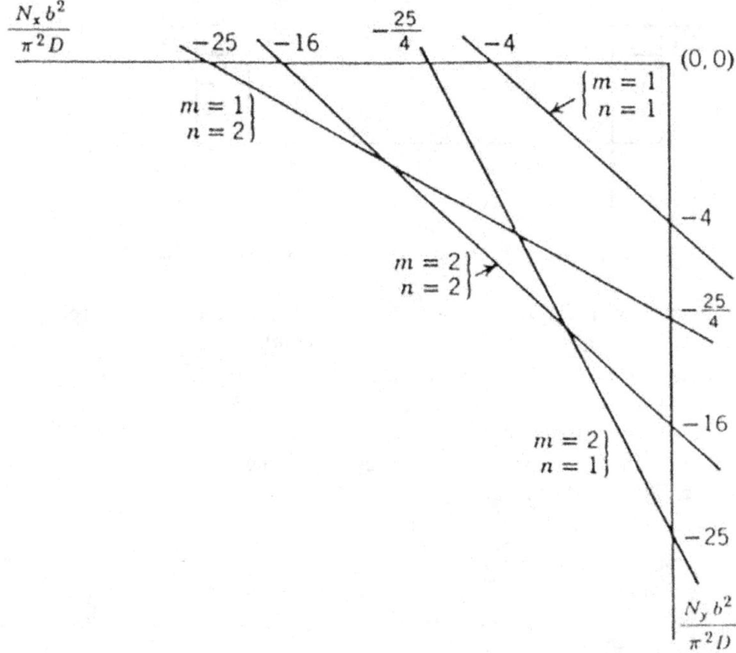

Figure 6.5. Values of biaxial loads causing buckling for square simply supported isotropic plate.

geometry r, what values of m and n provide the lowest buckling loads. One can make a plot such as Figure 6.5 above from manipulating Equation (6.22) (which is not shown to scale) for a square plate ($a = b$, $r = 1$).

It is seen from Figure 6.5 that for such a square plate, simply supported on all four edges, the plate will always buckle into a half sine wave ($m = n = 1$) under any combination of N_x and/or N_y, since that line is always closest to the origin, hence, the lowest buckling load situation.

Next consider a plate under an in-plane load in the x direction only, so $N_y = 0$, and $\alpha = 0$. In this case, Equation (6.21) can be written as

$$N_{x\,cr} = -\frac{D\pi^2 a^2}{m^2}\left[\frac{m^2}{a^2} + \frac{n^2}{b^2}\right]^2. \tag{6.24}$$

The loaded plate is shown in Figure 6.6.

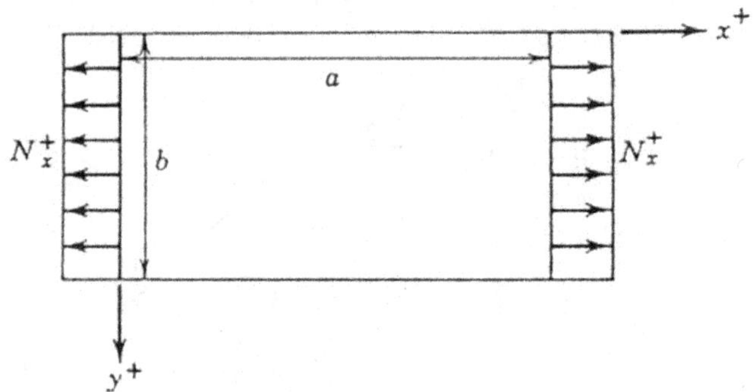

Figure 6.6. Plate subjected to in-plane load in the x direction.

Examination of Equation (6.24) shows that the first term is merely the Euler column load (6.13) for a column of unit width, including Poisson ratio effects. The second term clearly shows the buckle resisting effect providing by the simply supported side edges, and this effect diminishes as the plate gets wider, i.e., as b increases. In fact as $b \to \infty$, (6.24) shows that the plate acts merely as an infinity of unit width beams, simply supported at the ends, and because they are 'joined together', the Poisson ratio effect occurs, i.e., D instead of EI appears.

It is obvious from Equation (6.24) that the minimum values of N_x occurs when $n = 1$, since n appears only in the numerator. Thus for an isotropic plate, simply supported on all four edges, subjected only to an uniaxial in-plane load the buckling mode given by (6.19) will always be one half sine wave [sin(y/b)] across the span, regardless of the length or width of the plate.

Thus, since $n = 1$, Equation (6.24) can be written as

$$N_{x\,cr} = -\frac{D\pi^2}{b^2}\left(\frac{m}{r}+\frac{r}{m}\right)^2 \tag{6.25}$$

where it is remembered that $r = a/b$, termed the aspect ratio.

Now if $a < b$ (the plate wider than it is long), the second term is always less than the first, hence, the minimum value of N_x is always obtained by letting $m = 1$. Hence for $a \le b$, the buckling mode for the simply supported plate is always

$$w(x,y) = A_{11}\sin\left(\frac{\pi x}{a}\right)\sin\left(\frac{\pi y}{b}\right). \tag{6.26}$$

In that case,

$$N_{x\,cr} = -\frac{D\pi^2}{b^2}\left(\frac{1}{r}+r\right)^2. \tag{6.27}$$

To find out at what aspect ratio r, that N_x is truly a minimum, let

$$\frac{dN_{x\,cr}}{dr} = 0 = -\frac{2D\pi^2}{b^2}\left(\frac{1}{r}+r\right)\left(-\frac{1}{r^2}+1\right).$$

Therefore $r = 1$ provides that minimum value. Hence for $m = 1$, N_x is a minimum when $a = b$. Under that condition, from (6.27)

$$N_{x\,cr\,a=b} = -\frac{4D\pi^2}{b^2} = -\frac{4D\pi^2}{a^2}. \tag{6.28}$$

Comparing this with the Euler buckling load of (6.13) for a simply supported column, it is seen that the continuity of a plate and the support along the sides of the plate provide a factor of at least 4 over the buckling of a series of strips (columns) that are neither continuous nor supported along the unloaded edges.

Now as the length to width ratio increases, as a/b increases, the buckling load (6.27) will increase, and one can ask, will $m = 1$ always result in a minimum buckling load, or is there another value of m which will provide a lower buckling load as r increases (i.e., $\boldsymbol{N_{xcr}}$ **(m = 2)** $\le$ $\boldsymbol{N_{xcr}}$ **(m = 1)** for some value of r?)

Mathematically, this can be phrased as the following, using (6.25):

$$\left(\frac{m}{r}+\frac{r}{m}\right)^2 \overset{?}{\le} \left(\frac{m-1}{r}+\frac{r}{m-1}\right)^2.$$

This states the condition under which the plate of aspect ratio r will buckle in m half sine waves in the loaded direction rather than $m-1$ sine waves. Manipulating this inequality results in

$$m(m-1) \le r^2. \tag{6.29}$$

Equation (6.29) states that the plate will buckle in two half sine waves in the axial direction rather than one when $r \ge \sqrt{2}$. The plate will buckle in three half sine waves in the axial direction rather than 2, when $r \ge \sqrt{6}$, etc.

Again one can ask that when the plate buckles into $m = 2$ configuration, does a minimum buckling load occur, if so at what r and what is $N_{x\,cr\,(\min)}$?

From Equation (6.25)

$$\frac{dN_{x\,cr}}{dr}(m=2)=0=-\frac{2D\pi^2}{b^2}\left(\frac{2}{r}+\frac{r}{2}\right)\left(-\frac{2}{r^2}+\frac{1}{2}\right)=0$$

or $r^2 = 4$, $r = 2$.

$$N_{x\,cr\,(\min)} = -\frac{4D\pi^2}{b^2} \quad \text{for } m=2. \tag{6.30}$$

This is the same value as is given in Equation (6.28) for $m = 1$. Proceeding with all values of r and m, the following graph can be drawn, which clearly shows the results (Figure 6.7).

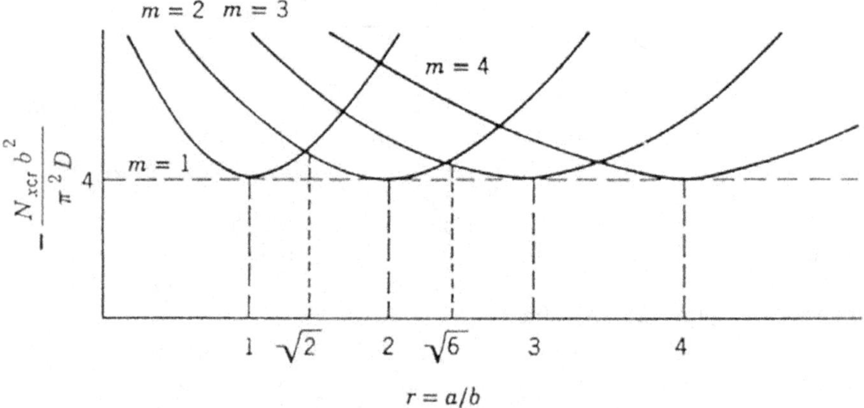

Figure 6.7. Buckling load as a function of aspect ratio for a simply supported isotropic plate.

Hence knowing the value of r, the figure provides the actual value of N_x and the corresponding value of the wave number m in the load direction. However, in practice for $r > 1$, universally one simply uses Equation (6.28) or (6.30) for the buckling load.

However, looking more closely at Equation (6.29), as m increases we see

$$m(m-1) \rightarrow m^2 = r^2 \quad \text{or} \quad m = r = a/b.$$

This means that for long plates, the number of half sine waves of the buckles have lengths approximately equal to the plate width. Another way to stating it is that a long plate simply supported on all four edges and subjected to a uniaxial compressive load attempts to buckle into a number of square cells.

Remembering that $\sigma_x = N_x/h$, Equation (6.28) or (6.30) can be written as the following for $a/b \geq 1$,

$$\sigma_{cr} = -\frac{\pi^2 E}{3(1-\nu^2)}\left(\frac{h}{b}\right)^2. \tag{6.31}$$

6.5 Buckling of Isotropic Plates with Other Loads and Boundary Conditions

The solution to the buckling of flat isotropic plates simply supported on all four sides subjected to uniaxial uniform compressive in-plane loads has been treated in detail. However, for many other boundary conditions, simple displacement functions like Equation (6.19) do not exist, and in some cases analytical, exact solutions analogous to Equations (6.21) and (6.31) have not been found. In those cases approximate solutions have been found using energy methods, which will be discussed in Chapters 8 and 9. These have been catalogued by Gerard and Becker [6.3] among others, and are presented in Figure 6.8 and k_c, given in the following equations:

$$\sigma_{x\,cr} = -\frac{k_c \pi^2 E}{12(1-\nu^2)}\left(\frac{h}{b}\right)^2; \quad N_{x\,cr} = -\frac{k_c \pi^2 D}{b^2} \tag{6.32}$$

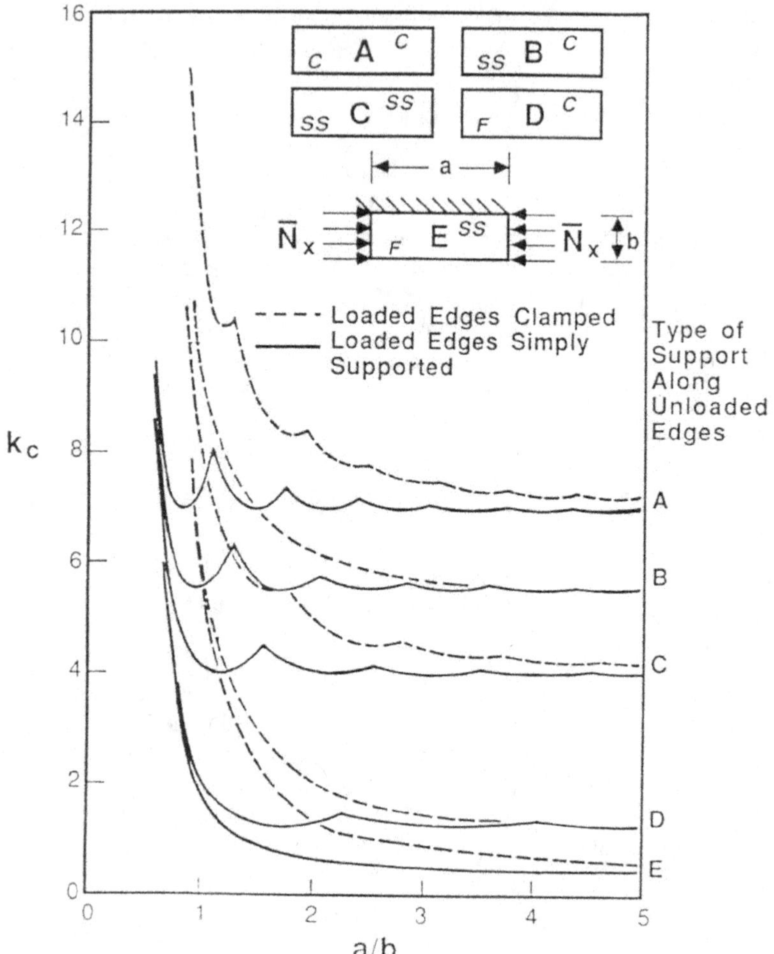

Figure 6.8. Compressive-buckling coefficients for flat rectangular isotropic plates.

In many practical applications, the edge rotational restraints lie somewhere between fully clamped and simply supported along the unloaded edges. For the case of the loaded edges simply supported, the buckling coefficient, k_c, of Equation (6.32) are also given by Gerard and Becker [6.3] as shown in Figure 6.9. The unloaded edge restraint, ε, is zero for simply supported edges and infinity for full clamping. Values in between these extremes require engineering judgment.

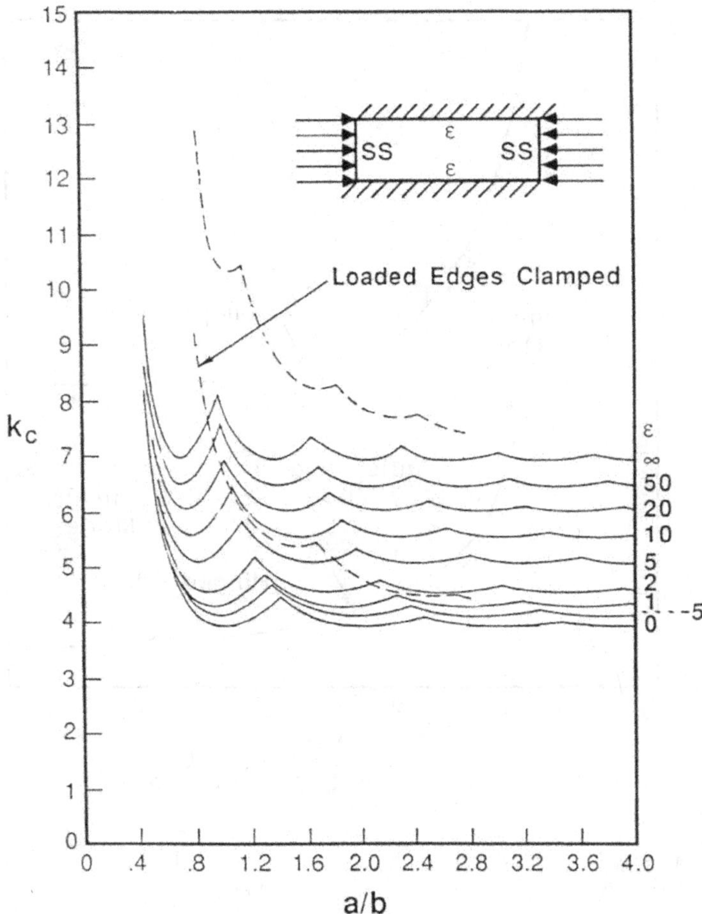

Figure 6.9. Compressive-buckling-stress coefficient of isotropic plates as a function of *a/b* for various amounts of edge rotational restraint.

For in-plane shear loading, the critical shear stress is given by the following equations:

$$\tau_{cr} = \frac{K_s \pi^2 E}{12(1-\nu^2)}\left(\frac{h}{b}\right)^2; \quad N_{xy_{cr}} = \frac{K_s \pi^2 D}{b^2} \tag{6.33}$$

where K_s is given in Figure 6.10 for various boundary conditions [6.3].

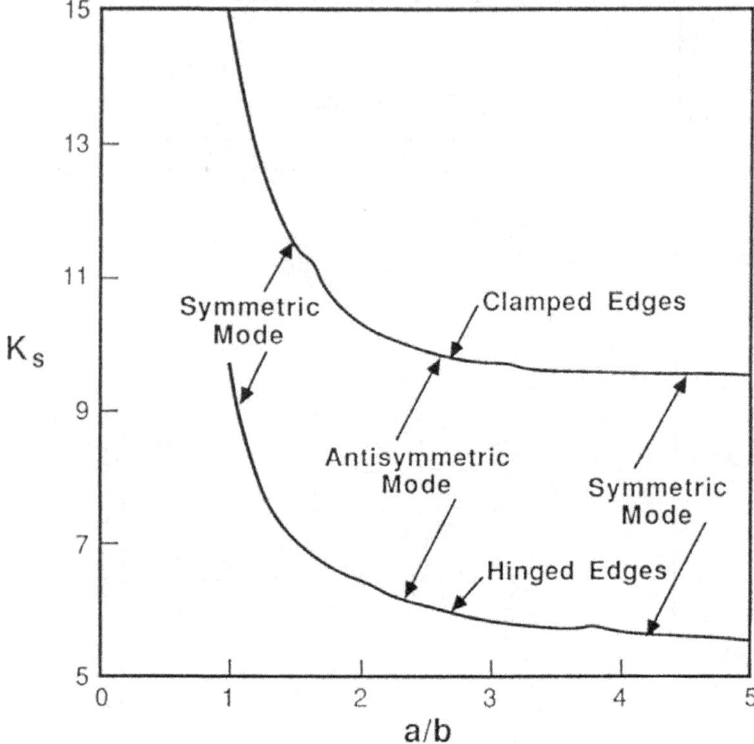

Figure 6.10. Shear-buckling-stress coefficient of isotropic plates as a function of a/b for clamped and hinged edges.

For rectangular plates subjected to in-plane bending loads, the following equation is used to determine the stress value for the buckling of the plate shown in Figure 6.11.

$$\sigma_B \quad \frac{k_b \pi^2 E}{12(1-\nu^2)}\left(\frac{h}{b}\right)^2 \tag{6.34}$$

where again ε is the value of the edge constraint as discussed previously.

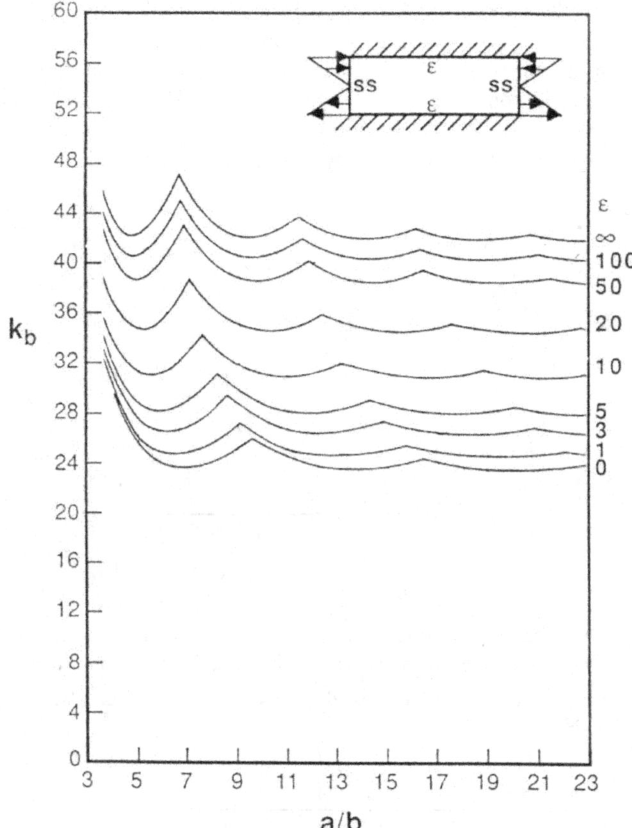

Figure 6.11. Bending-buckling coefficient of isotropic plates as a function of a/b for various amounts of edge rotational restraint.

6.6 The Buckling of an Isotropic Plate on an Elastic Foundation Subjected to Biaxial In-Plane Compressive Loads

It is important to consider that besides overall buckling of the entire plate, it is possible that a sandwich face plate may buckle, due to loads applied to the face. In this case the plate can be considered to be supported on a uniform elastic foundation, namely the core. In such a case the buckling equation for this phenomenon is

$$D\nabla^4 w + kw + \overline{N}_x \frac{\partial^2 w}{\partial x^2} + \overline{N}_y \frac{\partial^2 w}{\partial y^2} = 0 \tag{6.35}$$

where D is the flexural stiffness of the face plate, w is the lateral displacement of the face plate, k is the foundation modulus in force/unit area/unit deflection, and $\overline{N}_x, \overline{N}_y$ are the compressive loads per unit width in the subscripted direction (i.e., $\overline{N}_x = -N_x$, etc.) acting on that particular face plate.

Considering this localized buckling phenomenon, is has been found that the plate boundary conditions at the outer plate edges do not affect the buckling load. Therefore, for analytical simplicity, assume simply-supported edges on all four sides. Therefore, the Navier approach may be used for the solution, with the lateral deflection assumed to be

$$w(x,y) = A_{mn}\sin\left(\frac{m\pi x}{a}\right)\sin\left(\frac{n\pi y}{b}\right) \tag{6.36}$$

where A_{mn} is the deflection amplitude, a is the plate dimension in the x-direction, and b is the plate dimension in the y-direction.

For simplification, let $\phi = \overline{N}_y/\overline{N}_x$ and $r = a/b$. Substituting Equation (6.36) into (6.35) and using the above

$$\overline{N}_{x\,\mathrm{cr}} = \frac{(\pi^4 D/a^4)(m^2+n^2r^2)^2+k}{(\pi^2/a^2)(m^2+n^2r^2\phi)} \tag{6.37}$$

If $\phi = 1$, and $r = 1$, then the response is independent of direction. When the in-plane loads are caused by the cooling of a sandwich plate wherein the coefficients of thermal expansion between face and core cause the face to be compressed, then $\phi = 1$. Further because the buckling is a localized phenomenon, one can let $r = 1$. Then Equation (6.37) may be written as

$$\overline{N}_{x\,\mathrm{cr}} = \frac{(\pi^4 D/a^2)(m^2+n^2r^2)^2+ka^2}{\pi^2(m^2+n^2)} \tag{6.38}$$

First it is seen that the minimum value of $\overline{N}_x$ will occur when $m = n = 1$, therefore

$$\overline{N}_x = \frac{(4\pi^4 D/a^2)+ka^2}{2\pi^2} \tag{6.39}$$

To find the dimension a resulting in a minimum value of $\overline{N}_{x\,\mathrm{cr}}$, set $\partial\overline{N}_{x\,\mathrm{cr}}/\partial a = 0$, with the result that

$$a = 2^{1/2}\pi\left(\frac{D}{k}\right)^{\frac{1}{4}} \tag{6.40}$$

This is the half wavelength of the buckle that will occur, and it can be determined that this is a localized buckle in a reasonably sized face plate. Substituting Equation (6.40) into (6.39) results in

$$\overline{N}_{x\,\mathrm{cr}} = 2(kD)^{1/2}$$

As defined, $\overline{N}_{x\,\mathrm{cr}}$ is a compressive force per unit width and equal to $\overline{N}_{y\,\mathrm{cr}}$, since $\phi = 1$, or in the usual notation, where $\overline{N}_i = -N_i$,

$$N_{x\,\mathrm{cr}} = N_{y\,\mathrm{cr}} = -2(kD)^{1/2} \tag{6.41}$$

The buckling stress in the face plate is therefore

$$\sigma_{\mathrm{cr}} = \frac{-2}{h}(kD)^{1/2} \tag{6.42}$$

It has been found that in the fabrication of some sandwich plates, because of the cooling down subsequent to joining, the faces to the core in a rolling operation, differential thermal contractions caused sufficiently high compressive stresses in the faces to cause thermal buckling of the sandwich faces.

6.7 References

6.1. Timoshenko, S. and Gere, J. (1961) *Elastic Stability*, McGraw-Hill Book Co., Inc., 2nd Edition.

6.2. Bleich, H.H. (1952) *Buckling of Metal Structures*, McGraw-Hill Book Co., Inc.

6.3. Gerard, G. and Becker, H. (1957) *Handbook of Structural Stability, Part 1 – Buckling of Flat Plates*, NACA TN 3781.

6.4. Jones, R.M. (2004) *Buckling of Bars, Plates and Shells*, R.T. Edwards, Flourtown, PA.

6.8 Problems

6.1. In a plate clamped on all four edges, $\nu = 0.25$ and loaded in the x direction the critical buckling stress is given by (from Reference 7.1)

$$\sigma_{\mathrm{cr}} = \frac{k\pi^2 D}{b^2 h} = \frac{k\pi^2 E}{12(1-\nu^2)}\left(\frac{h}{b}\right)^2$$

where D is the flexural stiffness, b is the plate width, a is the plate length, and h is the plate thickness. k_c is given by

a/b	0.75	1.0	1.5	2.0	2.5	3.0
k	11.69	10.07	8.33	7.88	7.57	7.37

(a) Part of a support fixture for a missile launcher measure $45'' \times 15''$, and must support 145,000 lbs in axially compressive load. Its edges are all clamped. If the plate is composed of aluminum with $\boldsymbol{E} = \mathbf{10} \times \mathbf{10^6}$ psi, allowable 30,000 psi (both the tensile and compressive allowable stress is of magnitude 30,000 psi) and $\nu = 0.25$. What thickness is required to prevent buckling? What thickness is required to prevent overstressing?

(b) Suppose a steel plate of the same dimensions were used instead of the aluminum with the following properties: E_{steel} □ 30×10^6 psi, $\nu = 0.25$ and $\sigma_{\text{allowable}} = \pm 100{,}000$ psi. What thickness is needed to prevent buckling? Will the steel plate be overstressed?

(c) The density of steel is $0.283\ \text{lbs/in}^3$, the density of aluminum is $0.100\ \text{lbs/in}^3$. Which plate will be lighter?

6.2. A structural component in the interior of an underwater structure consists of a square plate of dimension a, simply supported on all four sides. If the component is subjected to in-plane compressive loads in both the x and y directions of equal magnitude, find N_x.

6.3. An aluminum support structure consists of a rectangular plate simply supported on all four edges is subjected to an in-plane uniaxial compressive load. If the length of the plate in the load direction is 4 feet, the width 3 feet, determine the minimum plate thickness to insure that the plate would buckling in the elastic range, if the material properties are $\boldsymbol{E} = \mathbf{10} \times \mathbf{10^6}$ psi, $\nu = 0.3$ and the compressive yield stress, $\sigma_y = 30{,}000$ psi.

6.4. A rectangular plate 4 feet $\times$ 2 feet is subjected to an in-plane compressive load N_x in the longer direction as shown in Figure 7.6. How much weight of plate can be saved by using a plate clamped on all four edges rather than having the plate simply supported on all four edges to resist the same compressive load $N_{x\,\text{cr}}$? Express the answer as a percentage.

6.5. An aluminum plate measure 6 feet $\times$ 3 feet, of thickness 0.1 inch is clamped on all four edges. Use the material properties in Problem 6.3 above.

(a) If it is subjected to a compressive in-plane load in the longer direction, what is the buckling stress?

(b) How much higher is the buckling stress compared to the same plate simply supported on all four edges?

CHAPTER 7

VIBRATIONS OF ISOTROPIC BEAMS AND PLATES

7.1 Introduction

Through the previous chapter, the static behavior of beams, rods, columns and plates has been treated to determine displacements, stresses, and buckling loads. This is important because many structures are stiffness critical (maximum deflections are limited) or strength critical (maximum stresses are limited). In Chapter 6, the elastic stability of these structures was treated because that is a third way in which structures can be rendered useless. In most cases when a structure becomes elastically unstable, it cannot fulfill its structural purpose.

In this chapter, the vibration of beams and plates is studied in some detail. Many textbooks have been written dealing with this subject, but here, only an introduction is made to show how one approaches and deals with such problems.

In linear vibrations, both natural vibration and forced vibrations are important. The former deals with natural characteristic of any elastic body, and these natural vibrations occur at discrete frequencies, depending on the geometry and material systems only. Such problems (like buckling) are *eigenvalue* problems, the natural frequencies are the *eigenvalues*, and the displacement field associated with each natural frequency are the *eigenfunctions*. One remembers that in a simple spring-mass system, there is one natural frequency and mode shape; in a system of two springs and two masses, there are two natural frequencies and two mode shapes. In a continuous elastic system, theoretically there are an infinite number of natural frequencies, and a mode shape associated with each.

Forced vibrations occur when an elastic body is subjected to a time dependent force or forces. In that case the response to the forced vibrations can be viewed as a linear superposition of all the eigenfunctions (vibrations modes), each with an amplitude determined by the form of the forcing function. In forced vibrations, the forces can by cyclic (harmonic vibration) or non-cyclic, including shock loads (those which occur over very small times).

7.2 Natural Vibrations of Beams

Consider again the beam flexure equation discussed previously.

$$\mathrm{EI}\frac{\mathrm{d}^4 w}{\mathrm{d}x^4} = q(x). \tag{7.1}$$

It is seen that the forcing function $q(x)$ is written in terms of force per unit length. Using d'Alembert's Principle for vibration, an inertial term can be written which is the mass times the acceleration per unit length. Also the forcing function can be a function of time, and of course the lateral deflections will be a function of both spatial and temporal coordinates. The result is that (7.1) becomes, for the flexural vibration,

$$\mathrm{EI}\frac{\partial^4 w}{\partial x^4} = q(x,t) - \rho_m A\frac{\partial^2 w}{\partial t^2}. \tag{7.2}$$

In the above ρ_m is the *mass* density of the beam material, and A is the beam cross-sectional area, both of which are taken here as constants for simplicity.

As stated previously, natural vibrations are functions of the beam material properties and geometry only, and are inherent properties of the elastic body – independent of any load. Thus, for natural vibrations, $q(x, t)$ is set equal to zero, and (7.2) becomes

$$\mathrm{EI}\frac{\partial^4 w}{\partial x^4} + \rho_m A\frac{\partial^2 w}{\partial t^2} = 0. \tag{7.3}$$

To solve this equation to obtain $w(x, t)$, in general, one can assume $w(x, t) = X(x)T(t)$, a separable solution, use separation of variables to obtain a spatial function $X(x)$ which satisfies all of the boundary conditions, and an harmonic function for $T(t)$, and thus arrive at a characteristic set of variables to satisfy (7.3) and its boundary conditions. In that process the natural frequencies and mode shapes are determined.

By way of a specific example, consider the beam to be simply supported at each end. Then the spatial function is a sine function such that

$$w(x,t) = \sum_{n=1}^{\infty} A_n \sin\frac{n\pi x}{L}\sin\omega_n t \tag{7.4}$$

where A_n is the amplitude, and ω_n is the natural circular frequency in radians per unit time for the nth vibrational mode.

Substituting (7.4) into (7.3) results in:

$$\sum_{n=1}^{\infty} A_n\left[\frac{n^4\pi^4}{L^4}\mathrm{EI} - \omega_n^2\rho A\right]\sin\frac{n\pi x}{L}\sin\omega_n t = 0. \tag{7.5}$$

For this to be an equation, then for each value of *n*, requires that

$$\omega_n = \frac{n^2\pi^2}{L^2}\sqrt{\frac{EI}{\rho_m A}}. \tag{7.6}$$

It is seen that for each integer *n*, there is a different natural frequency, and from (7.4) a corresponding mode shape (i.e., $n = 1$, a one half sine wave; $n = 2$, two half sine waves, etc.).

Unlike in buckling where one looks for the lowest buckling load only, in vibrations each natural frequency is important, because if a beam were subjected to an oscillating load coinciding with any one natural frequency, little energy would be needed to cause the amplitude to grow until failure occurs.

The lowest natural frequency, $n = 1$ in this case is called the *fundamental* frequency. Theoretically *n* could increase to infinity. However, at some point the governing equation (7.3) does not apply, and thus the resulting frequencies given by (7.6) become meaningless. For a beam of an isotropic material the classical beam equation (7.3) ceases to apply when the vibration half wave length approaches the beam depth, *h*, because then transverse shear deformation effects ($\varepsilon_{xz} \neq 0$) become important and (7.3) must be modified.

It is noted that for beams with boundary conditions other than simply supported at each end, the eigenfunctions (vibration modes) are not as simple as a sine wave. These are treated in detail in any of many fine texts on vibration. However, the natural frequencies, ω_n, are catalogued below for use in analysis and design. In this case, Equation (7.6) is modified slightly for general use.

$$\omega_n = \alpha_n^2\sqrt{\frac{EI}{\rho_m AL^4}} \tag{7.7}$$

where α_n^2 values are given by the following:

Simple Support-Simple Support Beam: $n^2\pi^2$
Cantilevered Beam: $\alpha_1^2 = 3.52$, $\alpha_2^2 = 22.6$, $\alpha_3^2 = 61.7$
Clamped-Clamped Beam: $\alpha_1^2 = 22.4$, $\alpha_2^2 = 61.7$, $\alpha_3^2 = 121.0$

In all of the above, classical beam theory is used, i.e., no transverse shear deformation effects. Also remember that in 90+% of the errors made by students calculating natural frequencies, the student used the weight density of the material rather than the mass density.

7.3 Natural Vibrations of Isotropic Plates

Consider again the equation for the bending of a rectangular plate subjected to a lateral load, $p(x, y)$, given by Equation (2.57).

$$D\nabla^4 w = p(x, y). \tag{7.8}$$

If d'Alembert's Principle were used to accommodate the motion, one would add a term to the right hand side equal to the negative of the product of the mass per unit area and the acceleration in the z direction. In that case, the right hand side of (7.8) becomes:

$$p(x, y, t) - \rho_m h \frac{\partial^2 w}{\partial t^2}(x, y, t) \tag{7.9}$$

where both p and w are functions of time as well as space, ρ is the *mass* density of the material and h is the plate thickness. For forced vibration $p(x, y, t)$ causes the dynamic response, and can vary from a harmonic oscillation to an intense one time impact.

As discussed in the previous section, to study the natural vibrations $p(x, y, t)$ is set equal to zero, and the governing equation becomes the following homogeneous equation:

$$D\left[\frac{\partial^4 w}{\partial x^4} + 2\frac{\partial^4 w}{\partial x^2 \partial y^2} + \frac{\partial^4 w}{\partial y^4}\right] + \rho_m h \frac{\partial^2 w}{\partial t^2} = 0. \tag{7.10}$$

As done previously, one can assume a solution for the lateral deflection, which spatially satisfies the boundary condition, is harmonic in time and satisfies (7.10) above. For the case of a plate simply-supported on all four edges, such a function is

$$w(x, y, t) = \sum_{m=1}^{\infty}\sum_{n=1}^{\infty} A_{mn} \sin\frac{m\pi x}{a} \sin\frac{n\pi y}{b} \sin\omega_n t \tag{7.11}$$

where a and b are the plate dimensions, A_{mn} is the vibration amplitude for each value of the integers m and n, and ω_{mn} is the natural circular frequency in radians per unit time.

$$\omega_{mn} = \left[\frac{\pi^4 D}{\rho_m h}\left(\frac{m^2}{a^2} + \frac{n^2}{b^2}\right)^2\right]^{1/2}. \tag{7.12}$$

In this case the fundamental natural frequency occurs for $m = n = 1$. Again, the amplitude A_{mn} cannot be determined from this linear eigenvalue problem, in which the eigenvalues are the natural circular frequencies of Equation (7.12) and the corresponding eigenfunctions are the mode shapes, given in (7.11).

To obtain the natural frequency of vibration in cycles per second, called Hertz (Hz), denoted by f_{mn},

$$f_{mn} = \frac{\omega_{mn}}{2\pi} \tag{7.13}$$

As in buckling, plates with other boundary conditions comprise more complicated problems, often most difficult to solve analytically. In many cases appropriate solutions are obtained using energy methods (Chapters 8 and 9).

Ma and Lin [7.1] have provided graphic descriptions of the first six vibration modes for an aluminum (isotropic) square plate, giving both the experimental observation and the mode shapes obtained by numerical calculation.

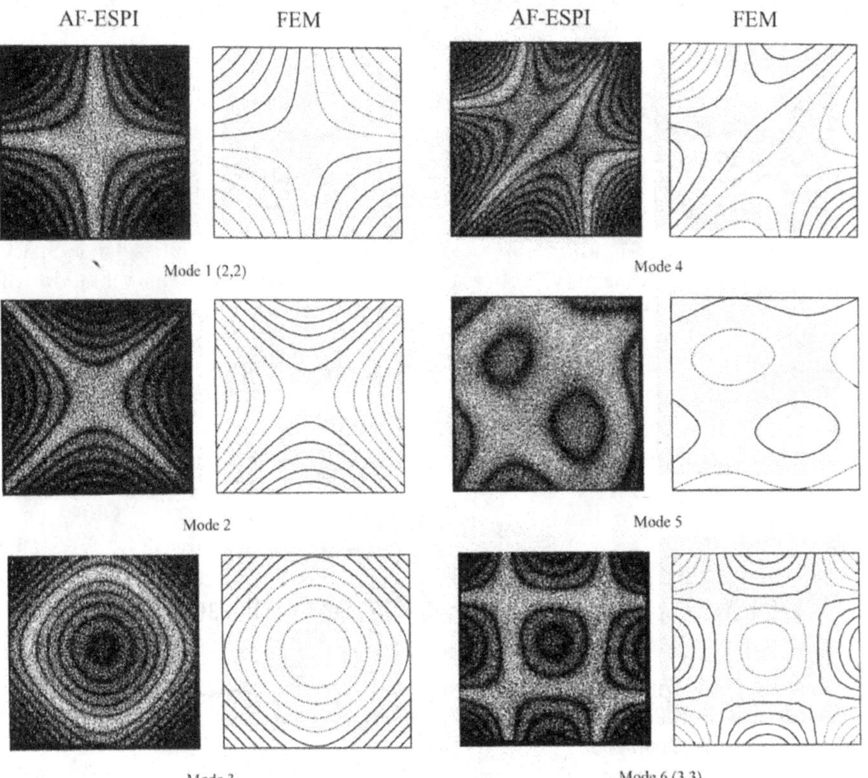

Figure 7.1. First six mode shapes for isotropic aluminum square plate obtained from experimental observation and numerical calculation.

7.4 Forced Vibrations of Beams and Plates

These will not be treated here. There are many fine texts at a basic level and voluminous literature dealing with this subject. Thomson [7.2] is such a text, and Leissa [7.3] provides solutions to numerous problems. Vibration damping is treated in the text by Nashif, Jones and Henderson [7.4].

One excellent paper by Dobyns [7.5] also given by Vinson and Sierakowski [1.7], and repeated here in Chapter 13, provide solutions to the dynamic response of anisotropic plates subjected to a variety of impact loads of practical value, and are treated herein in Section 13.4. Those solutions are easily simplified to treat plates of isotropic materials.

7.5 References

7.1. Ma, C-C and Lin, C-C (2001) Experimental Investigation of Vibrating Laminated Composite Plates by Optical Interferometry Method, *AIAA Journal*, Vol. 38, No. 3, March, pp. 491-497.

7.2. Thomson, W.T. (1965) *Vibration Theory and Applications*, Englewood Cliffs: Prentice Hall Publishers.

7.3. Leissa, A.W. (1969) *Vibration of Plates*, National Aeronautics and Space Administration Special Publication, also (1973) The Free Vibration of Rectangular Plates, *Journal of Sound and Vibration*, Vol. 31, No. 3, pp. 257-293.

7.4. Nashif, A.D., Jones, D.I.G. and Henderson, J. (1985) *Vibration Damping*, Wiley Interscience.

7.5. Dobyns, A.L. (1981) The Analysis of Simply-Supported Orthotopic Plates Subjected to Static and Dynamic Loads, *AIAA Journal*, May, pp. 642-650.

7.6 Problems

7.1. A beam is 30 inches long, 1 inch wide and made of steel ($\mathbf{E = 30 \times 10^6}$ psi, $\nu = 0.3$, weight density $\rho_w = 0.283$ lbs/in^3), simply supported at each end. What must the thickness h be to insure that the lowest natural frequency is not lower than 30 Hz?

7.2. For the beam of Problem 7.1 dimensions and material, what is the fundamental frequency if the beam is cantilevered?

7.3. For the beam of Problem 7.1 dimensions and material, what is the fundamental frequency if the beam is clamped at each end?

CHAPTER 8

THEOREM OF MINIMUM POTENTIAL ENERGY, HAMILTON'S PRINCIPLE AND THEIR APPLICATIONS

8.1 Introduction

Many structures involve complicated shapes and numerous or unusual loads for which solutions of the governing differential equations and/or the boundary conditions are difficult or impossible. For instance, a rectangular plate with a hole somewhere, or a plate with discontinuous boundary conditions poses a major difficulty in finding an analytical solution.

For preliminary design and analysis one needs simplified, easy to use analyses analogous to those that have been presented earlier. However, for final design, quite often transverse shear deformation and thermal effects must be included. Thermal effects have been described in Chapter 4. Analytically they cause considerable difficulty, because with their inclusion few boundary conditions are homogeneous, hence separation of variables, used throughout the plate solutions to this point, cannot be utilized in a straightforward manner. Only through the laborious process of transformation of variables can the procedures discussed herein be used [1.1]. Therefore, energy principles are much more convenient for use in design and analyses of plate structures when thermal effects are present.

In solving plate problems it is seen that in order to obtain an analytical solution one must solve the differential equations and satisfy the boundary conditions; if that cannot be accomplished, there is no solution. With energy methods, one can always obtain a good approximate solution, no matter what the structural complexities, the loads or the boundary condition complications may be, using a little ingenuity.

In structural mechanics three energy principles are used: Minimum Potential Energy, Minimum Complementary Energy and Reissner's Variational Theorem [8.1]. The first two are discussed at length in Sokolnikoff [1.1] and many other references. The Reissner Variational Theorem, likewise, is widely referenced. In solid mechanics, Minimum Complementary Energy is rarely used, because it requires assuming functions that insure that the stresses satisfy boundary conditions and equilibrium. It is usually far easier to make assumptions about functions that can represent displacements.

Minimum Potential Energy is widely used in solutions to problems involving plate structures. In fact, the more complicated the loading, the more complicated the geometry and the more complicated the boundary conditions (e.g., discontinuous or concentrated boundary conditions), the more desirable it is to use Minimum Potential Energy to obtain an approximate solution, compared to attempting to solve the governing differential equations and to satisfy the boundary conditions exactly.

In addition, in many cases energy principles can be useful for eigenvalue problems such as in the buckling and vibration problems as shall be shown.

There are numerous books dealing with energy theorems and variational methods. One of the more recent is that by Mura and Koya [8.2].

8.2 Theorem of Minimum Potential Energy

For any generalized elastic body, the potential energy of that body can be written as follows:

$$V \quad \int_R W \mathrm{d}R - \int_{S_T} T_i u_i \mathrm{d}S - \int_R F_i u_i \mathrm{d}R$$

(8.1)

where

W strain energy density function, defined in Equation (8.4) below
R volume of the elastic body
T_i *i*th component of the surface traction
u_i *i*th component of the deformation
F_i *i*th component of a body force
S_T □ portion of the body surface over which tractions are prescribed

One sees that the first term on the right-hand side of Equation (8.1) is the strain energy of the elastic body. The second and third terms are the work done by the surface tractions; and the body forces, respectively. The Theorem of Minimum Potential Energy can be stated as described in [1.1]: "Of all the displacements satisfying compatibility and the prescribed boundary conditions, those that satisfy the equilibrium equations make the potential energy a minimum."

Mathematically, the operation is simply stated as,

$$\delta V \;\square\; 0 \qquad (8.2)$$

The lowercase delta is a mathematical operation known as a variation. Operationally, it is analogous to partial differentiation. To employ variational operations in structural mechanics, only the following three operations are usually needed (where y is any dependent variable):

$$\frac{\mathrm{d}(\delta y)}{\mathrm{d}x} \quad \delta\left(\frac{\mathrm{d}y}{\mathrm{d}x}\right), \quad \delta\left(y^2\right) \quad 2y\delta y, \quad \int \delta y \mathrm{d}x \quad \delta \int y \mathrm{d}x \qquad (8.3)$$

In Equation (8.1) the strain energy density function, W, is defined as follows in a Cartesian coordinate frame:

$$W = \frac{1}{2}\sigma_{ij}\varepsilon_{ij} = \frac{1}{2}\sigma_x\varepsilon_x + \frac{1}{2}\sigma_y\varepsilon_y + \frac{1}{2}\sigma_z\varepsilon_z + \sigma_{xy}\varepsilon_{xy} + \sigma_{xz}\varepsilon_{xz} + \sigma_{yz}\varepsilon_{yz} \tag{8.4}$$

To utilize the Theorem of Minimum Potential Energy, the stress-strain relations for the elastic body are employed to change the stresses in Equation (8.4) to strains, and the strain-displacement relations are employed to change all strains to displacements. Thus, it is necessary for the analyst to select the proper stress-strain relations and strain-displacement relations for the problem being solved.

Although this text is dedicated to plate and panel structures, it is best to introduce the subject using isotropic monocoque beams, a much simpler structural component, to first illustrate the energy principles.

8.3 Analysis of a Beam In Bending Using the Theorem of Minimum Potential Energy

As the simplest example of the use of Minimum Potential Energy, consider a beam in bending, shown in Figure 8.1. In this section, Minimum Potential Energy methods are used to show that if one makes beam assumptions, one obtains the beam equation. However, the most useful employment of the Minimum Potential Energy Theorem is through making assumptions for the dependent variables (the deflection) and using the Theorem to obtain approximate solutions, as will be illustrated later.

From Figure 8.1 it is seen that the beam is of length L, in the x-direction, width b and height h. It is subjected to a lateral distributed load, $q(x)$ in the positive z-direction, in units of force per unit length. The modulus of elasticity of the isotropic beam material is E, and the stress-strain relation is simply

$$\sigma_x = \mathrm{E}\varepsilon_x \tag{8.5}$$

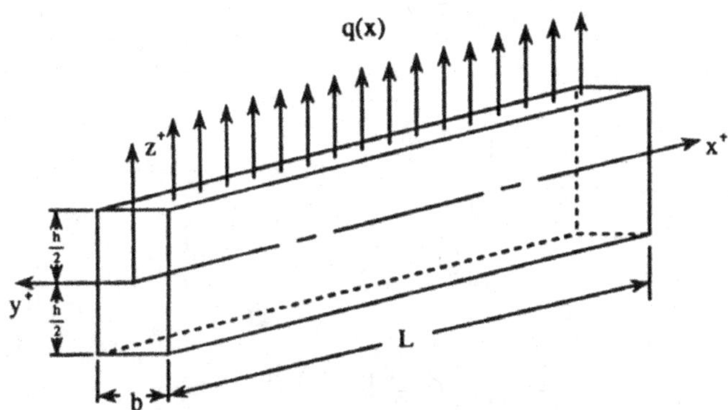

Figure 8.1. Beam in bending

The corresponding strain-displacement relation for a beam in bending only is, from (1.16), (2.1) and (2.27),

$$\varepsilon_x = \frac{\mathrm{d}u}{\mathrm{d}x} = -\frac{\mathrm{d}^2 w}{\mathrm{d}x^2} z \tag{8.6}$$

since in the bending of beams, $u = -z(\mathrm{d}w/\mathrm{d}x)$ only.

Looking at Equations (8.4) through (8.6) and remembering that in elementary beam theory

$$\sigma_y = \sigma_z = \sigma_{xy} = \varepsilon_{xz} = \varepsilon_{yz} = \sigma_{xy} = 0$$

then if the beam is subjected to bending only

$$W = \frac{1}{2}\sigma_x \varepsilon_x = \frac{1}{2}\mathrm{E}\varepsilon_x^2 = \frac{1}{2}\mathrm{E}\left(\frac{\mathrm{d}^2 w}{\mathrm{d}x^2}\right)^2 z^2 \tag{8.7}$$

Therefore, the strain energy, U, which is the volume integral of the strain energy density function, W, is

$$U = \int_0^L \int_{-b/2}^{+b/2} \int_{-h/2}^{+h/2} \frac{1}{2}\mathrm{E}\left(\frac{\mathrm{d}^2 w}{\mathrm{d}x^2}\right)^2 z^2 \mathbf{dzdydx} = \frac{\mathrm{EI}}{2}\int_0^L \left(\frac{\mathrm{d}^2 w}{\mathrm{d}x^2}\right)^2 \mathrm{d}x \tag{8.8}$$

where, $\mathbf{I} = bh^3/12$, the flexural stiffness for a beam of rectangular cross-section.

Similarly, from the surface traction work term in Equation (8.1) it is seen that

$$\int_{S_T} T_i u_i \mathrm{d}s = \int_0^L q(x)\, w(x)\mathrm{d}x$$

Equation (8.1) then becomes

$$V = \frac{\mathrm{EI}}{2}\int_0^L \left(\frac{\mathrm{d}w^2}{\mathrm{d}x^2}\right)^2 \mathrm{d}x - \int_0^L q(x)\, w(x)\mathrm{d}x \tag{8.9}$$

Following Equation (8.2) and remembering Equation (6.3) then

$$\delta V = 0 = \frac{\mathrm{EI}}{2}\int_0^L \delta\left(\frac{\mathrm{d}^2 w}{\mathrm{d}x^2}\right)^2 \mathrm{d}x - \int_0^L q(x)\, \delta w(x)\mathrm{d}x \tag{8.10}$$

The variation δ can be included under the integral, because the order of variation and integration can be interchanged. Also, there is no variation of E, I or $q(x)$ because they are all specified quantities.

Integrating by parts the first term on the right-hand side of Equation (8.10).

$$
\begin{aligned}
\frac{\mathrm{EI}}{2}\int_0^L \delta\left(\frac{\mathrm{d}^2 w}{\mathrm{d}x^2}\right)^2 \mathrm{d}x &= \mathrm{EI}\int_0^L \left(\frac{\mathrm{d}^2 w}{\mathrm{d}x^2}\right)\delta\left(\frac{\mathrm{d}^2 w}{\mathrm{d}x^2}\right)\mathrm{d}x \\
&= \mathrm{EI}\int_0^L \frac{\mathrm{d}^2 w}{\mathrm{d}x^2}\frac{\mathrm{d}^2(\delta w)}{\mathrm{d}x^2}\mathrm{d}x \\
&= \left[\mathrm{EI}\frac{\mathrm{d}^2 w}{\mathrm{d}x^2}\delta\left(\frac{\mathrm{d}w}{\mathrm{d}x}\right)\right]_0^L - \mathrm{EI}\int_0^L \frac{\mathrm{d}^3 w}{\mathrm{d}x^3}\frac{\mathrm{d}(\delta w)}{\mathrm{d}x}\mathrm{d}x \\
&= \left[\mathrm{EI}\frac{\mathrm{d}^2 w}{\mathrm{d}x^2}\delta\left(\frac{\mathrm{d}w}{\mathrm{d}x}\right)\right]_0^L - \left[\mathrm{EI}\frac{\mathrm{d}^3 w}{\mathrm{d}x^3}\delta w\right]_0^L \\
&\quad + \mathrm{EI}\int_0^L \frac{\mathrm{d}^4 w}{\mathrm{d}x^4}\delta w \mathrm{d}x
\end{aligned}
\tag{8.11}
$$

Substituting Equation (8.11) into (8.10) and rearranging, it is seen that:

$$
\begin{aligned}
\delta V = 0 = &\left[\mathrm{EI}\frac{\mathrm{d}^2 w}{\mathrm{d}x^2}\delta\left(\frac{\mathrm{d}w}{\mathrm{d}x}\right)\right]_0^L - \left[\mathrm{EI}\frac{\mathrm{d}^3 w}{\mathrm{d}x^3}\delta w\right]_0^L \\
&+ \int_0^L \left[\mathrm{EI}\frac{\mathrm{d}^4 w}{\mathrm{d}x^4} - q(x)\right]\delta w \mathrm{d}x
\end{aligned}
\tag{8.12}
$$

For this to be true, the following equation must be satisfied for the integral above to be zero:

$$
\mathrm{EI}\frac{\mathrm{d}^4 w}{\mathrm{d}x^4} = q(x) \tag{8.13}
$$

This is obviously the governing equation for the bending of a beam under a lateral load. So, it is seen that if one considers a beam-type structure, uses beam assumptions, and uses proper stress-strain relations and strain-displacement relations, the result is the beam bending equation. However, it can be emphasized that if a nonclassical-shaped elastic structure were being analyzed, by using physical intuition, experience or some other reasoning to formulate stress-strain relations, and strain-displacement relations for the body, then through the Theorem of Minimum Potential Energy one can formulate the governing differential equations for the structure and load analogous to Equation (8.13). Incidentally, the resulting governing differential equations derived from the Theorem of Minimum Potential Energy are called the Euler-Lagrange equations.

Note also for Equation (8.12) to be true, each of the first two terms must be zero. Hence, at $x = 0$ and $x = L$ (at each end) either $EI\left(d^2w/dx^2\right) - M_x = 0$ or $\left(dw/dx\right)$ must be specified (that is, its variation must be zero), also either $EI\left(d^3w/dx^3\right) - V_x = 0$ or w must be specified. These are the natural boundary conditions. All of the classical boundary conditions, including simple supported, clamped and free edges are contained in the above "natural boundary conditions." This is a nice byproduct from using the variational approach for deriving governing equations for analyzing any elastic structure.

The above discussion shows that if in using The Theorem of Minimum Potential Energy one makes all of the assumptions of classical beam theory, the resulting Euler-Lagrange equation is the classical beam equation (8.13) and the natural boundary conditions given in (8.12) as discussed above.

Equally or more important the Theorem of Minimum Potential Energy provides a means to obtain an approximate solution to practical engineering problems by assuming good deflection functions which satisfy the boundary conditions. As the simplest example consider a beam simply supported at each end subjected to a uniform lateral load per unit length $q(x) = -q_0$, a constant.

Here, an example, assume a deflection which satisfies the boundary conditions for a beam simply supported at each end, where A is a constant to be determined.

$$w(x) = A\sin\frac{\pi x}{L} \tag{8.14}$$

This is not the exact solution, but should lead to a good approximation because (8.14) is a continuous single valued function which satisfies the boundary conditions of the problem.
Proceeding,

$$w' = A\frac{\pi}{L}\cos\frac{\pi x}{L}, \qquad w'' = -\frac{A\pi^2}{L^2}\sin\frac{\pi x}{L}$$
$$(w'')^2 = A^2\frac{\pi^4}{L^4}\sin^2\frac{\pi x}{L} \tag{8.15}$$

Substituting (8.14) into (8.9) results in

$$
\begin{aligned}
V &= \frac{\mathrm{EI}}{2}\int_0^L \frac{A^2\pi^4}{L^4}\sin^2\frac{\pi x}{L}\,\mathrm{d}x - \int_0^L(-q_0)A\sin\frac{\pi x}{L}\,\mathrm{d}x \\
&= \frac{\mathrm{EI}}{2}\frac{A^2\pi^4}{L^4}\left(\frac{L}{2}\right) - (-q_0)A\frac{L}{\pi}\left[-\cos\frac{\pi x}{L}\right]_0^L \\
&= \frac{\pi^4}{4L^3}\mathrm{EI}\,A^2 + q_0 A\frac{L}{\pi}\left[-\cos\pi + 1\right]
\end{aligned}
\tag{8.16}
$$

$$
\delta V = 0 = \frac{\pi^4}{4L^3}\mathrm{EI}\,2A\,\delta A + \frac{2q_0 L}{\pi}\delta A = \delta A\left[\frac{\pi^4}{2L^3}\mathrm{EI}\,A + \frac{2q_0 L}{\pi}\right]
$$

Therefore,

$$
A = -\frac{4q_0L^4}{\pi^5\mathrm{EI}} = w(L/2)
\tag{8.17}
$$

The exact solution is

$$
w(L/2) = -\frac{5}{384}\frac{q_0L^4}{\mathrm{EI}}
\tag{8.18}
$$

The difference is seen to be 0.386%. So the Minimum Potential Energy solution is seen to be almost exact in determining the maximum deflection.

In determining maximum stresses the accuracy of the energy solution is less, because bending stresses are proportional to second derivatives of deflection. By taking derivatives the errors increase (conversely, integrating is an averaging process and errors decrease) so the stresses from the approximate solution differ more from the exact solution than do the deflections.

To continue this example for a one lamina composite beam, simply supported at each end, subjected to a constant uniform lateral load per unit length of $-q_0$, it is clear that the maximum stress occurs at $x = L/2$. From classical beam theory, the exact value of the maximum stress is

$$
\sigma_{\max} \equiv \sigma_x\left(\frac{L}{2}, \pm\frac{h}{2}\right) = \pm\frac{q_0L^2}{8}
\tag{8.19}
$$

Likewise, for the Minimum Potential Energy solution, using (8.15)

$$\sigma_{\max} = \sigma_x\left(\frac{L}{2},\pm\frac{h}{2}\right) = -\mathrm{EI}\,w''\left(\frac{L}{2},\pm\frac{h}{2}\right) = \pm\frac{4q_0L^2}{\pi^3} \tag{8.20}$$

The difference between the two is 3.2%, so the energy solution is quite accurate for many applications.

If one wishes to increase the accuracy, instead of using (8.14) one could use

$$w(x) = \sum_{n=1}^{N} A_n \sin\frac{n\pi x}{L} \tag{8.21}$$

If N were chosen to be three, for example, the expression for $w(x)$ is given by $w(x) = A_1\sin\frac{\pi x}{L} + A_2\sin\frac{2\pi x}{L} + A_3\sin\frac{3\pi x}{L}$ and one would proceed as before, taking variations with respect to A_1, A_2 and A_3 which provides three algebraic equations for determining the three A_n. Of course as N increases, the accuracy of the solution increases until as N approaches infinity it is another form of the exact solution.

As a second example, examine the same beam, this time subjected to a concentrated load P at the mid-length, $x = L/2$. To obtain an exact solution, one must divide the beam into two parts, so that the load discontinuity can be accommodated, with the result that there are two fourth order differential equations and eight boundary conditions. Not so with the case of Minimum Potential Energy to obtain an approximate solution, as follows. Again assume (8.14) as the approximate deflection because it is single valued, continuous and satisfies the boundary conditions at the end of the beam. There,

$$V = \frac{1}{2}\int_0^L \mathrm{EI}\left(\frac{d^2w}{dx^2}\right)^2 dx - Pw(L/2)$$

$$= \frac{\pi^4}{4L^3}\mathrm{EI}\,A^2 - PA$$

$$\delta V = 0 = \frac{\pi^4\mathrm{EI}\,A\delta A}{2L^3} - P\delta A$$

$$\text{or,}\quad A = \frac{2PL^3}{\pi^4\mathrm{EI}} = w(L/2) = w_{\max}$$

Again, instead of (8.14) one could have chosen (8.21) as the trial function to use in solving this problem.

Thus, the Theorem of Minimum Potential Energy can be used easily for complicated laterally distributed loads, concentrated lateral loads, any boundary conditions, and/or variable or discontinuous beam thicknesses. One only needs to select a form of the lateral displacement such as the following examples.

Clamped Clamped Beam

$$w(x) = A\left[1-\cos\frac{2\pi x}{L}\right] \tag{8.22}$$

Clamped-Simple Beam

$$w(x) = A\left[\mathrm{L}^3x-3Lx^3+2x^4\right] \tag{8.23}$$

Cantilevered Beam

$$w(x) = Ax^2 \tag{8.24}$$

8.4 The Buckling of Columns

In this case the strain energy is again given by Equation (8.8), where neglecting body forces F_i, the work done by surface tractions is given as follows:

$$\int_{S_T} \boldsymbol{T}_i\boldsymbol{u}_i\mathrm{d}\boldsymbol{S} = -\int_0^L \left\{P\left[\frac{\mathrm{d}u_0}{\mathrm{d}x}+\frac{1}{2}\left(\frac{\mathrm{d}w}{\mathrm{d}x}\right)^2\right]\right\}\mathrm{d}x.$$

This equation incorporates the more comprehensive theory employed in Chapter 6 to include buckling, and as discussed previously, to calculate buckling loads, $\boldsymbol{u}_0 = 0$, because at incipient buckling the arc length of the buckled column is equal to the original length. Also, in the above, P is the tensile load, considered constant to make the problem linear. Therefore, for column buckling,

$$V = \frac{EI}{2}\int_0^L \left(\frac{\mathrm{d}^2w}{\mathrm{d}x^2}\right)^2\mathrm{d}x+\frac{P}{2}\int_0^L\left(\frac{\mathrm{d}w}{\mathrm{d}x}\right)^2\mathrm{d}x. \tag{8.25}$$

Taking the variation of the potential energy, one obtains the following Euler-Lagrange equation analogous to Equation (8.13)

$$EI\frac{\mathrm{d}^4w}{\mathrm{d}x^4}-P\frac{\mathrm{d}^2w}{\mathrm{d}x^2} = 0 \tag{8.26}$$

as well as the natural boundary conditions discussed previously. However, assuming a form of $w(x)$, which satisfies the boundary conditions for the column, which approximates the exact buckled shape will provide an approximation to the exact buckling load.

Consider a column simply supported at each end, if one uses (8.14) in (8.25) and takes variation of A, the result is:

$$V = \left[\frac{\pi^4 EI}{4L^3} + \frac{P\pi^2}{4L}\right] A^2$$

(8.27)

$$\delta V = \left[\frac{\pi^4 EI}{4L^3} - \frac{P\pi^2}{4L}\right] 2A\delta A = 0$$

so the bracket must equal zero, or

$$P = -\pi^2 \frac{EI}{L^2}.$$

It is seen that this is the exact buckling load, because the exact buckling mode (8.14) was utilized. Some other approximate displacement functions satisfying the boundary conditions would give an approximate buckling load. It can be proven that such an approximate buckling load will always be greater than the exact buckling load. However, as long as the assumed displacement satisfies the boundary conditions, the error is never more than a very few percent of the exact value.

8.5 Vibration of Beams

The energy principle to utilize in dynamic analysis is Hamilton's Principle which employs the functional

$$I = \int_{t_1}^{t_2} (T - V)\mathrm{d}t \,. \qquad (8.28)$$

Hamilton's Principle states that in a conservative system

$$\delta I = 0 \,. \qquad (8.29)$$

In the above, the potential energy, V, is given by Equation (8.1), and T is the kinetic energy of the body. In a beam undergoing flexural vibration, the kinetic energy would be

$$T = \frac{1}{2}\int_0^L \rho_m A\left(\frac{\partial w}{\partial t}\right)^2 dx \tag{8.30}$$

where ρ_m is the *mass* density of the material, A is the beam cross-sectional area, and $\partial w/\partial t$ is the velocity of the beam.

Using Hamilton's Principle in the same way that was done before for Minimum Potential Energy, the resulting Euler-Lagrange equation is

$$EI\frac{\partial^4 w}{\partial x^4} + \rho_m A\frac{\partial^2 w}{\partial t^2} = 0 \tag{8.31}$$

which is identical to Equation (7.3). Also resulting are the natural boundary conditions, discussed previously.

Considering a beam simply supported at each end, if Equation (8.14) is modified to include a harmonic motion with time, such as

$$w(x,t) = C\sin\frac{n\pi x}{L}\cos\omega_n t$$

where C is a constant.
The result is an Euler-Lagrange equation of

$$\omega_n = \frac{n^2\pi^2}{L^2}\sqrt{\frac{EI}{\rho_m A}} \tag{8.32}$$

which is the exact solution for the natural circular frequency, ω_n, in radians/unit time [see Equation (7.6)] because the exact mode shape was assumed. Again, if the assumed displacement function is approximate, then approximate natural frequencies will be obtained; are higher than the exact frequencies, but the error will be at most a few percent. In any case the natural frequency, f (in Hz), is found by $\omega_n/2\pi$.

Note that in assuming mode shape functions in both buckling and vibration problems (eigenvalue problems), the closer the assumed approximate function is to the exact mode shape, the lower the resulting eigenvalue will be, and of course it will be closer to the exact eigenvalue, since the exact eigenvalue is always lower than any approximated value.

8.6 Minimum Potential Energy for Rectangular Isotropic Plates

The strain energy density function, W, for a three dimensional solid in rectangular coordinates is given by Equation (8.4). The assumptions associated with the classical plate theory of Chapter 2 are employed to modify (8.4) for a rectangular plate. If transverse shear deformation is neglected, then $\varepsilon_{xz} = \varepsilon_{yz} = 0$. If there is no plate thickening, then $\varepsilon_z = 0$. From Equations (1.9), (1.10), and (1.12), stresses are written in terms of strains, such that for the classical plate,

$$\sigma_x = \frac{E}{(1-\nu^2)}[\varepsilon_x + \nu\varepsilon_y]; \quad \sigma_y = \frac{E}{(1-\nu^2)}[\varepsilon_y + \nu\varepsilon_x]; \quad \sigma_{xy} = \frac{E}{(1-\nu^2)}\varepsilon_{xy}. \tag{8.33}$$

Therefore, (8.4) becomes

$$W = \frac{E\varepsilon_x}{2(1-\nu^2)}(\varepsilon_x + \nu\varepsilon_y) + \frac{E\varepsilon_y}{2(1-\nu^2)}(\varepsilon_y + \nu\varepsilon_x) + \frac{E}{(1+\nu)}\varepsilon_{xy}^2. \tag{8.34}$$

If the plate is subjected to bending and stretching, the deflection functions are given by Equations (2.24) through (2.28). Substituting these into (8.34) results in the following:

$$\begin{aligned} W = {} & \frac{1}{2}\frac{E}{(1-\nu^2)}\left\{\left(\frac{\partial u_0}{\partial x}\right)^2 + \left(\frac{\partial v_0}{\partial y}\right)^2 + 2\nu\left(\frac{\partial u_0}{\partial x}\right)\left(\frac{\partial v_0}{\partial y}\right) + \frac{1+\nu}{2}\left[\frac{\partial u_0}{\partial y} + \frac{\partial v_0}{\partial x}\right]^2\right\} \\ & + \frac{Ez^2}{2(1-\nu^2)}\left[\left(\frac{\partial^2 w}{\partial x^2}\right)^2 + \left(\frac{\partial^2 w}{\partial y^2}\right)^2 + 2\nu\left(\frac{\partial^2 w}{\partial x^2}\right)\left(\frac{\partial^2 w}{\partial y^2}\right)\right] \\ & + \frac{Ez^2}{(1+\nu)}\left(\frac{\partial^2 w}{\partial x \partial y}\right)^2. \end{aligned} \tag{8.35}$$

From this the strain energy $U(= \int_R W dR)$ is found.

$$\begin{aligned} U = {} & \frac{K}{2}\int_0^a\int_0^b\left\{\left(\frac{\partial u_0}{\partial x} + \frac{\partial v_0}{\partial y}\right)^2 - 2(1-\nu)\frac{\partial u_0}{\partial x}\frac{\partial v_0}{\partial y} + \frac{1-\nu}{2}\left(\frac{\partial u_0}{\partial y} + \frac{\partial v_0}{\partial x}\right)^2\right\}dxdy \\ & + \frac{D}{2}\int_0^a\int_0^b\left\{\left(\frac{\partial^2 w}{\partial x^2} + \frac{\partial^2 w}{\partial y^2}\right)^2 - 2(1-\nu)\left[\left(\frac{\partial^2 w}{\partial x^2}\right)\left(\frac{\partial^2 w}{\partial y^2}\right) - \left(\frac{\partial^2 w}{\partial x \partial y}\right)^2\right]\right\}dxdy. \end{aligned} \tag{8.36}$$

It is seen that the first term is the extensional or in-plane strain energy of the plate, and the second is the bending strain energy of the plate. In the latter, it is seen that the first term is proportional to the square of the average plate curvature, while the second term is known as the Gaussian curvature.

For the plate the total work term due to surface traction is seen to be

$$\int_{S_T} T_i u_i \,dS = \int_0^a \int_0^b p(x,y)\, w(x,y)\, dxdy - \int_0^a \int_0^b \left\{ N_x \left[\frac{\partial u_0}{\partial x} + \frac{1}{2}\left(\frac{\partial w}{\partial x}\right)^2 \right] \right.$$
$$\left. + N_y \left[\frac{\partial v_0}{\partial y} + \frac{1}{2}\left(\frac{\partial w}{\partial y}\right)^2 \right] + N_{xy}\left[\left(\frac{\partial u_0}{\partial y} + \frac{\partial v_0}{\partial x}\right) + \left(\frac{\partial w}{\partial x}\right)\left(\frac{\partial w}{\partial y}\right) \right] \right\} dxdy. \tag{8.37}$$

Hence, in (8.36) and (8.37) if one considers a plate subjected only to a lateral load $p(x, y)$, one assumes $u_0 = v_0 = N_x = N_y = N_{xy} = 0$. If one is considering in-plane loads only (except for buckling) assume $w(x, y) = p(x, y) = 0$. If one is looking for buckling loads, assume $\boldsymbol{p(x,y) = u_0 = v_0} = 0$. The rationale for all of this has been discussed previously.

8.7 The Buckling of an Isotropic Plate Under a Uniaxial Load, Simply Supported on Three Sides, and Free on an Unloaded Edge.

The most beneficial use of the Minimum Potential Energy Theorem occurs when one cannot formulate a suitable set of governing differential equations, and/or when one cannot ascertain a consistent set of boundary conditions. In that case one can make a reasonable assumption of the displacements, and then solves for an approximate solution using the Theorem of Minimum Potential Energy. This is illustrated in the following example.

Consider the plate shown below in Figure 8.2. The governing differential equation for this problem is obtained from Equation (6.7).

$$\frac{\partial^4 w}{\partial x^4} + 2\frac{\partial^4 w}{\partial x^2 \partial y^2} + \frac{\partial^4 w}{\partial y^4} = \frac{N_x}{D}\frac{\partial^2 w}{\partial x^2}. \tag{8.38}$$

To solve for the buckling load directly, a Levy type solution may be assumed:

$$w(x,y) = \sum_{m=1}^{\infty} \psi_m(y) \sin\frac{m\pi x}{a}. \tag{8.39}$$

Substituting (8.39) into (8.38) results in the following ordinary differential equation to solve:

$$\lambda_m^4 \psi - 2\lambda_m^2 \psi'' + \psi^{IV} = -\frac{N_x}{D}\lambda_m^2 \psi \tag{8.40}$$

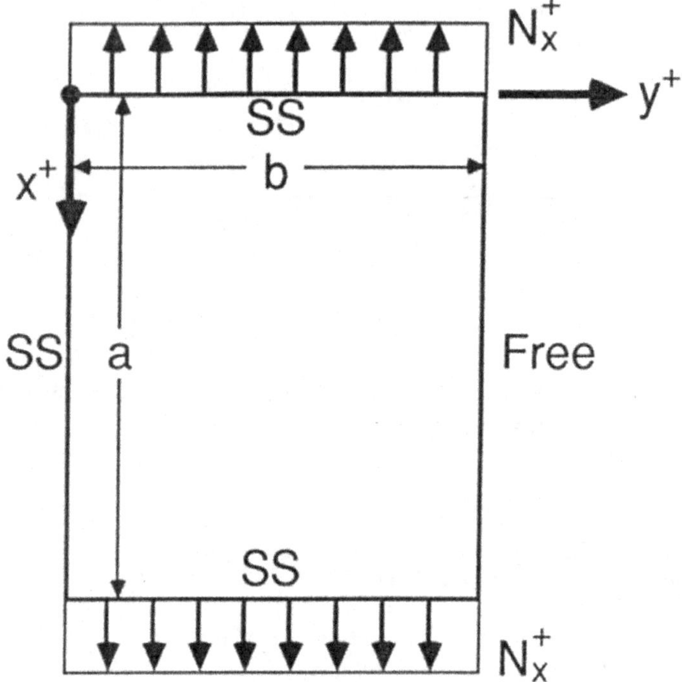

Figure 8.2. Plate studied in Section 8.7.

where

$$\lambda_m = \frac{m\pi}{a}, \quad (\ )'' = \frac{d^2(\ )}{dy^2}, \quad \text{and} \quad (\ )^{IV} = \frac{d^4(\ )}{dy^4} \quad .$$

Letting $\overline{N}_x = -N_x$, Equation (8.40) can be solved with the result that

$$\psi_m(y) = A\cosh\alpha\, y + B\sinh\alpha\, y + C\cos\beta\, y + E\sin\beta\, y \tag{8.41}$$

where

$$\alpha = \left[\lambda_m^2 + \lambda_m\sqrt{\frac{\overline{N}_{xm}}{D}}\right]^{1/2}$$

$$\beta = \left[-\lambda_m^2 + \lambda_m\sqrt{\frac{\overline{N}_{xm}}{D}}\right]^{1/2} .$$

The boundary conditions on the $y = 0$ and b edges are

$$\begin{aligned}
&w(x,0) = 0 \rightarrow \psi(0) = 0 \\
&M_y(x,0) = 0 \rightarrow \psi''(0) = 0 \\
&M_y(x,b) = 0 \rightarrow \psi''(b) - \nu\lambda_m^2\psi(b) = 0 \\
&V(x,b) = 0 \rightarrow \psi'''(b) - (2-\nu)\lambda_m^3\psi'(b) = 0.
\end{aligned} \tag{8.42}$$

It is clear that the first two boundary conditions require that $A = C = 0$. Satisfying the other two boundary conditions results in the following relationship for the eigenvalues (i.e., the buckling load $\overline{N}_x = -N_x$).

$$-\beta \tanh\alpha\, b[\alpha^2 - \nu\lambda_m^2]^2 + \alpha \tan\beta\, b[\beta^2 + \nu\lambda_m^2]^2 = 0. \tag{8.43}$$

Thus, knowing the plate geometry and the material properties, one can solve for the buckling loads for each value of m. It can be shown that the minimum buckling load will occur for $m = 1$, thus a one-half sine wave in the longitudinal direction. However, note the complexity both in obtaining Equation (8.43), and then using that equation to obtain the buckling load, compared to the relative simplicity of Section 6.4 for solving the simpler problem of the plate completely simple supported on all four edges. The solutions of this problem have been catalogued in Reference 6.1 and are given below:

$$N_x = -\frac{k\pi^2 D}{b^2} \quad \text{and} \quad \sigma_{cr} = -\frac{k\pi^2 E}{12(1-\nu^2)}\left(\frac{h}{b}\right)^2.$$

For $\nu = 0.25$

a/b	0.50	1.0	2.0	3.0	4.0	5.0
k	4.40	1.44	0.698	0.564	0.516	0.506

Now to solve the same problem using Minimum Potential Energy. However, before doing so a brief discussion regarding boundary conditions is in order. They can be divided into two categories: geometric and stress. Geometric boundary conditions involve specifications on the displacement function and the first derivative, such as specifying the lateral displacement w or the slope at the boundary, $\partial w/\partial x$ or $\partial w/\partial y$, stress boundary conditions involve the specifications of the second and third derivative of the displacement function, such as the stress couples, M_x, M_y, M_{xy}, or the transverse shear resultants Q_x, Q_y, or the effective transverse shear resultants V_x, V_y, discussed in Chapter 2.

In using the Minimum Potential Energy Theorem, one must choose a deflection function that *at least* satisfies the geometric boundary conditions specified on the boundaries. This suitable function will give a reasonable approximate solution. Better yet, by assuming a deflection function that satisfies all specified boundary conditions, one can achieve a very good approximate solution. If one could choose a deflection function that satisfies all boundary conditions and the governing differential equation for the problem also, that is the exact solution! Finally, if one chose a deflection function that did not satisfy even the geometric boundary conditions, the solution would be inaccurate because in effect the solution would not be for the problem to be solved, but for some other problem for which the assumed deflection does satisfy the geometric boundary conditions.

In this example, the following function is assumed for the lateral deflection:

$$w(x,y) = Ay\sin\frac{\pi x}{a} \tag{8.44}$$

This satisfies all boundary conditions on the $x = 0$, a edges. It satisfies the geometric boundary condition that $w(x,0) = 0$, but does not satisfy the stress boundary conditions that $M(x,0) = M(x,b) = V(x,b) = 0$. Substituting Equation (8.44) and its derivatives into Equations (8.1) using (8.36) and (8.37), where of course $N_y = N_{xy} = p(x,y) = 0$ produces

$$\begin{aligned} V = {} & \frac{D}{2}\int_0^a\int_0^b\left\{\left[-Ay\frac{\pi^2}{a^2}\sin\frac{\pi x}{a}\right]^2 + 2(1-\nu)\left[-A\frac{\pi}{a}\cos\frac{\pi x}{a}\right]^2\right\}dxdy \\ & + \frac{N_x}{2}\int_0^a\int_0^b A^2y^2\frac{\pi^2}{a^2}\cos^2\frac{\pi x}{a}\,dxdy. \end{aligned} \tag{8.45}$$

Integrating Equation (8.45) gives

$$V = A^2D\left[\frac{\pi^4}{a^3}\frac{b^3}{3} + 2(1-\nu)\frac{\pi^2 b}{a}\right] + N_xA^2\frac{\pi^2b^3}{3a}.$$

Setting $\delta V = 0$, where the only variable with which to take a variation is A, produces the requirement that

$$N_x = -\left[\frac{\pi^2D}{a^2} + \frac{6(1-\nu)D}{b^2}\right]. \tag{8.46}$$

To compare this approximate result with the exact solution shown previously, let $a/b = 1$, and $\nu = 0.25$. From Equation (8.46)

$$N_{x_{cr}} = -1.456\frac{\pi^2D}{b^2}. \tag{8.47}$$

In the exact solution, the coefficient is 1.440. Hence, the difference between the approximate solution and the exact solution is approximately 1%, yet the three stress boundary conditions on the y = constant edges were not satisfied.

8.8 Functions for Displacements in Using Minimum Potential Energy for Solving Beam, Column, and Plate Problems

In the use of Minimum Potential Energy methods to solve beam, column, and plate problems, one usually needs to assume an expression for the lateral deflection $w(x)$ for the beam or column, and $w(x, y)$ for the plate. These must be single valued, continuous functions that satisfy all the boundary conditions, or at least the geometric ones. Below are a few functions useful in the solutions of beam and column problems.

Simple-simple

$$w(x) = \sum_{n=1}^{\infty} A_n \sin\frac{n\pi x}{L} \tag{8.48}$$

Simple-free

$$w(x) = Ax \tag{8.49}$$

Clamped-clamped

$$w(x) = A\left[1 - \cos\frac{2m\pi x}{a}\right] \tag{8.50}$$

Clamped-free

$$w(x) = Ax^2 \tag{8.51}$$

Clamped-simple

$$w(x) = A[L^3 x - 3Lx^3 + 2x^4] \tag{8.52}$$

Free-free

$$w = A. \tag{8.53}$$

In the case of a plate with varied boundary conditions, let $w(x, y) = f(x)g(x)$ where for $f(x)$ and $g(y)$ use the appropriate beam functions above. For example, consider a plate clamped on edges $y = 0$ and $y = b$, and clamped at $x = 0$ and simply supported at $x = a$. Assume the function:

$$w(x,y) = A_m[L^3x - 3Lx^3 + 2x^4]\left[1 - \cos\frac{2m\pi y}{b}\right]. \qquad (8.54)$$

Keep in mind none of the above functions are unique, and thus the engineer may use his ingenuity to conceive functions best for the solution of that particular problem. For instance, suppose a plate had one edge simply supported at $y = 0$, $0 \le x \le a/2$, and clamped from $a/2 \le x \le a$. No analytical solution could be obtained but an approximate solution using energy methods is always attainable.

Perhaps the most complete and useful tabulation of functions, their derivatives and their integrals, to use in energy methods are those of Warburton [8.3] and Young and Felgar [3.1, 3.2].

8.9 References

8.1. Reissner, E. (1950) On a Variational Theorem in Elasticity, *J. Math. Phys.*, Vol. 29, pg. 90.

8.2. Mura, T. and Koya, T. (1992) *Variational Methods in Mechanics*, Oxford University Press, March.

8.3. Warburton, G. (1968) The Vibration of Rectangular Plates, *Proceedings of the Institute of Mechanical Engineers*, pp. 371-384.

8.10 Problems

8.1. Consider a steel plate ($E = 30\times10^6$ psi, $\nu = 0.25$, $\sigma_y = 30{,}000$ psi) used as a portion of a bulkhead on a ship. The bulkhead is 60″ long and 30″ wide subjected to an in-plane compressive load in the longer direction. What thickness must the plate be to have a buckling stress equal to the yield stress if:
(a) the plate is simply supported on all four edges?
(b) the plate is simply supported on three edges and free on one unloaded edge?

8.2. Given a column of width b, height h, and length L, simply supported at each end, use the principle of Minimum Potential Energy to determine the buckling load, if one assumes the deflection to be

(a) $w(x) = A\dfrac{x}{L}(L-x)$

(b) $w(x) = \dfrac{A}{L^3}[2Lx^3 - x^4 - L^3x]$

where in each case A is an amplitude.
Do the deflections assumed above satisfy the geometric boundary conditions? Do they satisfy the stress boundary conditions?

8.3. Consider the plates below, each subjected to a uniform axial compressive load per inch of width, $\overline{N}_x = -N_x$ (lbs./in.) in the x direction. Determine a suitable deflection function $w(x, y)$ for each case for subsequent use in the Principle of Minimum Potential Energy to determine the critical load $\overline{N}_x$.

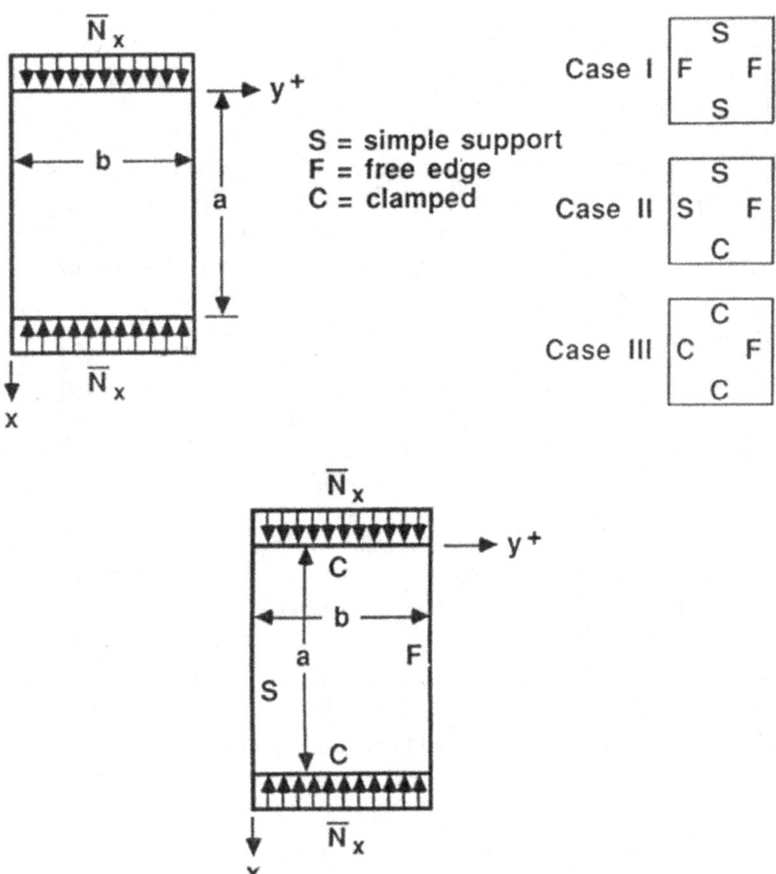

8.4. For an end plate in a support structure with the following boundary conditions, use the Principle of Minimum Potential Energy to determine the buckling load, if one assumes the deflection function to be $w = A[1-\cos(2\pi x/a)]$, where A is the unknown amplitude.

8.5. Consider a rectangular plate of $0 \le x \le a$, $0 \le y \le b$, $-h/2 \le z \le h/2$. If the lateral deflection $w(x, y)$ is assumed to be in a separable form $w = f(x)g(y)$, and if $w = 0$ on all boundaries, determine the amount of strain energy due to the terms comprising the Gaussian curvature. See (8.36).

8.6. The base of a missile launch platform consists in part of vertical rectangular plates of height a, and width b, where $a > b$. They are tied into the foundation below and the platform above such that those edges are considered clamped.

However, on their vertical edges they are tied into I-beams, such that those edges can only be considered simply supported. Using the Theory of Minimum Potential Energy, derive the equation for the buckling load per inch of edge distance, $N_{x_{cr}}$, for these plates, using a suitable deflection function, so that the plates can be designed to resist buckling.

8.7. An alternative to the design of Problem 8.6 would be to 'beef up' the vertical support beams such that the plate members can be considered to have their vertical edges clamped. Thus the plates have all four edges clamped. Employing a suitable deflection function, use the Theorem of Minimum Potential Energy to determine an expression for the critical buckling load per unit edge distance, $N_{x_{cr}}$, to use in designing the plates. Is the plate with all edges clamped thicker or thinner than the one with the sides simply supported in Problem 8.6, to have the same buckling load?

8.8. The legs of a water tower consist of three columns of length *a*, constant flexural stiffness *EI*, simply supported at one end and clamped at the other end. Using the Theorem of Minimum Potential Energy, and a suitable function for the lateral deflection, calculate the buckling load P_{cr} for each leg, in order that they may be properly designed.

8.9. Consider a beam of length *L*, and constant cross-section, i.e., *EI* is a constant. The beam is subjected to a load $q(x) = a + c(x/L)$, (lbs./in.) applied laterally where *a* and *c* are constants. The beam is simply supported on both ends. Using Minimum Potential Energy, and assuming $w(x) = B\sin(\pi x/L)$, determine the maximum deflection, *w*, and the maximum bending stress, σ_x. Consider the beam to be of unit width, i.e., $b = 1$.

8.10. A beam of length *L*, and constant cross-section (EI = constant) is subjected to a lateral load $q(x) = (q_0 x)/L$, where q_0 is a constant, and is simply supported at each end. Using Minimum Potential Energy, and assuming $w(x) = A\sin(\pi x/L)$, where *A* is a constant to be determined, determine the maximum deflection, *w*, and the maximum stress, σ_x, in the beam.

8.11. Consider the beam of Section 8.3 to be simply supported at each end and subjected to a uniform lateral load q_0 (lbs./in.). Assuming the deflection to be $w(x) = A\sin(\pi x/L)$, use the Principle of Minimum Potential Energy to determine *A*.

8.12. Consider a beam-column simply supported at one end and clamped at the other. Using the Theorem of Minimum Potential Energy, and assuming an admissible form for the lateral deflection, *w*(*x*), calculate the in-plane load, P_{cr} (lbs.), to buckle the column.

8.13. Consider a beam of stiffness *EI*, length *L*, width *b*, height *h*, simply supported at each end, subjected to a uniform lateral load, q_0 (lbs./in.). Use Minimum Potential Energy, employing a deflection function

$$w(x) = \sum_{n=1}^{N} A_n \sin\frac{n\pi x}{L}$$

where $N = 3$, to determine the maximum deflection and maximum stress. Compare the answer with the exact solution.

8.14. Consider a column of length L, clamped at one end and simply supported at the other end. Using a buckling mode shape of

$$w(x) = A[L^3 x - 3Lx^3 + 2x^4]$$

where A is the buckle amplitude. Use Minimum Potential Energy to determine the axial critical buckling load, P_{cr}.

8.15. Consider a beam of constant flexural stiffness EI, of length L, clamped at each end. Using Hamilton's Principle, and an assumed deflection of

$$w(x,t) = A[1 - \cos(2\pi x / L)] \sin \omega_n t,$$

determine the fundamental natural frequency, and compare it with the exact solution.

CHAPTER 9

REISSNER'S VARIATIONAL THEOREM AND ITS APPLICATIONS

9.1 Introduction

A general discussion of Reissner's Variational Theorem is presented, followed by a treatment of the theory of moderately thick beams which represents a striking example of the power of this technique. The first application is the development of the governing equations for the static deformations of moderately thick rectangular beams, including the effects of transverse shear deformation and transverse normal stress. The second application involves the use of the theorem, together with Hamilton's Principle, to develop a theory of beam vibrations including rotatory inertia, in addition to the other effects listed above.

The Calculus of Variations has long been recognized as a powerful mathematical tool in many branches of mathematical physics and engineering. Variational principles are found to constitute the central core of many of the most useful techniques in such fields as dynamics, optics and continuum mechanics. The utility of such principles is two-fold: first, they provide a very convenient method for the derivation of the governing equations and natural boundary conditions for complex problems and, second, they provide the mathematical foundation required to produce consistent approximate theories. It is in this second role that variational methods have been most useful in theory of elasticity. There are two variational principles in the classical theory of elasticity, namely the Principle of Minimum Potential Energy, treated in Chapter 8, and the Principle of Minimum Complementary Energy. It will be useful to discuss these two principles very briefly here, because it was certain of their features which led E. Reissner, in 1950, to propose a third, more general, variational theorem.

The Principle of Minimum Potential Energy was discussed in Chapter 8. It was noted that, in carrying out the variations to minimize the potential energy, V, the class of admissible variations are displacements satisfying the boundary conditions, and the appropriate stress-strain relations have to be obtained separately. The resulting Euler-Lagrange equations of the variational problem are then equilibrium equations, written in terms of displacements. When the principle is used to formulate approximate theories, i.e., beam, plate or shell theory, it can therefore only yield appropriate equilibrium equations and the stress-strain or stress-displacement relations must be obtained independently, as stated above

The Principle of Minimum Complementary Energy may be stated as follows: of all the stress systems satisfying equilibrium and the stress boundary conditions, that which satisfies the compatibility conditions corresponds to a minimum of the complementary energy V^* defined as,

$$V^* = \int_R W \mathrm{d}R - \int_{S_u} T_i u_i \mathrm{d}S \tag{9.1}$$

where S_u denotes that part of the boundary S on which displacements are prescribed.

It is emphasized that, in the Principle of Minimum Complementary Energy, the class of admissible variations are stresses which must satisfy equilibrium everywhere, as well as the stress boundary conditions. The Euler-Lagrange equations of the variational problem are here compatibility equations or stress-displacement relations which insure the satisfaction of the compatibility requirements. Thus, when this principle is used in developing approximate theories, only the stress-displacement relations may be obtained and the equilibrium relations must be derived independently.

It should be pointed out that, in the language of structural analysis, the Principle of Minimum Potential Energy corresponds to the Principle of Virtual Displacements, while the Principle of Minimum Complementary Energy corresponds to the Theorem of Castigliano.

So in these two theories, one must either satisfy the stress-strain relations exactly and formulate approximate equilibrium conditions or vice-versa. As a result, any approximate theory obtained by such means runs the risk of inconsistency. These considerations led Reissner in 1950 to propose a third variational theorem of elasticity which would yield as its Euler-Lagrange equations both the equilibrium equations and the stress-displacement relations. Clearly, if such a principle could be developed, its use would yield approximate theories which would satisfy both requirements to the same degree and would, therefore, have the advantage of consistency. The result of this investigation is the Reissner Variational Theorem, which may be stated as follows:

Of all the stress and displacement states satisfying the boundary conditions, those which also satisfy the equilibrium equations and the stress-displacement relations correspond to a minimum of functional ψ defined as,

$$\psi = \int_R H \mathrm{d}R - \int_R F_i u_i \mathrm{d}R - \int_{S_t} T_i u_i \mathrm{d}S \tag{9.2}$$

where S_t = portion of S on which stresses are prescribed. Again, F_i and T_i are body forces and surface tractions prescribed by the problem considered.

$$H = \sigma_{ij}\varepsilon_{ij} - W(\sigma_{ij})$$

$W(\sigma_{ij})$ = strain energy density function in terms of stresses only.

In a rectangular coordinate system $W(\sigma_{ij})$ in general is written as follows for an isotropic material

$$\begin{aligned} W(\sigma_{ij}) = \frac{1}{2E}\Big[&\sigma_x^2 + \sigma_y^2 + \sigma_z^2 - 2\nu(\sigma_x\sigma_y + \sigma_y\sigma_z + \sigma_z\sigma_x) \\ &+ 2(1+\nu)(\sigma_{xy}^2 + \sigma_{zx}^2 + \sigma_{yz}^2)\Big] \end{aligned} \tag{9.3}$$

The proof of the Theorem is now presented. The tensor notation is standard, and hence will not be explained here. Also, the operations used herein in taking variations are identical to those of partial differentiation, such as $\delta(\sigma_{ij}^2) = 2\sigma_{ij}\delta\sigma_{ij}$, $\delta(\sigma_{ij}\varepsilon_{ij}) = \sigma_{ij}\delta\varepsilon_{ij} + \varepsilon_{ij}\delta(\sigma_{ij})$, and $\frac{\partial}{\partial x_j}(\delta\sigma_{ij}) = \delta\left(\frac{\partial \sigma_{ij}}{\partial_{xj}}\right)$.

Taking the variation of ψ and equating it to zero, one obtains:

$$\delta\psi = \int_R \left[\sigma_{ij}\delta\varepsilon_{ij} + \varepsilon_{ij}\delta\sigma_{ij} - \frac{\partial W}{\partial \sigma_{ij}}\delta\sigma_{ij}\right]\mathrm{d}R - \int_R F_i\delta u_i \mathrm{d}R - \int_{S_t} T_i\delta u_i \mathrm{d}S = 0 \tag{9.4}$$

where $\varepsilon_{ij} = \frac{1}{2}\left(\frac{\partial u_i}{\partial x_j} + \frac{\partial u_j}{\partial x_i}\right)$.

It should be noted that all stress and strain components have been varied, while F_i and T_i, which are prescribed functions, are not. Rearranging the above expression, one obtains,

$$\begin{aligned}\delta\psi = {} & \int_R \left\{\left[\varepsilon_{ij} - \frac{\partial W}{\partial \sigma_{ij}}\right]\delta\sigma_{ij} + \frac{1}{2}\sigma_{ij}\left[\frac{\partial}{\partial x_j}(\delta u_i) + \frac{\partial}{\partial x_i}(\delta u_j)\right]\right\}\mathrm{d}R \\ & - \int_R F_i\delta u_i \mathrm{d}R - \int_{S_t} T_i u_i \mathrm{d}S = 0.\end{aligned} \tag{9.5}$$

The terms $\sigma_{ij}\frac{\partial}{\partial x_j}(\delta u_i)$ and $\sigma_{ij}\frac{\partial}{\partial x_i}(\delta u_j)$ are symmetric with respect to *i* and *j*, and we may, therefore, interchange these indices and obtain,

$$\begin{aligned}\delta\psi = {} & \int_R \left\{\left[\varepsilon_{ij} - \frac{\partial W}{\partial \sigma_{ij}}\right]\delta\sigma_{ij} + \sigma_{ij}\frac{\partial}{\partial x_j}(\delta u_i)\right\}\mathrm{d}R - \int_R F_i\delta u_i \mathrm{d}R \\ & - \int_{S_t} T_i\delta u_i \mathrm{d}S = 0\end{aligned} \tag{9.6}$$

Note that $\frac{\partial}{\partial x_j}(\sigma_{ij}\delta u_i) = \sigma_{ij}\frac{\partial}{\partial x_j}(\delta u_i) + \frac{\partial \sigma_{ij}}{\partial x_j}\delta u_i$, so that Equation (9.6) may be written,

$$\begin{aligned}\delta\psi = {} & \int_R \left\{\left[\varepsilon_{ij} - \frac{\partial W}{\partial \sigma_{ij}}\right]\delta\sigma_{ij} + \frac{\partial}{\partial x_j}(\sigma_{ij}\delta u_i) - \frac{\partial \sigma_{ij}}{\partial x_j}\delta u_i\right\}\mathrm{d}R - \int_R F_i\delta u_i \mathrm{d}R \\ & - \int_{S_t} T_i\delta u_i \mathrm{d}S = 0.\end{aligned}$$

Applying the Green-Gauss divergence theorem and remembering that $\boldsymbol{\delta u_i} = 0$, on all surfaces where displacements are prescribed, one obtains:

$$\int_R \frac{\partial}{\partial x_j}(\sigma_{ij}\delta u_i)\mathrm{d}R = \int_{S_t} \sigma_{ij}\nu_j \delta u_i \mathrm{d}S = \int_{S_t} T_i \delta u_i \mathrm{d}S \tag{9.7}$$

where here ν_j is the direction cosine.

Finally, substituting Equation (9.7) into (9.6) yields the equations,

$$\delta\psi = \int_R \left\{ \left[\varepsilon_{ij} - \frac{\partial W}{\partial \sigma_{ij}} \right] \delta\sigma_{ij} - \left[\frac{\partial \sigma_{ij}}{\partial x_j} + F_i \right] \delta u_i \right\} \mathrm{d}R = 0. \tag{9.8}$$

Since $\delta\sigma_{ij}$ and δu_i are arbitrary variations, Equation (9.8) is satisfied only if the stresses σ_{ij} and strains ε_{ij} satisfy the equations,

$$\frac{\partial \sigma_{ij}}{\partial x_j} + F_i = 0 \tag{9.9}$$

$$\varepsilon_{ij} = \frac{\partial W(\sigma_{ij})}{\partial \sigma_{ij}} \tag{9.10}$$

Equations (9.9) and (9.10) are the equilibrium and stress-strain displacement relations of elasticity. Thus, the Reissner Variational Theorem is found to be equivalent to the three-dimensional equations of elasticity and is, therefore, established. Now consider typical applications of the Theorem to the static and dynamic deformations of beams, because beams permit the simplest example.

9.2 Static Deformation of Moderately Thick Beams

As a first illustration, consider the development of a theory for the static deformations of moderately thick beams in which the effects of transverse shear deformation and transverse normal stress are taken into account. Consider a beam of rectangular cross-section of width b, height h and length L, as shown in Figure 9.1 subjected to a distributed load $q(x)$ acting on the surface $z = +h/2$.

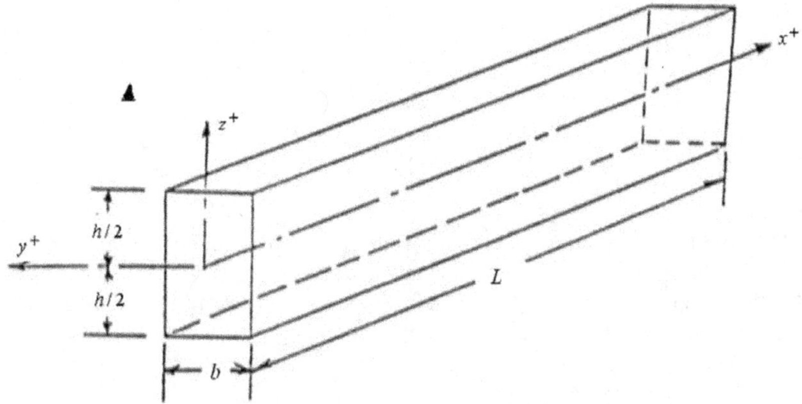

Figure 9.1. Rectangular Beam.

In order to apply the variational theorem, one must first assume functions for the stresses in the beams. In this case the following are assumed.

$$\sigma_x = \frac{Mz}{I}, \text{ where } I = \frac{bh^3}{12},$$

$$\sigma_{xz} = \frac{3Q}{2A}\left[1-\left(\frac{z}{h/2}\right)^2\right], \text{ where } A = bh,$$

$$\sigma_z = \frac{3q}{4b}\left[\frac{z}{h/2}+\frac{2}{3}-\frac{1}{3}\left(\frac{z}{h/2}\right)^3\right] \tag{9.11}$$

$$\sigma_y = \sigma_{yx} = \sigma_{yz} = 0.$$

It should be noted that the form of the stress components σ_x and σ_{xz} is identical to that of classical theory. The form of the transverse normal stress σ_z may easily be derived from the stress equation of equilibrium in the thickness directions, as a consequence of the assumptions made above for σ_x and σ_{xz}. The expression shown in (9.11) is derived for

$$\sigma_z(+h/2) = q \text{ and } \sigma_z(-h/2) = 0.$$

An analogous expression can be derived easily if there were a normal stress on the lower surface.

The stress-couple M and shear resultant Q are defined in the usual manner by the equations,

$$M(x) = \int_{-h/2}^{h/2} b\sigma_x z\mathrm{d}z$$

$$Q(x) = \int_{-h/2}^{h/2} b\sigma_{xz}\mathrm{d}z. \tag{9.12}$$

It should be noted that Equations (9.11) satisfy all of the stress boundary conditions.

As in classical beam bending theory, one assumes that beam cross-sections undergo translation and rotation but no deformation in the plane of the cross-section. Such displacements are of the following form for bending only (no stretching), which is the simplest case:

$$u = z\bar{\alpha}(x)$$

$$w = w(x) \tag{9.13}$$

It should be noted that the cross-sections will not be assumed to remain normal to the deformed middle surface; this assumption, made in classical beam theory, is equivalent to the neglect of transverse shear deformation, and will not be made here.

The appropriate strain displacement relations may be written

$$\varepsilon_x = \frac{\partial u}{\partial x} = z\bar{\alpha}'(x)$$

$$\varepsilon_{xz} = \frac{1}{2}\left(\frac{\partial u}{\partial z} + \frac{\partial w}{\partial x}\right) = \frac{1}{2}(\bar{\alpha} + w') \tag{9.14}$$

$$\varepsilon_z = \frac{\partial w}{\partial z} = 0$$

where the primes denote differentiation with respect to x.

For the present case, the functional ψ, Equation (9.2), takes the form,

$$\psi = \int_0^L \int_{-h/2}^{h/2} b\left\{\sigma_x z\bar{\alpha}' + \sigma_{xz}(\bar{\alpha} + w') - \frac{1}{2E}\left[\sigma_x^2 + \sigma_z^2 - 2\nu\sigma_x\sigma_z + 2(1+\nu)\sigma_{xz}^2\right]\right\}\mathrm{d}z\mathrm{d}x - \int_0^L wq\mathrm{d}x. \tag{9.15}$$

Substituting Equations (9.11) into Equation (9.15) and carrying out the integration with respect to z, one obtains,

$$\psi = \int_0^L \left\{ M\bar{\alpha}' + Q(\bar{\alpha} + w') - \frac{M^2}{2EI} + \frac{6\nu qM}{5EA} - \frac{3Q^2}{5GA} - qw \right\} \mathrm{d}x + \int_0^L \int_{-h/2}^{h/2} \frac{\sigma_z^2}{2E} b\mathrm{d}z\mathrm{d}x \tag{9.16}$$

It should be noted that the integration of the term in σ_z^2 has not been carried out because this term depends only on q and not on the basic unknown stresses and displacements, $\bar{\alpha}$, W, M, and Q. Thus, when variations to minimize ψ are taken, the term in σ_z^2 will not contribute to the result. One may now obtain the governing equations by minimizing the functional ψ of Equation (9.16). Taking the variation of this equation gives,

$$\delta\psi = \int_0^L \left\{ M\delta(\bar{\alpha}') + \bar{\alpha}'\delta M + Q(\delta\bar{\alpha} + \delta(w')) + (\bar{\alpha} + w')\delta Q - \frac{M}{EI}\delta M + \frac{6\nu q}{5EA}\delta M - \frac{6Q}{5GA}\delta Q - q\delta w \right\} \mathrm{d}x = 0 \tag{9.17}$$

Integrating by parts and rearranging, Equation (9.17) may be written in the form,

$$\delta\psi = \left[M\delta\bar{\alpha} + Q\delta w \right]_0^L + \int_0^L \left\{ [Q - M']\delta\bar{\alpha} - [Q' + q]\delta w + \left[\bar{\alpha}' - \frac{M}{EI} + \frac{6\nu q}{5EA} \right]\delta M + \left[\bar{\alpha} + w' - \frac{6Q}{5GA} \right]\delta Q \right\} \mathrm{d}x = 0 \tag{9.18}$$

Setting the first term equal to zero yields the natural boundary conditions for the beam. It is seen that, either $M = 0$ or $\bar{\alpha}$ must be prescribed at $x = 0$ and L and either $Q = 0$ or w must be prescribed at $x = 0$ and L.

Finally, since the variations $\delta\bar{\alpha}, \delta w, \delta M,$ and δQ are all independent arbitrary functions of x, the only way in which the definite integral of Equation (9.18) can be made to vanish is by requiring the unknowns M, Q, $\bar{\alpha}$ and w to satisfy the equations,

$$-\frac{\mathrm{d}M}{\mathrm{d}x} + Q = 0 \tag{9.19}$$

$$\frac{\mathrm{d}Q}{\mathrm{d}x} + q = 0 \tag{9.20}$$

$$\frac{\mathrm{d}\bar{\alpha}}{\mathrm{d}x} - \frac{M}{EI} + \frac{6\nu q}{5EA} = 0 \tag{9.21}$$

$$\bar{\alpha} + \frac{\mathrm{d}w}{\mathrm{d}x} - \frac{6Q}{5GA} = 0 \tag{9.22}$$

Note that Equations (9.19) and (9.20) are identical to the equilibrium equations of classical beam theory. This is as expected since no new stress resultants or stress couples were introduced. Considering Equation (9.22), it is seen that the quantity $\alpha + w'$ is precisely the change in the angle between the beam cross-section and the middle surface occurring during the deformation; Equation (9.22) shows that this angular change, which is a measure of the shear deformation, is proportional to Q/A which is the average shear stress. In addition, note that as $G \to \infty$, the shear deformation tends to vanish as assumed in classical beam theory. Finally, observe that the third term in Equation (9.21) depends on the lateral load q and the Poisson's ratio ν; this term would vanish if one assumed $\nu = 0$ as in classical beam theory. It is identified as the effect of the transverse normal stress σ_z which is proportional to q, according to the initial assumptions (see Equation (9.11)).

Solutions of Equations (9.19) through (9.22) may easily be obtained for typical loading and boundary conditions. These solutions reveal that for beams of isotropic materials, the effects of transverse shear deformation and transverse normal stress are negligible for sufficiently large values of L/h and become important as L/h decreases and becomes of order unity.

9.3 Flexural Vibrations of Moderately Thick Beams

As a second example, a theory of free vibrations for moderately thick beams of rectangular cross-section is treated; this will include the effects of transverse shear deformation and rotatory inertia.

In order to derive the equations of motion, one now applies Hamilton's Principle in conjunction with the Reissner Variational Theorem. It will be remembered that Hamilton's Principle is nothing but a variational statement of Newton's Laws of Motion. Thus, one may state that the motion of the beam of Figure 9.1 will be such as to minimize the integral

$$\Phi = \int_{t_1}^{t_2} (T - \psi)\mathrm{d}t \tag{9.23}$$

where T = kinetic energy of the system, ψ = Reissner functional, t = time. The quantity $(T - \psi) = L$ is often called the Lagrangian.

The equations of motion will now be obtained from the condition,

$$\delta\Phi = 0 \tag{9.24}$$

and it must be remembered that all stresses, strains and displacements are now functions of time, as well as the space coordinates x and z. Equations (9.23) and (9.24) are general, for any elastic body.

The kinetic energy for the beam of Figure 9.1 may be written in the form,

$$T = \int_0^L \int_{-h/2}^{h/2} \frac{\rho_m b}{2}\left[\left(\frac{\partial u}{\partial t}\right)^2 + \left(\frac{\partial w}{\partial t}\right)^2\right] \mathrm{d}z\mathrm{d}x \tag{9.25}$$

where ρ_m is the mass density of the beam material. Substituting Equations (9.13) into Equation (9.25) and integrating with respect to z gives,

$$T = \int_0^L \frac{\rho_m}{2}\left[I\left(\frac{\partial \bar{\alpha}}{\partial t}\right)^2 + A\left(\frac{\partial w}{\partial t}\right)^2\right] \mathrm{d}x \tag{9.26}$$

where I and A are the area moment of inertia and the cross-sectional area of the beam, respectively.
The substitution of Equations (9.16) and (9.26) into Equation (9.23) then yields,

$$\begin{aligned}\Phi = \int_{t_1}^{t_2} \int_0^L \Bigg\{ & \frac{\rho_m}{2}\left[I\left(\frac{\partial \bar{\alpha}}{\partial t}\right)^2 + A\left(\frac{\partial w}{\partial t}\right)^2\right] - M\frac{\partial \bar{\alpha}}{\partial x} - Q\left(\bar{\alpha} + \frac{\partial w}{\partial x}\right) \\ & + \frac{M^2}{2EI} - \frac{6\nu\, qM}{5EA} + \frac{3Q^2}{5GA} + qw\Bigg\} \mathrm{d}x\mathrm{d}t\end{aligned} \tag{9.27}$$

where the term in σ_z^2 has been dropped since it will not contribute to the variations (as explained previously).

The governing equations are then obtained by taking the variation of Equation (9.27) and setting the result equal to zero. It is found that the natural boundary conditions are the same as for the static case, while the initial deflection and velocity must also be specified. The equations of motion are obtained in the form,

$$Q - \frac{\partial M}{\partial x} + \rho_m I c\frac{\partial^2 \bar{\alpha}}{\partial t^2} = 0 \tag{9.28}$$

$$\frac{\partial Q}{\partial x} - \rho_m A\frac{\partial^2 w}{\partial t^2} + q(x,t) = 0 \tag{9.29}$$

$$\frac{\partial \bar{\alpha}}{\partial x} - \frac{M}{EI} + \frac{6\nu\, q}{5EA} = 0 \tag{9.30}$$

$$\bar{\alpha} + \frac{\partial w}{\partial x} - \frac{6Qk}{5GA} = 0 \tag{9.31}$$

In the above equations, two tracing constants c and k are introduced for the purpose of identifying terms. Note that Equation (9.28) is identical to the corresponding moment equilibrium condition of classical beam theory, except for the term $\rho_m Ic(\partial^2 \overline{\alpha}/\partial t^2)$ which represents the contribution of rotatory inertia. Thus, when $c = 1$ in the resulting solutions, rotatory inertia effects are included, when $c = 0$, the theory neglects the effect of rotatory inertia. Equation (9.29) is identical to the classical beam theory equation for transverse force equilibrium with the inertia term added. Equation (9.30) exhibits the term $6\nu q/5EA$, which is the contribution of transverse normal stress; since this is the only term in which ν appears explicitly, setting $\nu = 0$ is equivalent to neglecting the transverse normal stress. Equation (9.31) is nearly identical to the corresponding stress-strain relation of classical theory with the term $6Qk/5GA$ representing the effect of transverse shear deformation which is included when $k = 1$, and neglected when $k = 0$. Now consider a simple application of this theory.

9.3.1 Natural Frequencies of a Simple-Supported Beam

For free vibrations, Equations (9.28) through (9.31) reduce to,

$$\left.\begin{aligned} &Q - \frac{\partial M}{\partial x} + \rho_m Ic \frac{\partial^2 \overline{\alpha}}{\partial t^2} = 0 \\ &\frac{\partial Q}{\partial x} - \rho_m A \frac{\partial^2 w}{\partial t^2} = 0 \\ &\frac{\partial \overline{\alpha}}{\partial x} - \frac{M}{EI} = 0 \\ &\overline{\alpha} + \frac{\partial w}{\partial x} - \frac{6Qk}{5GA} = 0 \end{aligned}\right\} \tag{9.32}$$

It is convenient to reduce these equations to a system of two equations in the unknown displacements w and $\overline{\alpha}$.

From the first and the third of Equations (9.32), one obtains,

$$M = EI \frac{\partial \overline{\alpha}}{\partial x}$$

$$Q = EI \frac{\partial^2 \overline{\alpha}}{\partial x^2} - \rho_m Ic \frac{\partial^2 \overline{\alpha}}{\partial t^2}$$

and the substitution of these expressions in the second and fourth of Equations (9.32) yields,

$$
\left.\begin{aligned}
&\frac{\partial^3 \bar{\alpha}}{\partial x^3} - \frac{\rho_m c}{E}\frac{\partial^3 \bar{\alpha}}{\partial x \partial t^2} - \frac{\rho_m A}{EI}\frac{\partial^2 w}{\partial t^2} = 0 \\
&\bar{\alpha} + \frac{\partial w}{\partial x} - \frac{kh^2}{10}\left[\frac{E}{G}\frac{\partial^2 \bar{\alpha}}{\partial x^2} - \frac{\rho_m c}{G}\frac{\partial^2 \bar{\alpha}}{\partial t^2}\right] = 0
\end{aligned}\right\} \tag{9.33}
$$

For a simply supported beam of length L, the boundary conditions are

$$w = M = 0 \text{ for } x = 0, L,$$

and when the beam is oscillating in a normal mode, the motion is harmonic so that the solutions for $\bar{\alpha}$ and w may be taken in the form,

$$
\left.\begin{aligned}
w &= W_n \sin\frac{n\pi x}{L}\cos\omega_n t \\
\alpha &= \gamma_n \cos\frac{n\pi x}{L}\cos\omega_n t
\end{aligned}\right\} \tag{9.34}
$$

where W_n and γ_n are the amplitudes of the translation and rotation respectively, and ω_n is the natural circular frequency of the nth mode of vibration. It is easily verified that these expressions satisfy the boundary conditions. The substitution of Equation (9.34) in Equations (9.33) yields two simultaneous homogeneous algebraic equations for the amplitudes W_n and γ_n; these are,

$$
\left.\begin{aligned}
&\left[\left(\frac{n^3\pi^3}{L^3} - \frac{\rho_m c}{E}\left(\frac{n\pi}{L}\right)\omega_n^2\right)\right]\gamma_n + \frac{\rho_m A}{EI}\omega_n^2 W_n = 0 \\
&\left[1 + \frac{kh^2}{10}\left(\frac{E}{G}\frac{n^2\pi^2}{L^2} - \frac{\rho_m c}{G}\omega_n^2\right)\right]\gamma_n + \left(\frac{n\pi}{L}\right)W_n = 0
\end{aligned}\right\} \tag{9.35}
$$

Since (9.35) forms a homogeneous system, the condition for a non-trivial solution is that the determinant of the coefficients W_n and γ_n equal zero. That is termed an eigenvalue problem. Solving the determinant yields a frequency equation, from which solutions involve discrete values of the natural circular frequencies, which are termed eigenvalues.

The amplitude ratios are obtained from satisfying either of the two equations comprising (9.35). Thus, the amplitude ratio is given by,

$$\gamma_n = -\frac{(\rho_m A/EI)\omega_n^2}{(n\pi/L)^3 - (\rho_m c/E)\frac{n\pi}{L}\omega_n^2}W_n. \tag{9.36}$$

The frequency equation may be written in the form,

$$\omega_n^4 - \left[\frac{10G}{\rho_m h^2 kc} + \left(\frac{n\pi}{L}\right)^2 \left(\frac{E}{\rho_m c} + \frac{5}{6}\frac{G}{\rho_m k}\right)\right]\omega_n^2 + \frac{10}{h^2 kc}\left(\frac{G}{\rho_m}\right)\left(\frac{EI}{\rho_m A}\right)\left(\frac{n\pi}{L}\right)^4 = 0 \tag{9.37}$$

The natural frequencies for the case where both transverse shear deformation and rotatory inertia are included may be obtained by solving Equation (9.37) with $k = c = 1$. The frequency equation is then of the form,

$$\omega_n^4 - \left[\frac{10G}{\rho_m h^2} + \left(\frac{n\pi}{L}\right)^2 \left(\frac{E}{\rho_m} + \frac{5}{6}\frac{G}{\rho_m}\right)\right]\omega_n^2 + \frac{10}{h^2}\left(\frac{G}{\rho_m}\right)\left(\frac{EI}{\rho_m A}\right)\left(\frac{n\pi}{L}\right)^4 = 0 \tag{9.38}$$

To obtain a simplified theory neglecting the effect of rotatory inertia, but retaining transverse shear deformation, set $k = 1$ and $c = 0$ after multiplying the frequency equation (9.37) by c. The resulting simplified frequency equation may be written,

$$\left[\frac{10G}{\rho_m h^2} + \left(\frac{n\pi}{L}\right)^2 \left(\frac{E}{\rho_m}\right)\right]\omega_n^2 - \frac{10}{h^2}\left(\frac{G}{\rho_m}\right)\left(\frac{EI}{\rho_m A}\right)\left(\frac{n\pi}{L}\right)^4 = 0 \tag{9.39}$$

and the natural frequencies are given by,

$$\omega_n^2 = \frac{\left(\frac{EI}{\rho_m A}\right)\left(\frac{n\pi}{L}\right)^4}{1 + \frac{n^2\pi^2}{10}\left(\frac{E}{G}\right)\left(\frac{h}{L}\right)^2} \tag{9.40}$$

Finally, to obtain a frequency equation in which both shear deformation and rotatory inertia are neglected, multiply Equation (9.37) by kc and set $k = c = 0$; the frequencies are then given by,

$$\omega_n^2 = \left(\frac{EI}{\rho_m A}\right)\left(\frac{n\pi}{L}\right)^4 \tag{9.41}$$

Equation (9.41) is easily recognized to be the well-known solution of classical beam theory for a simply-supported beam. A few calculations using Equations (9.38) and (9.40) will show that most of the error (approximately 90%) of the classical theory is due to the neglect of transverse shear deformation, so that accurate results may be obtained by Equation (9.40) which still neglects rotatory inertia, but has the advantage of simplicity. Comparison of Equations (9.40) and (9.41) reveals that the effect of shear deformation is to reduce the square of the frequencies by a factor equal to $1+\frac{n^2\pi^2}{10}\left(\frac{E}{G}\right)\left(\frac{h}{L}\right)^2$. It is seen that this factor increases with increasing h/L, so that the errors through using classical theory tend to become larger as the beam becomes "stubbier". This factor also increases as n increases, indicating that classical theory is only adequate for lower modes of vibration, and it becomes increasingly inaccurate for higher modes.

CHAPTER 10

ANISOTROPIC ELASTICITY AND COMPOSITE LAMINATE THEORY

10.1 Introduction

As discussed in the first nine chapters, an isotropic material is one that has identical mechanical, physical, thermal and electrical properties in every direction. Isotropic materials involve only four elastic constants, the modulus of elasticity, E, the shear modulus, G, the bulk modulus K and Poisson's ratio, ν. However, only two are independent, and the following relationships exist: See Equation (1.3).

$$G = \frac{E}{2(1+\nu)}, \quad K = \frac{E}{3(1-2\nu)} \qquad \text{(Isotropic only)} \qquad (10.1)$$

Most engineers and material scientists are well schooled in the behavior and design of isotropic materials, which include the family of most metals and pure polymers. The rapidly increasing use of anisotropic materials such as composite materials has resulted in a materials revolution and requires a new knowledge base of anisotropic material behavior.

Before understanding the physical behavior of composite material structures and before being able to quantitatively determine the stresses, strains, deformations, natural frequencies, and buckling loads in such structures, a clear understanding of anisotropic elasticity is necessary. In general, isotropic materials are mathematical approximations to the true situation. For instance, in polycrystalline metals, the structure is usually made up of numerous anisotropic grains, wherein macroscopic isotropy exists in a statistical sense only because the anisotropic individual grains are randomly oriented. However, the same materials could be macroscopically anisotropic due to cold working, forging or spinning during a fabrication process. Other materials such as wood, human and animal bone, and most fiber reinforced materials are anisotropic.

Fiber reinforced composite materials are unique in application because the use of long fibers results in a material which has a higher strength-to-density ratio and/or stiffness-to-density ratio than any other material system at moderate temperatures, and there exists the opportunity to uniquely tailor the fiber orientations to a given geometry, applied load and environment. For short fiber composites, used mainly in high production, low cost systems, the use of fibers makes the composites competitive and superior to plastic and metal alternatives. Finally, the use of two or more kinds of dissimilar fibers within one matrix is termed a hybrid composite, where one fiber is stronger or stiffer while the other fiber is less expensive but desirable for less critical locations in an overall structural component. Other examples of a hybrid composite involve stronger and stiffer (but more brittle) fibers that are protected by outer plys of a tougher fiber composite to protect the composite from impact and other deleterious effects. Therefore through the use of composite materials, the engineer is not merely a materials selector, but is also a materials designer.

For small deflections, the linear elastic analysis of anisotropic composite material structures requires the use of the equilibrium equations, strain-displacement relations, and compatibility equations, which remain the same whether the structure is composed of an isotropic material or an anisotropic composite material. However, it is very necessary to drastically alter the stress-strain relations, also called the constitutive relations, to account for the anisotropy of the composite material structure.

A quantitative understanding of the virtues of using composite materials in a structure is found through deriving systematically the anisotropic elasticity tensor matrix, discussed below in Section 10.2.

10.2 Derivation of the Anisotropic Elastic Stiffness and Compliance Matrices

Consider an elastic solid body of any general shape, and assume it is composed of an infinity of *material points* within it. In order to deal with a *continuum*, one also assumes that the material points are infinitely large compared to the molecular lattice spacing of the particular material. If one assigns a Cartesian reference frame to the elastic body shown in Figure 10.1, one then calls this rectangular parallelepiped material point a *control element* or control volume of dimension dx, dy and dz in a Cartesian coordinate system. Figure 10.1 is identical to figure 1.1, but repeated here for continuity.

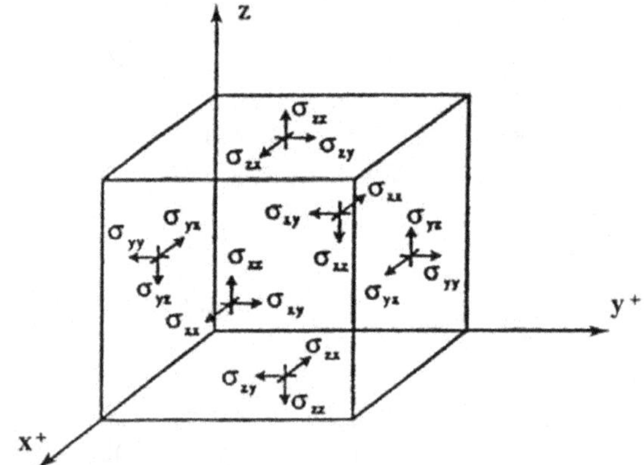

Figure 10.1. Positive Stresses on a Control Element of an Elastic Body.

On the surface of the control element there can exist both normal stresses (those perpendicular to the plane of the face) and shear stresses (those parallel to the plane of the face). On any one face the three mutually orthogonal stress components comprise a vector, which is called a *surface traction*.

It is important to note the sign convention and the meaning of the subscripts of these surface stresses. For a stress component on a face whose outward normal is in the direction of a positive axis, the stress component is positive when it is in the direction of a positive axis. Also, when a stress component is on a face whose outward normal is in the direction of a negative axis, the stress component is positive when it is in the direction of a negative axis. This can be seen clearly in Figure 10.1.

The first subscript of any stress component on any face of the control element signifies the axis to which the outward normal of the face is parallel; the second subscript refers to the axis to which that stress component is parallel. Again, see Figure 10.1.

The strains occurring in an elastic body have the same subscripts as the stress components but are of two types. Dilatational or extensional strains are denoted by ε_{ii}, where i = x, y, z, and are a measure of the change in dimension of the control volume in the subscripted direction due to normal stresses, σ_{ii}, acting on the control volume. Shear strains ε_{ij} $(i \neq j)$ are proportional to the change in angles of the control volume from 90°, changing the rectangular control volume into a parallelogram due to the shear stresses, σ_{ij}, $i \neq j$. For example, looking at the control volume *x-y* plane shown in Figure 10.2 below, shear stresses σ_{xy} and σ_{yx} cause the square control element with 90° corner angles to become a parallelogram with the corner angle ϕ as shown. Here, the change in angle γ_{xy} is

$$\gamma_{xy} = \frac{\pi}{2} - \phi \tag{10.2}$$

The shear strain ε_{xy}, a tensor quantity, is defined by

$$\varepsilon_{xy} = \gamma_{xy}/2 \tag{10.3}$$

Similarly, $\varepsilon_{xz} = \gamma_{xz}/2$, and $\varepsilon_{yz} = \gamma_{yz}/2$.

Having defined all of the elastic stress and strain tensor components, the stress-strain relations are now used to derive the anisotropic stiffness and compliance matrices.

The following derivation of the stress-strain relations for an anisotropic material parallels the derivation of Sokolnikoff [1.1], Vinson and Chou [2.8], Vinson [10.1, 10.2], and Vinson and Sierakowski [1.7]. Although the derivation is very formal mathematically to the reader who is primarily interested with the end result, the systematic derivation does provide confidence in the extended use of the results.

From knowledge of basic strength of materials [10.3], both the stresses, σ_{ij}, and the strains ε_{ij}, are second order tensor quantities, where in three dimensional space they have $3^2 = 9$ components. They are equated by means of the fourth order elasticity tensor, C_{ijkl}, which therefore has $3^4 = 81$ components, with the resulting constitutive equation:

$$\sigma_{ij} = C_{ijkl}\varepsilon_{kl} \tag{10.4}$$

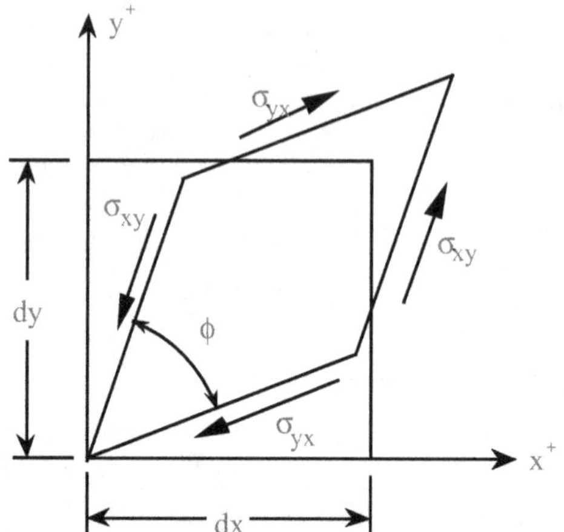

Figure 10.2. Shearing of a Control Element.

where *i*, *j*, *k* and *l* assume values of 1, 2, 3 or x, y, z in a Cartesian coordinate system. Fortunately, there is no actual material that has eighty-one elastic constants. Both the stress and strain tensors are symmetric, i.e., $\sigma_{ij} = \sigma_{ji}$ and $\varepsilon_{kl} = \varepsilon_{lk}$, and therefore the following shorthand notation may be used:

$$\begin{array}{llll} \sigma_{11} = \sigma_1 & \sigma_{23} = \sigma_4 & \varepsilon_{11} = \varepsilon_1 & \gamma_{23} = 2\varepsilon_{23} = \varepsilon_4 \\ \sigma_{22} = \sigma_2 & \sigma_{31} = \sigma_5 & \varepsilon_{22} = \varepsilon_2 & \gamma_{31} = 2\varepsilon_{31} = \varepsilon_5 \\ \sigma_{33} = \sigma_3 & \sigma_{12} = \sigma_6 & \varepsilon_{33} = \varepsilon_3 & \gamma_{12} = 2\varepsilon_{12} = \varepsilon_6 \end{array} \tag{10.5}$$

At the outset it is noted that ε_4, ε_5 and ε_6, which are quantities widely used in composite analyses, are not tensor quantities and therefore do not transform from one set of axes to another by affine transformation relationships. Care must also be taken regarding whether or not to use the factor of "two" when using shear strain relations, see (10.3) and (10.5). Using Equation (10.5), Equation (10.4) can be written:

$$\sigma_i = C_{ij}\varepsilon_j \tag{10.6}$$

It should be noted that the contracted C_{ij} quantities are also not tensor quantities, and therefore cannot be transformed as such.

Hence by the symmetry in the stress and strain tensors the elasticity tensor immediately reduces to the 36 components shown by Equation (10.6). In addition, if a strain energy density function, W, exists [1.1, 1.7, 2.8, 10.1, 10.2, 10.3], i.e.,

$$W = \frac{1}{2}\sigma_{ij}\varepsilon_{ij},$$

in such a way that

$$\frac{\partial W}{\partial \varepsilon_{ij}} = C_{ij} \ \ \varepsilon \ \ = \sigma_{ij}, \tag{10.7}$$

then the independent components of C_{ijkl} are reduced to 21 elastic constants, since $C_{ijkl} = C_{klij}$ and now it can be written $C_{ij} = C_{ji}$.

Next, to simplify the general mathematical anisotropy to the cases of very practical importance, consider the Cartesian coordinate system only. (However, the results are applicable to any curvilinear orthogonal coordinate system of which there are twelve, some of which are spherical, cylindrical, elliptical, etc. – see [1.7 and 10.2].

First, consider an elastic body whose properties are symmetric with respect to the X_1 - X_2 plane. The resulting symmetry can be expressed by the fact that the C_{ij}'s discussed above must be invariant under the transformation $x_1 = x_1'$, $x_2 = x_2'$ and $x_3 = -x_3'$, shown in Figure 10.3.

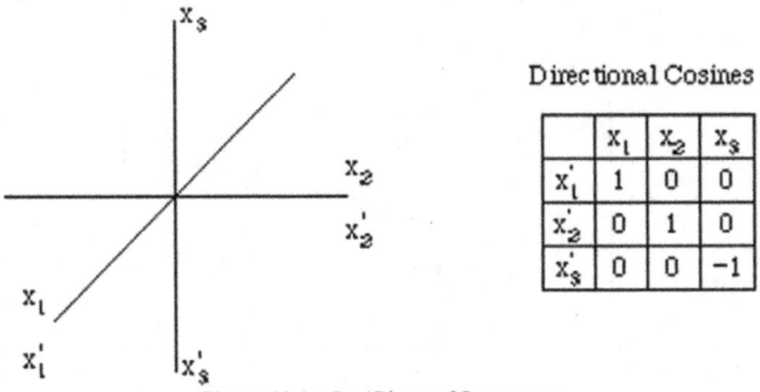

	x_1	x_2	x_3
x_1'	1	0	0
x_2'	0	1	0
x_3'	0	0	-1

Figure 10.3. One Plane of Symmetry.

Also shown in the table above are the direction cosines, t_{ij}, associated with this transformation. The stresses and strains of the primed coordinate system are related to those of the original (unprimed) coordinate system by the well-known relationships:

$$\sigma'_{\alpha\beta} = t_{\alpha i} t_{\beta j} \sigma_{ij} \text{ and } \varepsilon'_{\alpha\beta} = t_{\alpha i}\, t_{\beta j} \varepsilon_{ij}.$$

where, for $i = 1, 2, 3, 6$, $\sigma_i' = \sigma_i$ and $\varepsilon_i' = \varepsilon_i$, i.e., $\sigma_{11}' = t_{11} t_{11} \sigma_{11} = \sigma_{11}$. However, from the direction cosines, $\varepsilon_{23}' = -\varepsilon_{23}$ or $\varepsilon_4' = -\varepsilon_4$, and $\sigma_4' = -\sigma_4$; likewise $\varepsilon_{31}' = -\varepsilon_{31}$, hence $\varepsilon_5' = -\varepsilon_5$ and $\sigma_5' = -\sigma_5$. For example, $\sigma_{23}' = \sigma_4' = t_{22} t_{33} \sigma_{23} = (1)(-1)\sigma_{23} = -\sigma_{23} = -\sigma_4$.

If one looks in detail at Equation (10.6) then,

$$\sigma_4' = C_{41}\varepsilon_1' + C_{42}\varepsilon_2' + C_{43}\varepsilon_3' + C_{44}\varepsilon_4' + C_{45}\varepsilon_5' + C_{46}\varepsilon_6',$$
$$\sigma_4 = C_{41}\varepsilon_1 + C_{42}\varepsilon_2 + C_{43}\varepsilon_3 + C_{44}\varepsilon_4 + C_{45}\varepsilon_5 + C_{46}\varepsilon_6.$$

It is evident from these two equations that $C_{41} = C_{42} = C_{43} = C_{46} = 0$. From similar examinations of the other two axial symmetries, it can be seen that $C_{25} = C_{35} = C_{64} = C_{65} = 0$, $C_{51} = C_{52} = C_{53} = C_{56} = 0$, and $C_{14} = C_{15} = C_{16} = C_{24} = C_{34} = 0$.

So, for a material having only one plane of symmetry the number of elastic constants is now reduced to 13. Note that from a realistic engineering point of view this would still require thirteen independent physical tests (at each temperature and humidity condition!) - an almost impossible task both in manpower and budget.

Now, materials that have three mutually orthogonal planes of elastic symmetry are called "orthotropic" (a shortened term for orthogonally anisotropic). In such a case, other terms in the elasticity matrix are also zero, namely

$$C_{16} = C_{26} = C_{36} = C_{45} = 0.$$

Therefore, the elastic stiffness matrix for orthotropic materials is shown below, remembering that $C_{ij} = C_{ji}$.

$$C_{ij} = \begin{bmatrix} C_{11} & C_{12} & C_{13} & 0 & 0 & 0 \\ C_{21} & C_{22} & C_{23} & 0 & 0 & 0 \\ C_{31} & C_{32} & C_{33} & 0 & 0 & 0 \\ 0 & 0 & 0 & C_{44} & 0 & 0 \\ 0 & 0 & 0 & 0 & C_{55} & 0 \\ 0 & 0 & 0 & 0 & 0 & C_{66} \end{bmatrix} \tag{10.8}$$

Thus, for orthotropic elastic bodies, such as most composite materials in a three dimensional configuration, there are nine elastic constants.

Hence, using Equations (10.8) and (10.6), the explicit stress-strain relations for an orthotropic, three dimensional material are:

$$\begin{aligned}
\sigma_1 &= C_{11}\varepsilon_1 + C_{12}\varepsilon_2 + C_{13}\varepsilon_3 \\
\sigma_2 &= C_{21}\varepsilon_1 + C_{22}\varepsilon_2 + C_{23}\varepsilon_3 \\
\sigma_3 &= C_{31}\varepsilon_1 + C_{32}\varepsilon_2 + C_{33}\varepsilon_3 \\
\sigma_4 &= \sigma_{23} = C_{44}\varepsilon_4 = 2C_{44}\varepsilon_{23} = C_{44}\gamma_{23} \\
\sigma_5 &= \sigma_{31} = C_{55}\varepsilon_5 = 2C_{55}\varepsilon_{31} = C_{55}\gamma_{31} \\
\sigma_6 &= \sigma_{12} = C_{66}\varepsilon_6 = 2C_{66}\varepsilon_{12} = C_{66}\gamma_{12}
\end{aligned} \tag{10.9}$$

It should be noted that in the latter three relationships, which involve shear relations, the factor of two is present when one uses the tensor shear strains, ε_{23}, ε_{31} and ε_{12}.

If the Equations (10.9) are inverted, then, through standard matrix transformation:

$$\begin{aligned}
\varepsilon_1 &= a_{11}\sigma_1 + a_{12}\sigma_2 + a_{13}\sigma_3 \\
\varepsilon_2 &= a_{21}\sigma_1 + a_{22}\sigma_2 + a_{23}\sigma_3 \\
\varepsilon_3 &= a_{31}\sigma_1 + a_{32}\sigma_2 + a_{33}\sigma_3 \\
\varepsilon_4 &= 2\varepsilon_{23} = a_{44}\sigma_{23} = a_{44}\sigma_4 \\
\varepsilon_5 &= 2\varepsilon_{31} = a_{55}\sigma_{31} = a_{55}\sigma_5 \\
\varepsilon_6 &= 2\varepsilon_{12} = a_{66}\sigma_{12} = a_{66}\sigma_6.
\end{aligned} \tag{10.10}$$

The a_{ij} matrix, called the compliance matrix, involves the transpose of the cofactor (C_0) matrix of the C_{ij}'s divided by the determinant of the C_{ij} matrix with each term defined as

$$a_{ij} = \frac{[C_o C_{ij}]^T}{|C_{ij}|} \tag{10.11}$$

Again, the a_{ij} quantities are not tensors, and cannot be transformed as such. In fact, factors of 1, 2 and 4 appear in various terms when relating the tensor compliance quantities a_{ijkl} and the contracted compliance quantities a_{ij}.

It can be easily shown that $a_{ij} = a_{ji}$ and that

$$\varepsilon_i = a_{ij}\sigma_j \quad (\text{where } i, j = 1, 2, ..., 6). \tag{10.12}$$

Table 10.1 is useful for listing the number of elastic coefficients present in both two and three dimensional elastic bodies.

Table 10.1. Summary of the Number of Elastic Coefficients Involved for Certain Classes of Materials.

Class of Material	Number of nonzero coefficients	Number of independent coefficients
Three-Dimensional Case		
General Anisotropy	36	21
One-plane of symmetry	20	13
Two-planes of symmetry	12	9
Transverse isotropy	12	5
Isotropy	12	2
Two-Dimensional Case		
General anisotropy	9	6
One-plane of symmetry	9	6
Two-planes of symmetry	5	4
Transverse isotropy	5	4
Isotropy	5	2

10.3 The Physical Meaning of the Components of the Orthotropic Elasticity Tensor

So far, the components of both the stiffness matrix, C_{ij}, and the compliance matrix, a_{ij}, are mathematical symbols relating stresses and strains. By performing hypothetical simple tensile and shear tests all of the components above can be related to physical or mechanical properties.

Consider a simple, standard tensile test in the x_1 direction. The resulting stress and strain tensors are

$$\sigma_{ij} = \begin{bmatrix} \sigma_{11} & 0 & 0 \\ 0 & 0 & 0 \\ 0 & 0 & 0 \end{bmatrix}, \qquad \varepsilon_{ij} = \begin{bmatrix} \varepsilon_{11} & 0 & 0 \\ 0 & -\nu_{12}\varepsilon_{11} & 0 \\ 0 & 0 & -\nu_{13}\varepsilon_{11} \end{bmatrix} \tag{10.13}$$

where the Poisson's ratio, ν_{ij}, is very carefully defined as the negative of the ratio of the strain in the x_j direction to the strain in the x_i direction due to an applied stress in the x_i direction. In other words in the above it is seen that $\varepsilon_{22} = -\nu_{12}\varepsilon_{11}$ or $\nu_{12} = -\varepsilon_{22}/\varepsilon_{11}$. Care must be taken to not confuse ν_{ij} with ν_{ji}, because in some unidirectional composites, i.e.

a composite in which all of the fibers are aligned in one direction, one may be ten or more times larger than the other.

Also, the constant of proportionality between stress and strain in the 1 direction is denoted as E_{11} (or E_1), the modulus of elasticity in the x_i direction. Thus,

$$\varepsilon_1 = a_{11}\sigma_1 = \frac{\sigma_1}{E_1}$$

$$\varepsilon_2 = a_{21}\sigma_1 = -\nu_{12}\varepsilon_1 = -\left(\frac{\nu_{12}\sigma_1}{E_1}\right)$$

$$\varepsilon_3 = a_{31}\sigma_1 = -\nu_{13}\varepsilon_1 = -\left(\frac{\nu_{13}\sigma_1}{E_1}\right)$$

Therefore,

$$a_{11} = 1/E_1,\ a_{21} = -\nu_{12}/E_1,\ a_{31} = -\nu_{13}/E_1 \tag{10.14}$$

For a simple tensile test in the x_2 direction, it is found that

$$a_{12} = -\nu_{21}/E_2,\ a_{22} = 1/E_2,\ a_{32} = -\nu_{23}/E_2. \tag{10.15}$$

Likewise, a tensile test in the x_3 direction yields

$$a_{13} = -\nu_{31}/E_3,\ a_{23} = -\nu_{32}/E_3,\ a_{33} = 1/E_3. \tag{10.16}$$

From the fact that $a_{ij} = a_{ji}$, then

$$\frac{\nu_{ij}}{E_i} = \frac{\nu_{ji}}{E_j} \quad (i, j = 1, 2, 3) \tag{10.17}$$

The Equation (10.17) is most valuable and widely used in the analysis of all composite material bodies.

Next, consider a hypothetical simple shear test as shown in Figure 10.4. In this case the stress, strain, and displacement tensor components are:

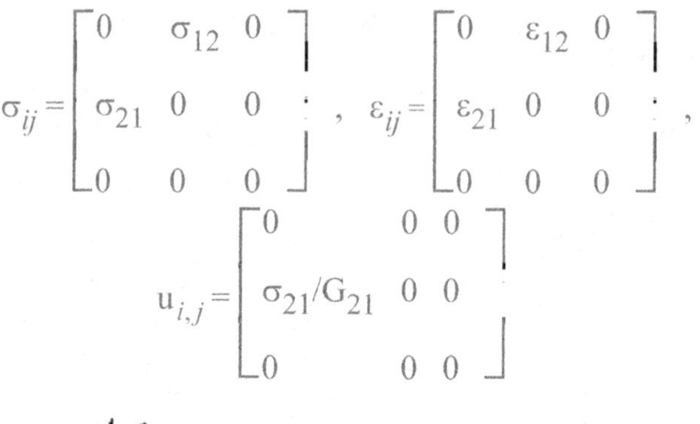

$$\sigma_{ij} = \begin{bmatrix} 0 & \sigma_{12} & 0 \\ \sigma_{21} & 0 & 0 \\ 0 & 0 & 0 \end{bmatrix}, \quad \varepsilon_{ij} = \begin{bmatrix} 0 & \varepsilon_{12} & 0 \\ \varepsilon_{21} & 0 & 0 \\ 0 & 0 & 0 \end{bmatrix},$$

$$u_{i,j} = \begin{bmatrix} 0 & 0 & 0 \\ \sigma_{21}/G_{21} & 0 & 0 \\ 0 & 0 & 0 \end{bmatrix}$$

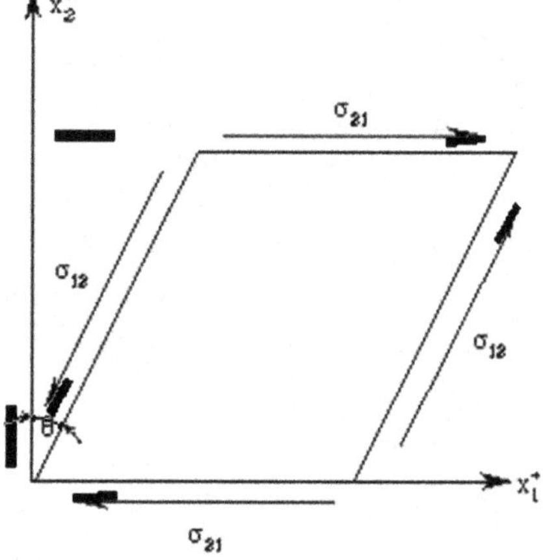

Figure 10.4. Shear Stresses and Strains.

In the above, u_i is the displacement and $u_{i,j} = (\partial u_i)/(\partial x_j)$. From elementary strength of materials the constant of proportionality between the shear stress σ_{21} and the angle θ is G_{21}, the shear modulus in the x_1 - x_2 plane.

From the theory of elasticity

$$\varepsilon_{12} = \frac{1}{2}(u_{1,2} + u_{2,1}) = \frac{\sigma_{21}}{2G_{21}} = \frac{\tan\theta}{2} . \qquad (10.18)$$

From Equation (10.10), $\varepsilon_6 = a_{66}\sigma_6$, or

$$\varepsilon_{12} = \frac{a_{66}\sigma_{21}}{2} = \frac{\sigma_{21}}{2G_{21}} .$$

Hence,

$$a_{66} = \frac{1}{G_{21}} = \frac{1}{G_{12}} \quad . \tag{10.19}$$

Similarly,

$$a_{44} = \frac{1}{G_{23}} \text{ and } a_{55} = \frac{1}{G_{13}} \quad . \tag{10.20}$$

Thus, all a_{ij} components have now been related to mechanical properties, and it is seen that to characterize a three dimensional orthotropic body, nine physical quantities are needed (that is E_1, E_2, E_3, G_{12}, G_{23}, G_{31}, ν_{12}, ν_{13}, ν_{21}, ν_{23}, ν_{31} and ν_{32}, and using Equation (10.17). However, because of (10.17) only six separate tests are needed to obtain the nine physical quantities. The standardized tests used to obtain these anistropic elastic constants are given in ASTM standards, and are described in a text by Carlsson and Pipes [10.4]. For convenience, the compliance matrix is given explicitly as:

$$a_{ij} \begin{bmatrix} \frac{1}{E_1} & -\frac{\nu_{21}}{E_2} & -\frac{\nu_{31}}{E_3} & 0 & 0 & 0 \\ -\frac{\nu_{12}}{E_1} & \frac{1}{E_2} & -\frac{\nu_{32}}{E_3} & 0 & 0 & 0 \\ -\frac{\nu_{13}}{E_1} & -\frac{\nu_{23}}{E_2} & \frac{1}{E_3} & 0 & 0 & 0 \\ 0 & 0 & 0 & \frac{1}{G_{23}} & 0 & 0 \\ 0 & 0 & 0 & 0 & \frac{1}{G_{13}} & 0 \\ 0 & 0 & 0 & 0 & 0 & \frac{1}{G_{12}} \end{bmatrix} \tag{10.21}$$

10.4 Methods to Obtain Composite Elastic Properties from Fiber and Matrix Properties

In order to minimize the time and expense to experimentally determine mechanical properties of the almost infinite varieties of composite materials, it is very useful to analytically predict the properties of a unidirectional (one in which all of the fibers are in one direction) composite, if the properties of the fibers and the properties of the matrix are known.

There are several sets of equations for obtaining the composite elastic properties from those of the fiber and matrix materials. These include those of Halpin and Tsai [10.5], Hashin [10.6], and Christensen [10.7]. In 1980, Hahn [10.8] codified certain results for fibers of circular cross section which are randomly distributed in a plane normal to the unidirectionally oriented fibers. For that case the composite is macroscopically, transversely isotropic, that is $()_{12} = ()_{13}$, $()_{22} = ()_{33}$ and $()_{55} = ()_{66}$, where in the parentheses the quantity could be E, G, or v; hence, the elastic properties involve only five independent constants, namely $()_{11}, ()_{22}, ()_{12}, ()_{23}$ and $()_{66}$.

For several of the elastic constants, Hahn states that they all have the same functional form:

$$P = \frac{(P_f V_f + \eta P_m V_m)}{(V_f + \eta V_m)} \qquad (10.22)$$

where for the elastic constant P, the P_f, P_m and η are given in Table 10.2 below, and where V_f and V_m are the volume fractions of the fibers and matrix respectively (and whose sum equals unity):

Table 10.2. Determination of Composite Properties From Fiber and Matrix Properties.

Elastic Constant	P	P_f	P_m	η
E_{11}	E_{11}	E_{11f}	E_m	1
v_{12}	v_{12}	v_{12f}	v_m	1
G_{12}	$1/G_{12}$	$1/G_{12f}$	$1/G_m$	η_6
G_{23}	$1/G_{23}$	$1/G_{23f}$	$1/G_m$	η_4
K_T	$1/K_T$	$1/K_f$	$1/K_m$	η_K

In Table 10.2 and Equation (10.26), unless the anisotropic properties are given specifically one must assume the fiber is isotropic.

The expressions for E_{11} and v_{12} in Table 10.2 are called the Rule of Mixtures. In the above K_T is the plane strain bulk modulus, $K_f = [E_f/2(1-v_f)]$ and $K_m = [E_m/2(1-v_m)]$. Also, the η's are given as follows:

$$\eta_6 = \frac{1 + G_m/G_{12f}}{2}, \quad \eta_4 = \frac{3 - 4v_m + G_m/G_{23f}}{4(1-v_m)}, \quad \eta_K = \frac{1 + G_m/K_f}{2(1-v_m)}.$$

The shear modulus of the matrix material, G_m, if isotropic, is given by $G_m = E_m/2(1+v_m)$.

The transverse moduli of the composite, $E_{22} = E_{33}$, are found from the following equation:

$$E_{22} = E_{33} = \frac{4K_T G_{23}}{K_T + mG_{23}} \quad , \tag{10.23}$$

where

$$m = 1 + \frac{4K_T \nu_{12}^2}{E_{11}} \quad .$$

The equations above have been written in general for composites reinforced with anisotropic fibers such as some graphite and aramid (Kevlar) fibers. If the fibers are isotropic, the fiber properties involve E_f, G_f and ν_f, where $G_f = \dfrac{E_f}{2(1+\nu_f)}$. In that case also η_K becomes

$$\eta_K = \frac{1 + (1\text{-}2\nu_f)G_m/G_f}{2(1\text{-}\nu_m)} \quad . \tag{10.24}$$

Hahn notes that for most polymeric matrix structural composites, $G_m/G_f < 0.05$. If that is the case then the η parameters are approximately:

$$\eta_6 \approx 0.5; \;\; \eta_4 = \frac{3\text{-}4\nu_m}{4(1\text{-}\nu_m)}; \;\; \eta_K = \frac{1}{2(1\text{-}\nu_m)} \quad . \tag{10.25}$$

Finally, noting that $\nu_m = 0.35$ for most epoxies, then $\eta_4 = .62$ and $\eta_K = 0.77$.

Also, the Poisson's ratio, ν_{23}, can be written as

$$\nu_{23} = \nu_{12f} V_f + \nu_m (1\text{-}V_f) \left[\frac{1 + \nu_m - \nu_{12}\left(\dfrac{E_m}{E_{11}}\right)}{1\text{-}\nu_m^{\,2} + \nu_m \nu_{12}\left(\dfrac{E_m}{E_{11}}\right)} \right] \quad . \tag{10.26}$$

where ν_{12f} is the fiber Poisson's ratio.

The above equations along with Equation (10.17) provide the engineer with the wherewithal to calculate the elastic constants for a unidirectional composite material if the constituent properties of the fibers and matrix, and the fiber volume fraction are known. In a few instances only the weight fraction of the fiber, W_f, is known. In that case the volume fraction is obtained from the following equation, where W_m is the weight fraction of the matrix, and ρ_m and ρ_f are the respective densities:

$$V_f = \frac{\rho_m W_f}{\rho_m W_f + \rho_f W_m} \tag{10.27}$$

For determining the composite elastic constants for short fiber composites, hybrid composites, textile composites, and very flexible composites, Chou [10.9] provides a comprehensive treatment.

Of course if the composite manufacturer has added a filler to the structural matrix to reduce cost, then (10.27) must be modified.

10.5 Thermal and Hygrothermal Considerations

In the previous two sections, the elastic relations developed pertain only to an anisotropic elastic body at one temperature, that temperature being the "stress free" temperature, i.e. the temperature at which the body is considered to be free of stress if it is under no mechanical static or dynamic loadings.

However, in both metallic and composite structures changes in temperature are commonplace both during fabrication and during structural usage. Changes in temperature result in two effects that are very important. First, most materials expand when heated and contract when cooled, and in most cases this expansion is proportional to the temperature change. If, for instance, one had a long thin bar of a given material then with change in temperature, the ratio of the change in length of the bar, ΔL, to the original length, L, is related to the temperature of the bar, T, as shown in Figure 10.5.

Mathematically, this can be written as

$$\varepsilon_{\text{thermal}} = \frac{\Delta L}{L} = \alpha \Delta T \tag{10.28}$$

where α is the coefficient of thermal expansion i.e., the proportionality constant between the "thermal" strain ($\Delta L/L$) and the change in temperature, ΔT, from some reference temperature at which there are no thermal stresses or thermal strains. For almost all materials, α is constant unless a phase change occurs in the material.

The second major effect of temperature change relates to stiffness and strength. Most materials become softer, more ductile, and weaker as they are heated. Typical plots of ultimate strength, yield stress and modulus of elasticity as functions of temperature are shown in Figure 10.6. In performing a stress analysis, determining the natural frequencies, or finding the buckling load of a heated or cooled structure one must use the strengths and the moduli of elasticity of the material at the temperature at which the structure is expected to perform.

In an orthotropic material, such as a composite, there can be up to three different coefficients of thermal expansion, and three different thermal strains, one in each of the orthogonal directions comprising the orthotropic material. Equation (10.28) would then have subscripts of 1, 2 and 3 on both the strains and the coefficients of thermal expansion. Notice that, for the primary material axes, all thermal effects are dilatational only; there are no thermal effects in shear.

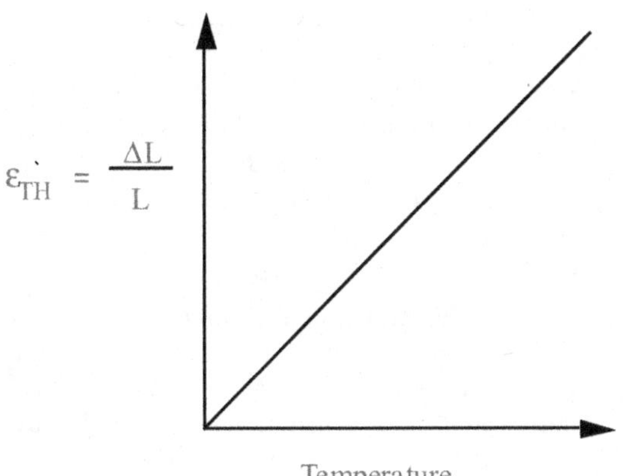

Figure 10.5. Change in Length of a Bar or Rod as a Function of Temperature.

Some general articles and monographs on thermomechanical effects on composite material structures include those by Tauchert [10.10], Argyris and Tenek [10.11], Turvey and Marshall [10.12], Noor and Burton [10.13] and Huang and Tauchert [10.14], Springer [10.15], Milke and Vizzini [10.16], and Sun and Chen [10.17].

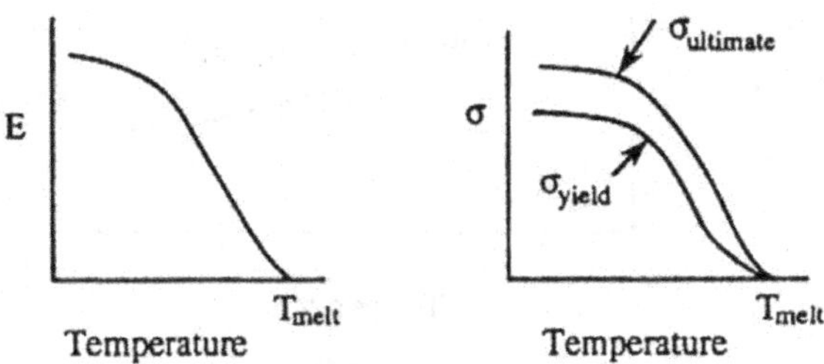

Figure 10.6. Modulus of Elasticity and Strengths as Functions of Temperature.

During the mid-1970's another physical phenomenon associated with polymer matrix composites was recognized as important. It was found that the combination of high temperature and high humidity caused a doubly deleterious effect on the structural performance of these composites. Engineers and material scientists became very concerned about these effects, and considerable research effort was expended in studying this new phenomenon. Conferences [10.18] were held which discussed the problem, and both short range and long range research plans were proposed. The twofold problem involves the fact that the combination of high temperature and high humidity results in the entrapment of moisture in the polymer matrix, with attendant weight increase ($\leq 2\%$)

and more importantly, a swelling of the matrix. It was realized [10.19] that the ingestion of moisture varied linearly with the swelling so that in fact

$$\varepsilon_{\text{hygrothemal}} = \frac{\Delta L}{L} = \beta \Delta m \tag{10.29}$$

where Δm is the increase from zero moisture measured in percentage weight increase, and β is the coefficient of hygrothermal expansion, analogous to the coefficient of thermal expansion, depicted in Equation (10.28). This analogy is a very important one because one can see that the hygrothermal effects are entirely analogous mathematically to the thermal effect. Therefore, if one has the solutions to a thermoelastic problem, merely substituting $\beta\Delta m$ for or adding it to the $\alpha\Delta T$ terms provides the hygrothermal solution. The test methods to obtain values of the coefficient of hygrothermal expansion β are given in [10.20].

The second effect (i.e. the reduction of strength and stiffness) is also similar to the thermal effect. This is shown qualitatively in Figure 10.7. Dry polymers have properties that are usually rather constant until a particular temperature is reached, traditionally called by polymer chemists the "glass transition temperature," above which both strength and stiffness deteriorate rapidly. If the same polymer is saturated with moisture, not only are the mechanical properties degraded at any one temperature but the glass transition temperature for that polymer is significantly lower.

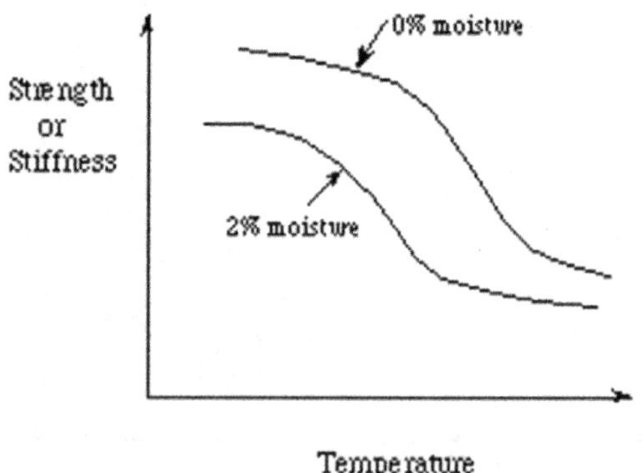

Figure 10.7. Mechanical Properties as a Function of Temperature and Moisture Absorption.

In 1981, Shen and Springer [10.21] investigated the hygrothermal effects on the tensile strength.

As a quantitative example, Figure 10.8 clearly shows the diminution in tensile and shear strength due to a long term hygrothermal environment. Short time tensile and shear tests were performed on random mat glass/polyester resin specimens. It is clearly seen that there is a significant reduction in tensile strength, and a 29.3% and a 37.1% reduction in ultimate shear strength of these materials over a 100^+ day soak period. If these effects are not accounted for in design analysis, catastrophic failures can and have occurred in such structures as the axial fans in waste disposal facilities.

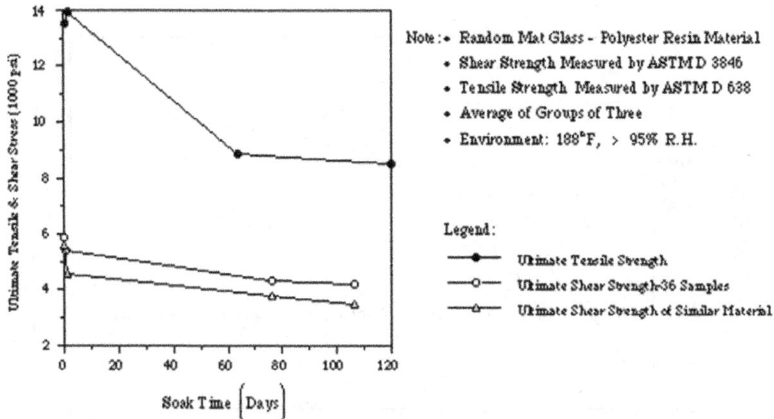

Figure 10.8. Strength as a Function of Soak Time in a Hot, Wet Environment.

Thus, for modern polymer matrix composites one must include not only the thermal effects but also the hygrothermal effects or the structure can be considerably under designed, resulting in potential failure.

Thus, to deal with the real world of polymer composites, Equation (10.12) must be modified to read

$$\varepsilon_i = a_{ij}\sigma_j + \alpha_i \Delta T + \beta_i \Delta m \quad (i = 1, 2, 3) \tag{10.30}$$

$$\varepsilon_i = a_{ij}\sigma_j \qquad (i = 4, 5, 6) \tag{10.31}$$

where in each equation $j = 1$ - 6.

Two types of equations are shown above because in the primary materials system of axes ($i,j = 1, 2, \ldots, 6$) both thermal and hygrothermal effects are dilatational only, that is, they cause an expansion or contraction, but do not affect the shear stresses or strains. This is important to remember.

Although the thermal and moisture effects are analogous, they have significantly different time scales. For a structure subjected to a change in temperature that would require minutes or at most hours to come to equilibrium at the new temperature, the same structure would require weeks or months to come to moisture equilibrium (saturation) if that dry structure were placed in a 95-100% relative humidity environment. Figure 10.9 illustrates the point, as an example. A 1/4" thick random mat glass polyester matrix material requires 49 days of soak time at 188°F and 95% relative humidity to become saturated.

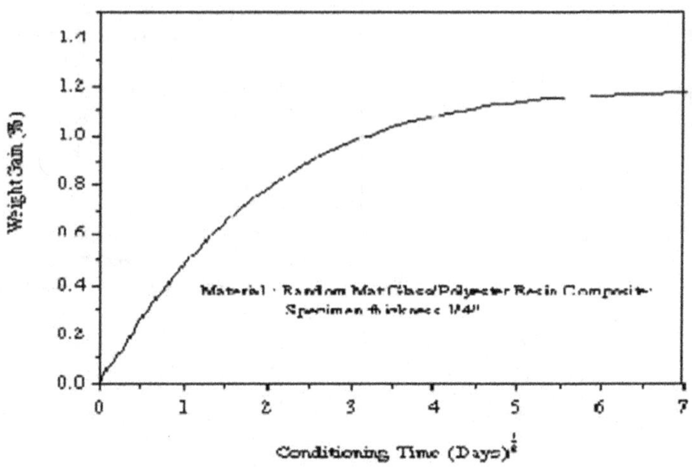

Figure 10.9. Moisture Absorption as a Function of Soak Time.

Woldesenbet [10.22] soaked a large number of IM7/8551-7 graphite epoxy unidirectional test pieces, some in room temperature water to saturation. For the 1/4" diameter by 3/8" long cylinders soaked at room temperature the time required to reach saturation was 55 weeks. Other test pieces were soaked at an elevated temperature to reduce soak time.

For additional reading on this subject, see Shen and Springer [10.21] and Zhu and Sun [10.23].

In Chapter 18 herein another effect is added to the hygrothermal equations (10.30) and (10.31). When piezoelectric materials are used in the structure an additional term is added to the right hand side of (10.30) and (10.31).

10.6 Time-Temperature Effects on Composite Materials

In addition to the effects of temperature and moisture on the short time properties discussed above, if a structure is maintained under a constant load for a period of time, then creep and viscoelastic effects can become very important in the design and analysis of that structure. The subject of creep is discussed in numerous materials science and strength of materials texts and will not be described here in detail.

Creep and viscoelasticity can become significant in any material above certain temperatures, but can be particularly important in polymer matrix materials whose operating temperatures must be kept below maximum temperatures of 250°F, 350°F, or in some cases 600°F for short periods of time, dependent upon the specific polymer material. See Christensen [10.24].

From a structural mechanics point of view, almost all of the viscoelastic effects occur in the polymer matrix, while little or no creep occurs in the fibers. Thus, the study of creep in the polymeric materials, which comprise the matrix, provides the data necessary to study creep in composites. Jurf [10.25] experimentally studied the effects of temperature and moisture (hygrothermal effects) on various epoxy materials (FM 73M and FM 300M adhesives). They established that at least for some epoxy materials it is possible to construct a master creep curve using a temperature shift factor, and established the fact that a moisture shift factor can also be employed. The importance of this is that by these experimentally determined temperature and moisture shift factors, for the shear modulus of the epoxy, the results of short time creep tests can be used for a multitude of time/temperature/moisture combinations over the lifetime and environment of a structure comprised of that material. Wilson [10.26, 10.27] studied the effects of viscoelasticity on the buckling of columns and rectangular plates and found that significant reductions of the buckling loads can occur. Wilson found that for the materials he studied, the buckling load diminished over the first 400 hours, then stabilized at a constant value. However that value may be a small fraction of the elastic buckling load if the composite properties in the load direction were matrix dominated properties (described later in the text). Wilson also established that for the problems studied it was quite satisfactory to bypass the complexities of a full-scale viscoelastic analysis using the Correspondence Principle and Laplace transformations. The use of the appropriate short time stiffness properties of the composite experimentally determined with specimens that have been held at the temperature and until the time for which the structural calculations are being made.

Hu and Sun have studied the equivalence of moisture and temperature in physical aging of polymer composites by using momentary creep tests [10.28]. Such equivalence may permit the substitution of a moisture creep test with tests under an equivalent temperature thus saving much time and expense.

10.7 High Strain Rate Effects on Material Properties

Another consideration in the analysis of all composite material structures is the effect of high strain rate on the strength and stiffness properties of the materials used. Most materials have significantly different strengths, moduli, and strains to failure at high strain rates compared to static values. However most of the major finite element codes such as those which involve 10^5-10^6 elements using 10^1-10^2 hours of computer time to describe underwater and other explosion effects on structures, still utilize static material properties. High strain rate properties of materials are sorely needed. Some dynamic properties have been found, and test techniques established. For more information see Lindholm [10.29], Daniel, La Bedz, and Liber [10.30], Nicholas [10.31], Zukas [10.32], and Sierakowski [10.33, 10.34], Rajapakse and Vinson [10.35], and Abrate [10.36, 10.37].

Vinson and his colleagues have found through testing over thirty various composite materials over the range of strain rates tested up to 1600/sec, that in comparing high strain rate values to static values, the yield stresses can increase by a factor up to 3.6, the yield strains can change by factors of 3.1, strains to failure can change by factors up to 4.7, moduli of elasticity can change by factors up to 2.4, elastic strain energy densities can change by factors up to 6, while strain energy densities to failure can change by factors up to 8.1. Thus the use of static material properties to analyze and design structures subjected to impact, explosions, crashes, or other dynamic loads should be carefully reviewed.

Most recently, Song experimentally determined the high strain rate mechanical properties of several polymer matrix composite materials over temperature ranges from room temperature to liquid nitrogen temperatures (-196°C). Most significantly, the strength of these composites increased as much as a factor of 5 at liquid nitrogen temperatures compared to the properties at room temperature [10.38 through 10.41].

10.8 Laminae of Composite Materials

Almost all practical composite material structures are thin in the thickness direction because the superior material properties of composites permit the use of thin walled structures. Many polymeric matrix composites are made in the form of a uniaxial set of fibers surrounded by a polymeric matrix in the form of a tape several inches wide termed as a "prepreg." The basic element in most long fiber composite structures is a lamina of fiber plus matrix, all fibers oriented in one direction, made by laying the prepreg tape of a certain length side by side. In the next section, 10.9, the stacking of various laminae to form a superior structure termed a laminate will be discussed.

In modern manufacturing methods, such as many liquid injection molding techniques, the fibers are placed in the mold as a “preform”. In that case the analyst must decide whether the molded composite can best be modeled as one lamina or a laminate.

Also if a composite lamina has a thermal gradient across the thickness such that the material properties vary significantly from one surface to the other, then the analyst could model the composite as a laminate with differing material properties in each lamina.

To describe this, consider a small element of a lamina of constant thickness h, wherein the principal material axes are labeled 1 and 2, that is the 1 direction is parallel to the fibers, the 2 direction is normal to them, and consider that the beam, plate or shell geometric axes are x and y as depicted in Figure 10.10. For the material axes 1 and 2, the 1 axis is always in the direction involving the stiffer and stronger material properties, compared to the 2 direction.

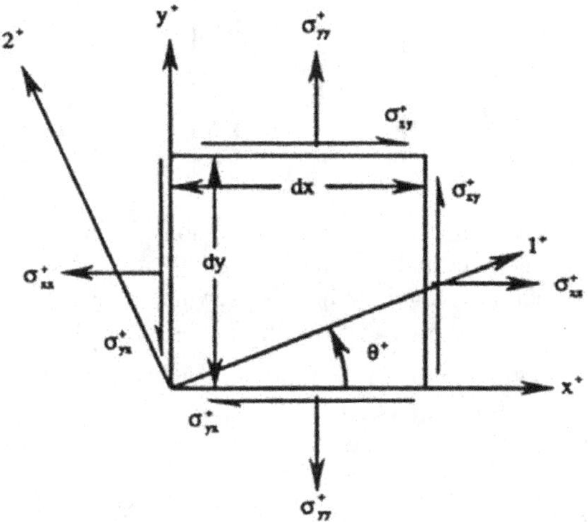

Figure 10.10. Lamina Coordinate System.

The element shown in Figure 10.10 has the stresses shown in the positive directions consistent with references [1.7, 2.7, 2.8, 3.3, 10.1, 10.2, 10.42]. If one performs a force equilibrium study to relate σ_x, σ_y and σ_{xy} to σ_1, σ_2, and σ_{12}, it is exactly analogous to the Mohr's circle analysis in basic strength of materials with the result that, in matrix form,

$$\begin{bmatrix} \sigma_1 \\ \sigma_2 \\ \sigma_6 \end{bmatrix} = [T]_{CL} \begin{bmatrix} \sigma_x \\ \sigma_y \\ \sigma_{xy} \end{bmatrix} \tag{10.32}$$

where

$$[T]_{CL} = \begin{bmatrix} m^2 & n^2 & +2mn \\ n^2 & m^2 & -2mn \\ -mn & mn & (m^2-n^2) \end{bmatrix} \tag{10.33}$$

and where here m = cos θ, n = sin θ, and θ is defined positive as shown in Figure 10.10, and the subscripts CL refer to the classical two-dimensional case only, that is, in the 1-2 plane or the *x-y* plane only.

Analogously, a strain relationship also follows for the classical isothermal case

$$\begin{bmatrix} \varepsilon_1 \\ \varepsilon_2 \\ \varepsilon_{12} \end{bmatrix} = [T]_{CL} \begin{bmatrix} \varepsilon_x \\ \varepsilon_y \\ \varepsilon_{xy} \end{bmatrix} \tag{10.34}$$

However, these classical two-dimensional relationships must be modified to treat a composite material to include thermal effects, hygrothermal effects, and the effects of transverse shear deformation treated in detail elsewhere [e.g., 1.7]. The effects of transverse shear deformation, shown through the inclusion of the σ_4 - ε_4 and σ_5 - ε_5 relations shown in Equations (10.32) and (10.34), must be included in composite materials, because in the fiber direction the composite has many of the mechanical properties of the fiber itself (strong and stiff) while in the thickness direction the fibers are basically ineffective and the shear properties are dominated by the weaker matrix material. Similarly, because quite often the matrix material has much higher coefficients of thermal and hygrothermal expansion (α and β), thickening and thinning of the lamina cannot be ignored in some cases. Hence, without undue derivation, the Equations (10.32) through (10.34) are modified to be:

$$\begin{bmatrix} \sigma_1 \\ \sigma_2 \\ \sigma_3 \\ \sigma_4 \\ \sigma_5 \\ \sigma_6 \end{bmatrix} [T] \begin{bmatrix} \sigma_x \\ \sigma_y \\ \sigma_z \\ \sigma_{yz} \\ \sigma_{xz} \\ \sigma_{xy} \end{bmatrix} \text{ and } \begin{bmatrix} \varepsilon_1 \\ \varepsilon_2 \\ \varepsilon_3 \\ \varepsilon_{4/2} \\ \varepsilon_{5/2} \\ \varepsilon_{6/2} \end{bmatrix} = [T] \begin{bmatrix} \varepsilon_x \\ \varepsilon_y \\ \varepsilon_z \\ \varepsilon_{yz} \\ \varepsilon_{xz} \\ \varepsilon_{xy} \end{bmatrix} \tag{10.35}$$

where

$$[T] = \begin{bmatrix} m^2 & n^2 & 0 & 0 & 0 & 2mn \\ n^2 & m^2 & 0 & 0 & 0 & -2mn \\ 0 & 0 & 1 & 0 & 0 & 0 \\ 0 & 0 & 0 & m & -n & 0 \\ 0 & 0 & 0 & n & m & 0 \\ -mn & mn & 0 & 0 & 0 & (m^2 - n^2) \end{bmatrix} \tag{10.36}$$

Please note the introduction of the factor of $\frac{1}{2}$ as noted in the strain expressions, because of the way ε_4, ε_5 and ε_6 are defined in Equation (10.5).

For completeness, the reverse transformations are given.

$$\begin{bmatrix} \sigma_x \\ \sigma_y \\ \sigma_z \\ \sigma_{yz} \\ \sigma_{xz} \\ \sigma_{xy} \end{bmatrix} = [T]^{-1} \begin{bmatrix} \sigma_1 \\ \sigma_2 \\ \sigma_3 \\ \sigma_4 \\ \sigma_5 \\ \sigma_6 \end{bmatrix} \text{ and } \begin{bmatrix} \varepsilon_x \\ \varepsilon_y \\ \varepsilon_z \\ \varepsilon_{yz} \\ \varepsilon_{xz} \\ \varepsilon_{xy} \end{bmatrix} = [T]^{-1} \begin{bmatrix} \varepsilon_1 \\ \varepsilon_2 \\ \varepsilon_3 \\ \left(\frac{\varepsilon_4}{2}\right) \\ \left(\frac{\varepsilon_5}{2}\right) \\ \left(\frac{\varepsilon_6}{2}\right) \end{bmatrix} = [T]^{-1} \begin{bmatrix} \varepsilon_{11} \\ \varepsilon_{22} \\ \varepsilon_{33} \\ \varepsilon_{23} \\ \varepsilon_{13} \\ \varepsilon_{12} \end{bmatrix} \tag{10.37}$$

where*

$$[T]^{-1} \begin{bmatrix} m^2 & n^2 & 0 & 0 & 0 & -2mn \\ n^2 & m^2 & 0 & 0 & 0 & 2mn \\ 0 & 0 & 1 & 0 & 0 & 0 \\ 0 & 0 & 0 & m & n & 0 \\ 0 & 0 & 0 & -n & m & 0 \\ mn & -mn & 0 & 0 & 0 & (m^2-n^2) \end{bmatrix} \tag{10.38}$$

Again, please note that the transformations can be made only with *tensor* strains. Hence, from Equation (10.5), it is necessary to divide ε_4, ε_5 and ε_6 by two.

If one systematically uses these expressions, and utilizes Hooke's Law relating stress and strain, and includes the thermal and hygrothermal effects, one can produce the following overall general equations for a lamina of a fiber reinforced composite material in terms of the principal material directions (1, 2, 3); see Equations (10.8) - (10.21).

*$[T]^{-1}$ can be found by replacing θ by $(-\theta)$ in [T].

$$\begin{Bmatrix} \sigma_1 \\ \sigma_2 \\ \sigma_3 \\ \sigma_4 \\ \sigma_5 \\ \sigma_6 \end{Bmatrix} = \begin{bmatrix} Q_{11} & Q_{12} & Q_{13} & 0 & 0 & 0 \\ Q_{12} & Q_{22} & Q_{23} & 0 & 0 & 0 \\ Q_{13} & Q_{23} & Q_{33} & 0 & 0 & 0 \\ 0 & 0 & 0 & Q_{44} & 0 & 0 \\ 0 & 0 & 0 & 0 & Q_{55} & 0 \\ 0 & 0 & 0 & 0 & 0 & Q_{66} \end{bmatrix} \begin{Bmatrix} \varepsilon_1 - \alpha_1\Delta T - \beta_1\Delta m \\ \varepsilon_2 - \alpha_2\Delta T - \beta_2\Delta m \\ \varepsilon_3 - \alpha_3\Delta T - \beta_3\Delta m \\ 2\varepsilon_{23} \\ 2\varepsilon_{31} \\ 2\varepsilon_{12} \end{Bmatrix} \quad (10.39)$$

In the above, the Q_{ij} quantities are used for the stiffness matrix quantities obtained directly from Equation (10.8) through (10.21). One should also remember that $\varepsilon_{23} = (1/2G_{23})\sigma_4$, $\varepsilon_{31} = (1/2G_{31})\sigma_5$ and $\varepsilon_{12} = (1/2G_{12})\sigma_6$, hence the coefficients of "two" appearing with the tensor shear strains ε_{23}, ε_{31}, and ε_{12} above. Using the notation of Sloan [10.42], the stiffness matrix quantities can be written as follows:

$$Q_{11} = E_{11}(1-\nu_{23}\nu_{32})/\Delta,\ Q_{22} = E_{22}(1-\nu_{31}\nu_{13})/\Delta$$

$$Q_{33} = E_{33}(1-\nu_{12}\nu_{21})/\Delta,\ Q_{44} = G_{23},\ Q_{55} = G_{13},\ Q_{66} = G_{12}$$

$$Q_{12} = (\nu_{21} + \nu_{31}\nu_{23})E_{11}/\Delta = (\nu_{12} + \nu_{32}\nu_{13})E_{22}/\Delta$$

$$Q_{13} = (\nu_{31} + \nu_{21}\nu_{32})E_{11}/\Delta = (\nu_{13} + \nu_{12}\nu_{23})E_{33}/\Delta \quad (10.40)$$

$$Q_{23} = (\nu_{32} + \nu_{12}\nu_{31})E_{22}/\Delta = (\nu_{23} + \nu_{21}\nu_{13})E_{33}/\Delta$$

$$\Delta = 1 - \nu_{12}\nu_{21} - \nu_{23}\nu_{32} - \nu_{31}\nu_{13} - 2\nu_{21}\nu_{32}\nu_{13}$$

Incidentally in the above expressions, if the lamina is transversely isotropic, i.e. has the same properties in both the 2 and 3 directions, then $\nu_{12} = \nu_{13}$, $G_{12} = G_{13}$, $E_{22} = E_{33}$ with resulting simplification.

For preliminary calculations in design or where great accuracy is not needed, simpler forms [1.7] for some of the expressions in Equation (10.40) can be used, as shown below, with little loss in numerical accuracy:

$$Q_{11} = E_{11}/(1 - \nu_{12}\nu_{21}), \quad Q_{22} = E_{22}/(1-\nu_{12}\nu_{21})$$

$$Q_{12} = Q_{21} = \nu_{21}E_{11}/(1-\nu_{12}\nu_{21}) = \nu_{12}E_{22}/(1-\nu_{12}\nu_{21}) \tag{10.41}$$

$$Q_{66} = G_{12}$$

If these simpler forms are used then one would use the classical form of the constitutive relations instead of Equation (10.39), neglecting transverse shear deformation and transverse normal stress, i.e., letting σ_3, σ_4 and σ_5 equal zero, thus obtaining

$$\begin{Bmatrix} \sigma_1 \\ \sigma_2 \\ \sigma_3 \end{Bmatrix} = \begin{bmatrix} Q_{11} & Q_{12} & 0 \\ Q_{12} & Q_{22} & 0 \\ 0 & 0 & Q_{66} \end{bmatrix} \begin{Bmatrix} \varepsilon_1 - \alpha_1 \Delta T - \beta_1 \Delta m \\ \varepsilon_2 - \alpha_2 \Delta T - \beta_2 \Delta m \\ 2\varepsilon_{12} \end{Bmatrix} \tag{10.42}$$

where one should remember also that $2\varepsilon_{12} = \varepsilon_6$, hence the appearance of the factor of two before ε_{12}. As stated above for many cases it is sufficient to use Equations (10.41) and (10.42) rather than Equations (10.39) and (10.40) for faster and easier calculation.

In the case where there are no 16 and 26 terms the anisotropic composite material is frequently referred to as "specially orthotropic".

When the structural axes, x, y and z, are not aligned with the principle materials axes, 1, 2, 3, as described in Figure 10.10, then a coordinate transformation is necessary. To relate these relationships to the *x-y-z* coordinate system, one utilizes Equations (10.37) through (10.39). The result is

$$\begin{Bmatrix} \sigma_x \\ \sigma_y \\ \sigma_z \\ \sigma_{yz} \\ \sigma_{xz} \\ \sigma_{xy} \end{Bmatrix} = \begin{bmatrix} \bar{Q}_{11} & \bar{Q}_{12} & \bar{Q}_{13} & 0 & 0 & \bar{Q}_{16} \\ \bar{Q}_{12} & \bar{Q}_{22} & \bar{Q}_{23} & 0 & 0 & \bar{Q}_{26} \\ \bar{Q}_{13} & \bar{Q}_{23} & \bar{Q}_{33} & 0 & 0 & \bar{Q}_{36} \\ 0 & 0 & 0 & \bar{Q}_{44} & \bar{Q}_{45} & 0 \\ 0 & 0 & 0 & \bar{Q}_{45} & \bar{Q}_{55} & 0 \\ \bar{Q}_{16} & \bar{Q}_{26} & \bar{Q}_{36} & 0 & 0 & \bar{Q}_{66} \end{bmatrix} \begin{Bmatrix} \varepsilon_x - \alpha_x\Delta T - \beta_x\Delta m \\ \varepsilon_y - \alpha_y\Delta T - \beta_y\Delta m \\ \varepsilon_z - \alpha_z\Delta T - \beta_z\Delta m \\ 2\varepsilon_{yz} \\ 2\varepsilon_{xz} \\ 2(\varepsilon_{xy} - \alpha_{xy}\Delta T - \beta_{xy}\Delta m) \end{Bmatrix} \tag{10.43}$$

where $[\bar{Q}] = [T]^{-1}[Q][T]$, or more explicitly,

$$\bar{Q}_{11} = Q_{11}m^4 + 2(Q_{12} + 2Q_{66})m^2n^2 + Q_{22}n^4$$

$$\bar{Q}_{12} = (Q_{11} + Q_{22} - 4Q_{66})m^2n^2 + Q_{12}(m^4 + n^4)$$

$$\bar{Q}_{13} = Q_{13}m^2 + Q_{23}n^2$$

$$\bar{Q}_{16} = -mn^3Q_{22} + m^3nQ_{11} - mn(m^2 - n^2)(Q_{12} + 2Q_{66})$$

$$\bar{Q}_{22} = Q_{11}n^4 + 2(Q_{12} + 2Q_{66})m^2n^2 + Q_{22}m^4$$

$$\bar{Q}_{23} = n^2Q_{13} + m^2Q_{23}$$

$$\bar{Q}_{33} = Q_{33}$$

$$\bar{Q}_{26} = -m^3nQ_{22} + mn^3Q_{11} + mn(m^2 - n^2)(Q_{12} + 2Q_{66}) \qquad (10.44)$$

$$\bar{Q}_{36} = (Q_{13} - Q_{23})mn$$

$$\bar{Q}_{44} = Q_{44}m^2 + Q_{55}n^2$$

$$\bar{Q}_{45} = (Q_{55} - Q_{44})mn$$

$$\bar{Q}_{55} = Q_{55}m^2 + Q_{44}n^2$$

$$\bar{Q}_{66} = (Q_{11} + Q_{22} - 2Q_{12})m^2n^2 + Q_{66}(m^2 - n^2)^2$$

$$\alpha_x = \alpha_1 m^2 + \alpha_2 n^2 \qquad \beta_x = \beta_1 m^2 + \beta_2 n^2$$

$$\alpha_y = \alpha_2 m^2 + \alpha_1 n^2 \qquad \beta_y = \beta_2 m^2 + \beta_1 n^2$$

$$\alpha_z = \alpha_3 \qquad \beta_z = \beta_3$$

$$\alpha_{xy} = (\alpha_1 - \alpha_2)mn \qquad \beta_{xy} = (\beta_1 - \beta_2)mn.$$

It should be remembered that although the coefficients of both thermal and hygrothermal expansion are purely dilatational in the material coordinate system 1-2, rotation into the structural coordinate system *x-y*, results in an α_{xy} and a β_{xy}, simply because the composite materials will probably have $\alpha_1 \neq \alpha_2$ and $\beta_1 \neq \beta_2$.

Again, for preliminary design purposes or for approximate but usually accurate calculations one can use the simpler classical form of

$$\begin{Bmatrix} \sigma_x \\ \sigma_y \\ \sigma_{xy} \end{Bmatrix} = \begin{bmatrix} \bar{Q}_{11} & \bar{Q}_{12} & \bar{Q}_{16} \\ \bar{Q}_{12} & \bar{Q}_{22} & \bar{Q}_{26} \\ \bar{Q}_{16} & \bar{Q}_{26} & \bar{Q}_{66} \end{bmatrix} \begin{Bmatrix} \varepsilon_x - \alpha_x \Delta T - \beta_x \Delta m \\ \varepsilon_y - \alpha_y \Delta T - \beta_y \Delta m \\ 2(\varepsilon_{xy} - \alpha_{xy} \Delta T - \beta_{xy} \Delta m) \end{Bmatrix} \tag{10.45}$$

where the $\bar{Q}_{ij}$ are defined in Equation (10.44), but one can use the Q_{ij} of Equation (10.41) instead of Equation (10.40) for consistency with the simpler expressions above.

One interesting variation of the above classical quantities of Equation (10.43) resulted when Tsai and Pagano [10.43] rewrote many of the quantities in terms of material invariants and trigonometric functions involving (2θ) and (4θ). Their method provides an alternative formulation or a check of numerical results.

At this point, given a lamina of a unidirectional composite of known elastic properties, if used in a plate or panel, (as well as a beam, ring or shell, discussed later) with the 1-2 material axis at an angle θ from the plate or panel x-y axes, all stiffness quantities Q_{ij} and $\bar{Q}_{ij}$ can be determined relating stresses and strains in either coordinate system.

10.9 Laminate Analysis

In the previous section the generalized constitutive equations for one lamina of a composite material were formulated. Many structures of composite materials including sandwich structures are composed of numerous laminae, which are bonded and/or cured together. In fact, over and above the superior properties in strength and stiffness that composites possess, the ability to stack laminae one on the other in a varied but unique fashion to result in the optimum laminate material properties for a given structural size and set of loadings is one of the major advantages that composites have over more conventional structures. Up to this point, the concentration has been on the stress-strain or constitutive relations. Now the other three sets of equations comprising the equations of elasticity will be considered: the strain-displacement relations, the equilibrium equations and the compatibility equations.

Consider a laminate composed of N laminae. For the kth lamina of the laminate, Equation (10.43) can be written as:

$$\begin{Bmatrix} \sigma_x \\ \sigma_y \\ \sigma_z \\ \sigma_{yz} \\ \sigma_{xz} \\ \sigma_{xy} \end{Bmatrix}_k = [\bar{Q}]_k \begin{Bmatrix} \varepsilon_x - \alpha_x \Delta T - \beta_x \Delta m \\ \varepsilon_y - \alpha_y \Delta T - \beta_y \Delta m \\ \varepsilon_z - \alpha_z \Delta T - \beta_z \Delta m \\ 2\varepsilon_{yz} \\ 2\varepsilon_{xz} \\ 2(\varepsilon_{xy} - \alpha_{xy} \Delta T - \beta_{xy} \Delta m) \end{Bmatrix}_k , \tag{10.46}$$

where all of the above matrices must have the subscript k due to the material and its orientation for each particular lamina with respect to the structural x-y coordinates and therefore its unique $[\bar{Q}]_k$, $[\alpha_i]_k$ and $[\beta_i]_k$.

For any elastic body the strain-displacement equations, i.e., those kinematic relations describing the functional relations between the elastic strains in the body and its displacements, are given by the following expression when considering linear elastic deformation:

$$\varepsilon_{ij} = \frac{1}{2}\left(u_{i,j} + u_{j,i}\right) \tag{10.47}$$

where i, j = x, y, z in a Cartesian coordinate frame, and the comma denotes partial differentiation with respect to the coordinate denoted by the symbol after the comma. Explicitly, the relations are:

$$\varepsilon_x = \frac{\partial u}{\partial x}, \quad \varepsilon_y = \frac{\partial v}{\partial y}, \quad \varepsilon_z = \frac{\partial w}{\partial z}$$

$$\varepsilon_{xz} = \frac{1}{2}\left(\frac{\partial u}{\partial z} + \frac{\partial w}{\partial x}\right), \quad \varepsilon_{yz} = \frac{1}{2}\left(\frac{\partial v}{\partial z} + \frac{\partial w}{\partial y}\right) \tag{10.48}$$

$$\varepsilon_{xy} = \frac{1}{2}\left(\frac{\partial u}{\partial y} + \frac{\partial v}{\partial x}\right)$$

In the above u, v, and w are the displacements in the structural x, y, and z directions, respectively. In linear elastic beam, plate, ring and shell theory, it is assumed that a lineal element extending through the thickness of a thin plate and perpendicular to the middle surface (that is, the x-y plane in Figure 10.11 below) prior to loading, upon the application of a load undergoes at most a translation and a rotation with respect to the original coordinate system. Based upon that one assumption the functional form of the displacements for a laminated plate is:

$$u(x, y, z) = u_0(x, y) + z\overline{\alpha}(x, y)$$

$$v(x, y, z) = v_0(x, y) + z\overline{\beta}(x, y) \tag{10.49}$$

$$w(x, y, z) = w(x, y)$$

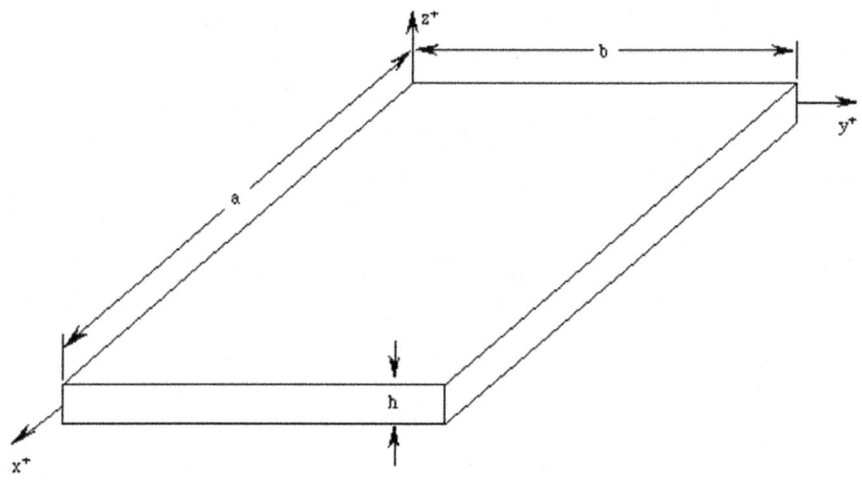

Figure 10.11. Typical Rectangular Plate.

where u_0, v_0 and w are the middle surface displacements, i.e., the translations of the lineal element, and the second terms in the first two equations are the rotations of a lineal element through the thickness. In classical beam and plate theory $\bar{\alpha}$ and $\bar{\beta}$ are the negative of the first derivative of the lateral displacement with respect to the x and y coordinates respectively (i.e., $\bar{\alpha} = -(\partial w/\partial x)$ and $\bar{\beta} = -(\partial w/\partial y)$, the negative of the slope), but if transverse shear deformation is included, $\bar{\alpha}$ and $\bar{\beta}$ are unknown dependent variables which must be solved for; this will be discussed later. Also, in classical theory, it is assumed that the lineal element across the thickness of the beam, plate or shell cannot extend nor shrink because at most it undergoes a translation and rotation, hence $w = w(x,y)$ only. Sloan [10.42] has shown that for many practical composites, plate thickening (where the lateral deflection is a function of z) is unimportant and can be neglected, hence it is not included here, but can be found in detail in [10.42].

Substituting Equation (10.49) into Equation (10.48) results in:

$$\varepsilon_x = \frac{\partial u_0}{\partial x} + z\frac{\partial \bar{\alpha}}{\partial x}, \quad \varepsilon_y = \frac{\partial v_0}{\partial y} + z\frac{\partial \bar{\beta}}{\partial y}, \quad \varepsilon_z = 0$$

$$\varepsilon_{xz} = \frac{1}{2}\left(\bar{\alpha} + \frac{\partial w}{\partial x}\right), \quad \varepsilon_{yz} = \frac{1}{2}\left(\bar{\beta} + \frac{\partial w}{\partial y}\right) \tag{10.50}$$

$$\varepsilon_{xy} = \frac{1}{2}\left(\frac{\partial u_0}{\partial y} + \frac{\partial v_0}{\partial x}\right) + \frac{z}{2}\left(\frac{\partial \bar{\alpha}}{\partial y} + \frac{\partial \bar{\beta}}{\partial x}\right)$$

The mid-surface strains can be written as:

$$\varepsilon_{x_0} = \frac{\partial u_0}{\partial x}, \quad \varepsilon_{y_0} = \frac{\partial v_0}{\partial y}, \quad \varepsilon_{xy_0} = \frac{1}{2}\left(\frac{\partial u_0}{\partial y} + \frac{\partial v_0}{\partial x}\right) \tag{10.51}$$

The curvatures can be written as

$$\kappa_x = \frac{\partial \bar{\alpha}}{\partial x}, \quad \kappa_y = \frac{\partial \bar{\beta}}{\partial y}, \quad \kappa_{xy} = \frac{1}{2}\left(\frac{\partial \bar{\alpha}}{\partial y} + \frac{\partial \bar{\beta}}{\partial x}\right) \tag{10.52}$$

The classical and first order shear deformation theories utilize displacements and strains to describe the strains and displacements of a laminated or sandwich structure composed of composite material, because all of the individual laminae are bonded together, therefore the same assumptions are made regarding the lineal element through the laminate thickness. Thus, a continuity of strains and displacements occurs across the laminated structure regardless of the orientation of individual laminae.

Substituting (10.50) through (10.52) into Equation (10.46) results in Equation (10.53); wherein because it is assumed that $\varepsilon_z = 0$, (because plate thickening is neglected), σ_z for a thin walled structure of composite material is usually negligible and will not be considered further.

$$\begin{Bmatrix} \sigma_x \\ \sigma_y \\ \sigma_{yz} \\ \sigma_{xz} \\ \sigma_{xy} \end{Bmatrix}_k = \left[\bar{Q}\right]_k \begin{Bmatrix} \varepsilon_{x_0} + z\kappa_x - \alpha_x \Delta T - \beta_x \Delta m \\ \varepsilon_{y_0} + z\kappa_y - \alpha_y \Delta T - \beta_y \Delta m \\ 2\varepsilon_{yz} \\ 2\varepsilon_{xz} \\ 2(\varepsilon_{xy_0} + \kappa_{xy} z - \alpha_{xy}\Delta T - \beta_{xy}\Delta m) \end{Bmatrix}_k \tag{10.53}$$

Note that without the hygrothermal terms, the strain-curvature matrix at the right in Equation (10.53) would suffice for the entire laminate independent of orientation, because the displacements, and strains are continuous over the thickness of the laminate. In that case the subscript k on that matrix would not be needed. However, even though there is continuity of the mid-surface strains and curvatures across the thickness of the laminate, the stresses are discontinuous across the laminate thickness because of the various orientations of each lamina, hence, the subscript k in the stress matrix above. It is seen from Equation (10.53) that if all quantities on the right hand side are known, one can easily calculate each stress component in each lamina comprising the laminate.

Consider a laminated plate or panel of thickness h as shown below, in Figure 10.12. It is seen that h_k is the *vectorial* distance from the panel mid-plane, z = 0, to the upper surface of the kth lamina, i.e., *any dimension below the midsurface is a negative dimension* and any dimension above the midsurface is positive. For example, consider a laminate 0.52 mm (0.020") thick, composed of four equally thick laminae, each being 0.13 mm (0.005") thick. Then h_0 = -0.26 mm (-0.010"), h_1 = -0.13 mm (-0.005"), h_2 = 0, h_3 = 0.13 mm (0.005") and h_4 = 0.26 mm (0.010").

As in classical beam, plate and shell theory [1.7, 2.7, 2.8, 3.3, 10.1, 10.2], one defines and uses stress resultants (N), stress couples (M), and transverse shear resultants (Q) per unit width, with appropriate subscript, for the overall structure regardless of the number and the orientation of the laminae, hence:

$$\begin{Bmatrix} N_x \\ N_y \\ N_{xy} \\ Q_x \\ Q_y \end{Bmatrix} = \int_{-h/2}^{+h/2} \begin{Bmatrix} \sigma_x \\ \sigma_y \\ \sigma_{xy} \\ \sigma_{xz} \\ \sigma_{yz} \end{Bmatrix} dz, \quad \begin{Bmatrix} M_x \\ M_y \\ M_{xy} \end{Bmatrix} = \int_{-h/2}^{+h/2} \begin{Bmatrix} \sigma_x \\ \sigma_y \\ \sigma_{xy} \end{Bmatrix} z\,dz \tag{10.54}$$

It must be emphasized that all of the above quantities are a force per unit width and a couple per unit width because in plate and shell structures these quantities vary in both the x and y directions.

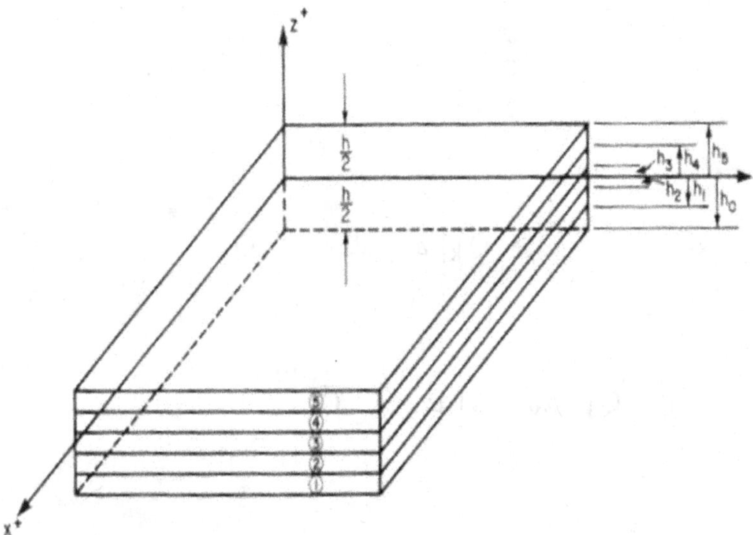

Figure 10.12. Nomenclature for the Stacking Sequence.

In the plate shown in Figure 10.13, the positive directions of all the stress resultants and stress couples are shown, consistent with the definitions of the quantities given in Equation (10.54).

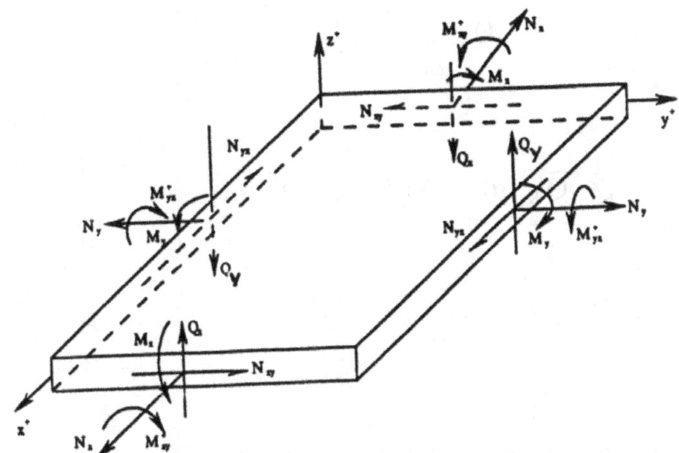

Figure 10.13. Positive directions for Stress Resultants and Stress Couples for a Plate.

For a laminated plate, the stress components can be integrated across each lamina, but must then be added together across the laminae as follows; employing Equations (10.47) and (10.50) through Equation (10.52):

$$\begin{Bmatrix} N_x \\ N_y \\ N_{xy} \end{Bmatrix} = \sum_{k=1}^{N} \int_{h_{k-1}}^{h_k} \begin{Bmatrix} \sigma_x \\ \sigma_y \\ \sigma_{xy} \end{Bmatrix} dz$$

$$\begin{Bmatrix} N_x \\ N_y \\ N_{xy} \end{Bmatrix} = \sum_{k=1}^{N} \left\{ \int_{h_{k-1}}^{h_k} [\bar{Q}]_k \begin{Bmatrix} \varepsilon_{x_0} \\ \varepsilon_{y_0} \\ \varepsilon_{xy_0} \end{Bmatrix} dz + \int_{h_{k-1}}^{h_k} [\bar{Q}]_k \begin{Bmatrix} \kappa_x \\ \kappa_y \\ \kappa_{xy} \end{Bmatrix} zdz \right.$$
$$\left. - \int_{h_{k-1}}^{h_k} [\bar{Q}]_k \begin{Bmatrix} \alpha_x \\ \alpha_y \\ \alpha_{xy} \end{Bmatrix}_k \Delta T dz - \int_{h_{k-1}}^{h_k} [\bar{Q}]_k \begin{Bmatrix} \beta_x \\ \beta_y \\ \beta_{xy} \end{Bmatrix}_k \Delta m dz \right\} \quad (10.55)$$

where only the pertinent portions of the $[\bar{Q}]_k$ matrix are used.

Since the derivatives of the mid-surface displacements (u_0 and v_0), the rotations ($\bar{\alpha}$ and $\bar{\beta}$) and the $\bar{Q}$'s are not functions of z, Equation (10.55) can be rewritten as:

$$\begin{Bmatrix} N_x \\ N_y \\ N_{xy} \end{Bmatrix} = \sum_{k=1}^{N} \left\{ [\bar{Q}]_k \begin{Bmatrix} \varepsilon_{x_0} \\ \varepsilon_{y_0} \\ \varepsilon_{xy_0} \end{Bmatrix}_k \int_{h_{k-1}}^{h_k} dz + [\bar{Q}]_k \begin{Bmatrix} \kappa_x \\ \kappa_y \\ \kappa_{xy} \end{Bmatrix}_k \int_{h_{k-1}}^{h_k} zdz \right.$$
$$\left. - \int_{h_{k-1}}^{h_k} [\bar{Q}]_k \begin{Bmatrix} \alpha_x \\ \alpha_y \\ \alpha_{xy} \end{Bmatrix}_k \Delta T dz - \int_{h_{k-1}}^{h_k} [\bar{Q}]_k \begin{Bmatrix} \beta_x \\ \beta_y \\ \beta_{xy} \end{Bmatrix}_k \Delta m dz \right\} \quad (10.56)$$

Finally, Equation (10.56) can be written succinctly as:

$$[N] = [A][\varepsilon_0] + [B][\kappa] - [N]^T - [N]^m, \quad (10.57)$$

where it is shown later in Equation (10.66) that a factor of 2 is necessary in some terms, and where

$$A_{ij} = \sum_{k=1}^{N} (\bar{Q}_{ij})_k [h_k - h_{k-1}], \quad [i,j = 1, 2, 6] \quad (10.58)$$

$$B_{ij} = \frac{1}{2}\sum_{k=1}^{N}\left(\overline{Q}_{ij}\right)_k\left[h_k^2 - h_{k-1}^2\right], \qquad [i,j = 1, 2, 6] \quad (10.59)$$

$$N_{ij}^T = \sum_{k=1}^{N}\int_{h_{k-1}}^{h_k}\left(\overline{Q}_{ij}\right)_k\left[\alpha_{ij}\right]_k \Delta T dz \qquad [i,j = 1, 2, 6] \quad (10.60)$$

$$N_{ij}^m = \sum_{k=1}^{N}\int_{h_{k-1}}^{h_k}\left(\overline{Q}_{ij}\right)_k\left[\beta_{ij}\right]_k \Delta m dz \qquad [i,j = 1, 2, 6] \quad (10.61)$$

where it is obvious from Equation (10.56) how the $[\alpha_{ij}]_k$ and $[\beta_{ij}]_k$ matrices are defined.

From Equation (10.57), it is seen that the in-plane stress resultants for a laminated thin walled structure are not only functions of the mid-plane strains ($\varepsilon_{x_0} = \partial u_0/\partial x$, etc.) as they are in a homogeneous beam, plate or shell, but they can also be functions of the curvatures and twists ($\kappa_x = \partial \alpha / \partial x$, etc.) as well. Therefore in-plane forces can cause curvatures or twisting deformations in composite laminated structures.

Similar to the above, but multiplying Equation (10.53) through by z first before integrating, as in Equation (10.55), the following can be found:

$$\begin{bmatrix} M_x \\ M_y \\ M_{xy} \end{bmatrix} = \sum_{k=1}^{N}\left\{\left[\overline{Q}\right]_k \begin{bmatrix} \varepsilon_{x_0} \\ \varepsilon_{y_0} \\ \varepsilon_{xy_0} \end{bmatrix}\int_{h_{k-1}}^{h_k} z dz + \left[\overline{Q}\right]_k \begin{bmatrix} \kappa_x \\ \kappa_y \\ \kappa_{xy} \end{bmatrix}\int_{h_{k-1}}^{h_k} z^2 dz \right.$$

$$\left. - \int_{h_{k-1}}^{h_k}\left[\overline{Q}\right]_k \begin{bmatrix} \alpha_x \\ \alpha_y \\ \alpha_{xy} \end{bmatrix} \Delta T z dz - \int_{h_{k-1}}^{h_k}\left[\overline{Q}\right]_k \begin{bmatrix} \beta_x \\ \beta_y \\ \beta_{xy} \end{bmatrix} \Delta m z dz \right\}$$

$$[M] = [B]\,[\varepsilon_0] + [D]\,[\kappa] - [M]^T - [M]^m \qquad (10.62)$$

where it will be shown in Equation (10.66) that factors of 2 are necessary in some terms, and where

$$D_{ij} = \frac{1}{3}\sum_{k=1}^{N}\left(\overline{Q}_{ij}\right)_k\left[h_k^3 - h_{k-1}^3\right], \qquad [i,j = 1, 2, 6] \quad (10.63)$$

$$M_{ij}^T = \sum_{k=1}^{N}\int_{h_{k-1}}^{h_k}\left(\overline{Q}_{ij}\right)_k\left[\alpha_{ij}\right]_k (\Delta T) z dz \qquad [i,j = 1, 2, 6] \quad (10.64)$$

$$M_{ij}^m = \sum_{k=1}^{N}\int_{h_{k-1}}^{h_k}\left(\overline{Q}_{ij}\right)_k\left[\beta_{ij}\right]_k (\Delta m) z dz \qquad [i,j = 1, 2, 6] \quad (10.65)$$

It is noted that in Equations (10.60), (10.61), (10.64) and (10.65) the Q_{ij}, α_{ij} and β_{ij} could have been placed outside of the integral sign since in a lamina, they are not functions of z.

Finally, the results of (10.57) and (10.62) can be written succinctly as follows to form perhaps the most important and most used equation in this text.

$$\begin{bmatrix} N_x \\ N_y \\ N_{xy} \\ --- \\ M_x \\ M_y \\ M_{xy} \end{bmatrix} = \begin{bmatrix} A_{11} & A_{12} & A_{16} & | & B_{11} & B_{12} & B_{16} \\ A_{12} & A_{22} & A_{26} & | & B_{12} & B_{22} & B_{26} \\ A_{16} & A_{26} & A_{66} & | & B_{16} & B_{26} & B_{66} \\ -- & -- & -- & & -- & -- & -- \\ B_{11} & B_{12} & B_{16} & | & D_{11} & D_{12} & D_{16} \\ B_{12} & B_{22} & B_{26} & | & D_{12} & D_{22} & D_{26} \\ B_{16} & B_{26} & B_{66} & | & D_{16} & D_{26} & D_{66} \end{bmatrix} \begin{bmatrix} \varepsilon_{x0} \\ \varepsilon_{y0} \\ 2\varepsilon_{xy0} \\ --- \\ \kappa_x \\ \kappa_y \\ 2\kappa_{xy} \end{bmatrix} - \begin{bmatrix} N_x^T \\ N_y^T \\ N_{xy}^T \\ --- \\ M_x^T \\ M_y^T \\ M_{xy}^T \end{bmatrix} - \begin{bmatrix} N_x^m \\ N_y^m \\ N_{xy}^m \\ --- \\ M_x^m \\ M_y^m \\ M_{xy}^m \end{bmatrix} \tag{10.66}$$

The [A] matrix represents the extensional stiffness matrix relating the in-plane stress resultants (N's) to the mid-surface strains (ε_0's) and the [D] matrix is the flexural stiffness matrix relating the stress couples (M's) to the curvatures (κ's). Since the [B] matrix relates the M's to ε_0's and N's to κ's, it is called the bending-stretching coupling matrix. It should be noted that a laminated structure can have bending-stretching coupling even if all laminae are isotropic. For example, a laminate composed of one lamina of steel and another of polyester will have bending-stretching coupling. In fact, only when the structure is exactly symmetric about its middle surface are all of the B_{ij} components equal to zero, and this requires symmetry in laminae properties, orientation, and distance from the middle surface.

It is seen that stretching-shearing coupling occurs when A_{16} and A_{26} are non-zero. Twisting-stretching coupling as well as bending-shearing coupling occurs when the B_{16} and B_{26} terms are non-zero, and bending-twisting coupling comes from non-zero values of the D_{16} and D_{26} terms. Usually the $(\,)_{16}$ and $(\,)_{26}$ terms are avoided by proper stacking sequences, but there could be some structural applications where these effects could be used to advantage, such as in aeroelastic tailoring.

Examples of these effects in several cross-ply laminates (i.e., combinations of 0° and 90° plies), angle ply laminates (combinations of $+\theta$ and $-\theta$ plies), and unidirectional laminates (all 0° plies) are involved in problems at the end of this chapter. It is seen in Equation (10.58) that $(h_k - h_{k-1})$ is always positive, and from Equation (10.63) $\left(h_k^3 - h_{k-1}^3\right)$

is always positive, hence, in all symmetric cross-ply laminates, all $(\ )_{16}$ and $(\ )_{26}$ terms are zero. If one describes a single layer isotropic plate one can see that, for example $A_{11} = A_{22} = Eh(1-\nu^2)$ and $D_{11} = D_{22} = \dfrac{Eh^3}{12(1-\nu^2)}$.

The inclusion of transverse shear deformation effects on the structural behavior, results in an improved theory as follows. To determine the transverse shear resultants Q_x and Q_y, defined in Equation (10.54), it is assumed that the transverse shear stresses are distributed parabolically across the laminate thickness. In spite of the discontinuities at the interface between laminae, a continuous function f(z) is used as a weighting function by some authors, which includes a factor of 5/4 so that the shear factor calculated for an orthotropic laminate is consistent with the established shear factor from the previous work of Reissner [10.44] and Mindlin [10.45] for the homogeneous case.

$$f(z) = \frac{5}{4}\left[1 - \left(\frac{z}{h/2}\right)^2\right] \tag{10.67}$$

Then from Equations (10.40), (10.43), (10.50), (10.54), and (10.67)

$$\sigma_{xz_k} = 2\bar{Q}_{55_k}\varepsilon_{xz} + 2\bar{Q}_{45_k}\varepsilon_{yz} \qquad \sigma_{yz_k} = 2\bar{Q}_{45_k}\varepsilon_{xz} + 2\bar{Q}_{44_k}\varepsilon_{yz} \tag{10.68}$$

where,

$$\bar{Q}_{44} = Q_{44}m^2 + Q_{55}n^2,\ \bar{Q}_{45} = (Q_{55}-Q_{44})mn,\ \bar{Q}_{55} = Q_{55}m^2 + Q_{44}n^2,$$

$$Q_{44} = G_{23},\ \text{and}\ Q_{55} = G_{13},$$

Hence, the transverse shear resultants Q_x and Q_y are as follows:

$$Q_x = 2(A_{55}\,\varepsilon_{xz} + A_{45}\,\varepsilon_{yz}) \tag{10.69}$$

$$Q_y = 2(A_{45}\varepsilon_{xz} + A_{44}\varepsilon_{yz}) \tag{10.70}$$

where,

$$A_{ij} = \frac{5}{4}\sum_{k=1}^{N}(\bar{Q}_{ij})_k\left[h_k - h_{k-1} - \frac{4}{3}(h_k^3 - h_{k-1}^3)\frac{1}{h^2}\right] \quad (i,j = 4,5 \text{ only}) \tag{10.71}$$

where h is the total thickness of the laminated plate. Some authors use other weighting functions. As an example if one considers the laminate to be only one lamina,

$-\frac{h}{2} \leq z \leq \frac{h}{2}$, then $A_{55} = \left(\frac{5}{6}\right) Q_{55} h$, illustrating clearly the weighting factor introduced by Reissner of $\frac{5}{6}$ [10.44].

At this point the stress-strain relations, or constitutive relations Equation (10.66) can be combined with the appropriate stress equations of equilibrium, and the strain-displacement relations to form an appropriate beam, plate or shell theory including thermal and hygrothermal effects as well as transverse shear deformation.

Another reference for the developments of this chapter is a recent text by Jones [10.46].

10.10 References

10.1. Vinson, J.R. (1999) *The Behavior of Sandwich Structures Composed of Isotropic and Composite Materials*, CRC Press.

10.2. Vinson, J.R. (1993) *The Behavior of Shells Composed of Isotropic and Composite Materials*, Kluwer Academic Publishers, Dordrecht, The Netherlands.

10.3. Shames, I.H. (1975) *Introduction to Solid Mechanics.* New York: Prentice-Hall, Inc.

10.4. Carlsson, L.A. and Pipes, R.B. (1996) *Experimental Characterization of Advanced Composite Materials*, Prentice-Hall Publishing Co., Inc., Englewood Cliffs, N.J., Second Edition.

10.5. Halpin, J.C. and Tsai, S.W. (1967) Environmental Factors in Composite Materials Design, *Air Force Materials Laboratory Technical Report*, pp. 67-423.

10.6. Hashin, Z. (1972) Theory of Fiber Reinforced Materials, *National Aeronautics and Space Administration Contractors Report 1974*.

10.7. Christensen, R.M. (1979) *Mechanics of Composite Materials.* New York: John Wiley and Sons, Inc.

10.8. Hahn, H.T. (1980) Simplified Formulas for Elastic Moduli of Unidirectional Continuous Fiber Composites, *Composites Technology Review*, Fall.

10.9. Chou, T.-W. (1992) *Microstructural Design of Fiber Composites*, Cambridge University Press.

10.10. Tauchert, T.R. (1991) Thermally Induced Flexure, Buckling and Vibration of Composite Laminated Plates, *Applied Mechanics Reviews*, Vol. 44, No. 8, pp. 347-360.

10.11. Argyris, J. and Tenek, L. (1997) Recent Advances in Computational Thermostructural Analysis of Composite Plate and Shells with Strong Nonlinearities, *Applied Mechanics Reviews*, Vol. 50, No. 5, pp. 285-305.

10.12. Turvey, O.J. and Marshall, I.H. (1995) *Buckling and Local Buckling of Composite Plates*, Chapman and Hall, London.

10.13. Noor, A.K. and Burton, W.S. (1992) Computational Models for High Temperature Multilayered Composite Plates and Shells, *Applied Mechanics Reviews*, Vol. 45, No. 10, pp. 414-446.

10.14. Huang, N.N. and Tauchert, T.R. (1988) Large Deformation of Antisymmetric Angle Ply Laminates Resulting from Non-uniform Temperature Loadings, *Journal of Themral Stresses*, Vol. 11, pp. 287-297.

10.15. Springer, G.S. (1984) Model for Predicting the Mechanical Properties of Composites at Elevated Temperatures, *Environmental Effects on Composite Materials*, Vol. 2, pp. 151-161.

10.16. Milke, J.A. and Vizzini, A.J. (1991) Thermal Response of Fire Exposed Composites, *Journal of Composites Technology and Research*, Vol. 13, pp. 145-151.

10.17. Sun, C.T. and Chen, J.K. (1982) Transient Thermal Stress Analysis in Graphite-Epoxy Composite Laminates, *Developments in Theoretical and Applied Mechanics*, Vol. 11, pp. 309-328.

10.18. Vinson, J.R., Walker, W.J., Pipes, R.B. and Ulrich, O.R. (1977) The Effects of Relative Humidity and Elevated Temperatures on Composite Structures, *Air Force Office of Scientific Research Technical Report 77-0030*, February.

10.19. Pipes, R.B., Vinson, J.R. and Chou, T.W. (1976) On the Hygrothermal Response of Laminated Composite Systems, *Journal of Composite Materials*, April, pp. 130-148.

10.20. ASTM Standard D5229-92 Test Method for Moisture Absorption Properties and Equilibrium Conditioning of Polymer Matrix Composite Materials.

10.21. Shen, C. and Springer, G.S. (1981) Environmental Effect on the Elastic Moduli of Composite Material, Ed. G.S. Springer, Technomic Publishing Co., Inc.

10.22. Woldesenbet, E. (1995) High Strain Rate Properties of Composites, *Ph.D. Dissertation*, Department of Mechanical Engineering, University of Delaware, January.

10.23. Zhu, R.P. and Sun, C.T. (2003) Effect of Fiber Orientation and Elastic Constants on Coefficients of Thermal Expansion in Laminates, *Mechanics of Advanced Materials and Structures*, Vol. 10, No. 2, April-June, pp. 99-107.

10.24. Christensen, R.M. (1981) Theory of Viscoelasticity: An Introduction, Academic Press, New York, 2nd Edition.

10.25. Jurf, R.A. and Vinson, J.R. (1985) Effect of Moisture on the Static and Viscoelastic Shear Properties of Epoxy Adhesives, *Journal of Materials Science*, Vol. 20, pp. 2979-2989.

10.26. Wilson, D.W. and Vinson, J.R. (1984) Viscoelastic Analysis of Laminated Plate Buckling, *AIAA Journal*, Vol. 22, No. 7, July, pp. 982-988.

10.27. Wilson, D.W. and Vinson, J.R. (1985). Viscoelastic Effects on the Buckling Response of Laminated Columns, *ASTM Special Publication, 864*, "Recent Advances in Composites in the United States and Japan.

10.28. Hu, H. and Sun, C.T. (2003) The Equivalence of Moisture and Temperature in Physical Aging of Polymeric Composites, Vol. 37, No. 10, pp. 913-928.

10.29. Lindholm, V.S. (1964) Some Experiments With the Split Hopkinson Pressure Bar, *Journal of Mechanics and Physics of Solids*, Vol. 12, pp. 317-335.

10.30. Daniel, I.M., La Bedz, R.H. and Liber, T. (1981) New Method for Testing Composites at Very High Strain Rates, *Experimental Mechanics*, February.

10.31. Nicholas, T. (1981) Tensile Testing of Materials at High Rates of Strain, *Experimental Mechanics*, pp. 177-185, May.

10.32. Zukas, J.A. (1982) *Impact Dynamics*, John Wiley and Sons, New York, N.Y.

10.33. Sierakowski, R.L. and Chaturvedi, S.K. (1997) *Dynamic Loading and Characterization of Fiber-Reinforced Composites*, Wiley Interscience.

10.34. Sierakowski, R.L. (1997) Strain Rate Effects in Composites, *Applied Mechanics Reviews*, Vol. 50, No. 12, pp. 741-761, December.

10.35. Rajapakse, Y. and Vinson, J.R. (1995) High Strain Rate Effects on Polymer, Metal and Ceramic Matrix Composites and Other Advanced Materials, *ASME Bound Volume AD-Vol. 48.*

10.36. Abrate, S. (1991) Impact on Laminated Composite Materials, *Applied Mechanics Reviews*, Vol. 44, No. 4, pp. 155-190, April.

10.37. Abrate, S. (1994) Impact on Laminated Composites: Recent Advances, *Applied Mechanics Reviews*, Vol. 47, No. 11, pp. 517-544, November.

10.38. Song, S., Vinson, J.R. and Crane, R.M. (2002) Low Temperature Effects on E-Glass/Urethane Composite Materials of High Strain Rates, *Proceedings of the 43rd Annual AIAA/ASME/ASCE/AHS/ASC Structures*, Structural Dynamics and Materials Conference, April.

10.39. Song, S. and Vinson, J.R. (2002) Low Temperature Effects on G30-500/EH-80 Composite Material Properties at High Strain Rates, Fourteenth U.S. National Congress on Theoretical and Applied Mechanics, June.

10.40. Song, S. and Vinson, J.R. (2002) Low Temperature Effects on the Ultimate Strength of IM7/977-3 Composites at High Strain Rates, *Proceedings of the 17th Annual Technical Conference of the American Society for Composites*, October.

10.41. Song, S. and Vinson, J.R. (2002) Low Temperature Effects on IM7/977-3 Cross-Ply Composite Material Properties at High Strain Rates, *Proceedings of the 10th Japan-U.S. Conference on Composite Materials*, September 16-18.

10.42. Sloan, J.G. (1979) The Behavior of Rectangular Composite Material Plates Under Lateral and Hygrothermal Loads, *MMAE Thesis*, University of Delaware (also AFOSR-TR-78-1477, July 1978).

10.43. Tsai, S.W. and Pagano, N.J. (1968) Invariant Properties of Composite Materials, Tsai, S.W. et al., eds., *Composite Materials Workshop*, Technomic Publishing Co., Inc., Lancaster, PA, pp. 233-253.

10.44. Reissner, E. (1950) On a Variational Theorem in Elasticity, *J. Math. Phys., 29*, pp. 90.

10.45. Mindlin, R.D. (1951) Influence of Rotatory Inertia and Shear on Flexural Motions of Isotropic Elastic Plates, *Journal of Applied Mechanics*, pg. 73.

10.46. Jones, R.M. (2002) *Mechanics of Composite Materials*, Second Edition, R.T. Edwards, Flourtown, PA.

10.11 Problems

10.1 Consider a laminate composed of boron-epoxy with the following properties:

$Q_{11} = 2.43 \times 10^5$ MPa (35.32 × 10^6 psi) $Q_{12} = 2.43 \times 10^4$ MPa (1.06 × 10^6 psi)

$Q_{22} = 2.43 \times 10^4$MPa (3.532 × 10^6 psi) $Q_{66} = 1.034 \times 10^3$ MPa (1.5 × 10^6 psi)

If the laminate is a cross–ply with [0°/90°/90°/0°], with each ply being 0.25 mm (0.11") thick, and if the laminate is loaded in tension in the x direction (i.e., the 0° direction):

(a) What percentage of the load is carried by the 0 plies? The 90° plies?

(b) If the strength of the 0° plies is 1.364 ×10^3 MPa (198,000 psi), and the strength of the 90° plies is 44.8 MPa (6,500psi) which plies will fail first?

(c) What is the maximum load, N_{max}, that the laminate can carry at incipient failure? What stress exists in the remaining two plies, at the failure load of the other two others?

(d) If the structure can tolerate failure of two plies, what is the maximum load, N_{max} that the other two plies can withstand to failure?

10.2. A laminate is composed of graphite epoxy (GY70/339) with the following properties: E_{11} = 2.89×105 MPa (42 ×10^6 psi), E_{22} = 6.063 × 10^3 MPa (0.88 × 10^6 psi), G_{12} = 4.134×10^3 MPa (0.6×106 psi) and ν_{12} = 0.31. Determine the elements of the A, B and D matrices for a two-ply laminate [+45°/–45°], where each ply is 0.15 mm (0.006") thick.

10.3. Consider a square panel composed of one ply with the fibers in the directions as shown in Figure 10.12.

Which of the orientations above would be the stiffest for the loads given in Figure 10.12?

10.4. For a panel consisting of boron-epoxy with the properties of Problem 1 above, and a stacking sequence of [0°/+45°/-45°/0°], and a ply thickness of 0.14 mm (0.005"), determine the elements of the elements of the A, B and D matrices.

10.5. The properties of graphite fibers and a polyimide matrix are as follows:

E = 2.756 × 105 MPa (40 × 106 psi) E = 2.756 × 103 MPa (0.4 × 106 psi)

ν = 0.2 ν = 0.33

(a) Find the modulus of elasticity in the fiber direction, E_{11}, of a laminate of graphite-polymide composite with 60% fiber volume ratio.

(b) Find the Poisson's ratio, ν_{12}?

(c) Find the modulus of elasticity normal to the fiber direction, E_{22}.

(d) What is the Poisson's ratio, ν_{21}?

10.6. Consider a laminate composed of GY70/339 graphite epoxy whose properties are given above in Problem 10.2. For a lamina thickness of 0.127 mm (0.005"), calculate the elements of the A, B and D matrices for the following:

(a) [0°, 0°, 0°, 0°] (unidirectional);

(b) [0°, 90°, 90°, 0°] (across-ply);

(c) $[\pm 45]_s$, i.e. [+45°/-45°/-45°/+45°] (an angle-ply);

(d) $[0°/+45°/-45°/90°]_s$ (a quasi-isotropic, 8 plies);
(e) Compare the various stiffness quantities for the four laminates above.

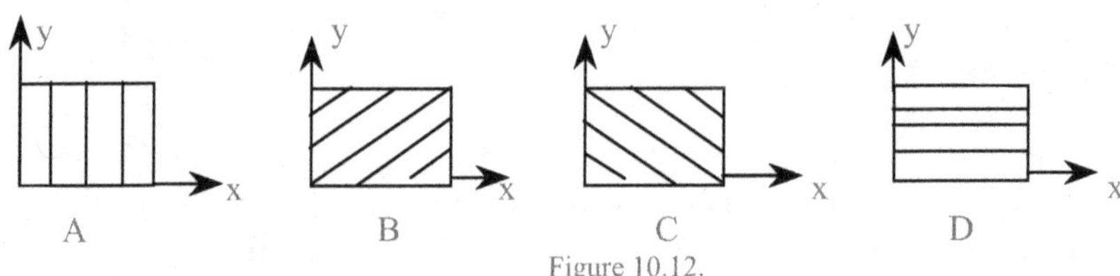

Figure 10.12.

10.7. Consider a laminate composed of GY70/339 graphite epoxy whose properties are given above in Problem 10.2. For a lamina thickness of 0.127 mm (0.005") cited in Problem 10.6, calculate the elements of the [A], [B] and [D] matrices for the following laminates:
(a) $[\pm(45)2]_s$, $[+45/-45/+45/-45]_s$
(b) $[\pm45]_s$, $[+45/-45/-45/+45]$
(c) $[\pm45]Q_s$, $[+45/-45/+45/-45]$
(d) $[\pm(45)2]_{Qs}$, $[+45/-45/+45/-45]_{Qs}$
Compare the forms of the A, B and D matrices between laminate type.

10.8. What type of coupling would you expect in the (B) matrix for (a) and (b) below:
(a) 0°/90° laminate
(b) $+\theta/-\theta$ laminate

10.9. Given a composite laminate composed of continuous fiber laminate laminae of High Strength Graphite/Epoxy with properties of Table 10.3, if the laminate architecture is [0°, 90°, 90°, 0°], determine A_{11} if ν_{12} □ 0.3, and each ply thickness is 0.006".

10.10. Consider a plate composed of a 0.01" thick steel plate joined perfectly to an aluminum plate, 0.01" thick. Using the properties of Table 10.3 calculate B_{11}, if the Poisson's Ratio of each material is $\nu = 0.3$.

10.11. Consider a unidirectional composite composed of a polyimide matrix and graphite fibers with properties given in Problem 10.5 above. In the fiber direction, what volume fraction is required to have a composite stiffness of $E_{11} = 10 \times 10^6$ psi to match an aluminum stiffness.

10.12. A laminate is composed of ultra high modulus graphite epoxy with properties given in Table 10.3 below. Determine the elements of the [A], [B] and [D] matrices for a two ply laminate [+45°/-45°], where each ply is 0.006" thick. For the material $\nu_{12} = 0.31$.

10.13. A laminate is composed of boron-epoxy with the properties of Problem 10.1 and a stacking sequence of [0/+45°/–45°/0°], and a ply thickness of 0.006″. Determine the elements of the A, B and D matrices.

Table 10.3. Unidirectional properties.

Material	Elastic moduli			Ultimate strength			Density
	Axial	Transverse	Shear	Axial tens.	Trans. tens.	Shear.tens.	
	E_{11}	E_{22}	G_{12}	σ_{11}	σ_{22}	τ_{12}	
High strength	20	1.0	0.65	220	6	14	0.057
GR/epoxy	(138)	(6.9)	(4.5)	(1517)	(41)	(97)	(1.57)
High Modulus	32	1.0	0.7	175	5	10	0.058
GR/epoxy	(221)	(6.9)	(4.8)	(1206)	(34)	(69)	(1.60)
Ultra high	44	1.0	0.95	110	4	7	0.061
modulus	(303)	(6.9)	(6.6)	(758)	(28)	(48)	(1.68)
GR/epoxy							
Kevlar49/	12.5	0.8	0.3	220	4	6	0.050
epoxy	(86)	(5.5)	(2.1)	(1517)	(28)	(41)	(1.38)
S glass	8	1.0	0.5	260	6	10	0.073
epoxy	(55)	(6.9)	(3.4)	(1793)	(41)	(69)	(2.00)
Steel	30	30	11.5	60	60	35	0.284
	(207)	(207)	(79)	(414)	(414)	(241)	(7.83)
Aluminum	10.5	10.5	3.8	42	42	28	0.098
6061-T6	(72)	(72)	(26)	(290)	(290)	(193)	(2.70)

Moduli in Msi (GPa); Stress in Ksi (MPa); Density in lb/in^3(g/cm^3)

10.14. Consider a composite laminae made up of continuous Boron fibers imbedded in an epoxy matrix. The volume fraction of the Boron fibers in the composite is 40%. Assuming that the modulus of elasticity of the Boron fiber is 5×10^7 psi and the epoxy is 5×10^5 psi, find:

(a) The Young's moduli of the composite in the 1 and 2 direction.

(b) Consider an identical second lamina to be glued to the first so that the fibers of the second lamina are parallel to the 2 direction. Assuming the thickness of each lamina to be 0.1" and neglecting Poisson's Ratio, what are the new moduli in the 1 and 2 directions.

10.15. The properties of graphite fibers and a polyimide matrix are as follows:

GRAPHITE	POLYIMIDE
$E = 40 \times 10^6$ psi	$E = 0.4 \times 10^6$ psi
ν ⌊ 0.2	ν □ 0.33

(a) Finder the modulus of elasticity in the fiber direction, E_{11}, of a lamina of graphite – polyimide composite with 70% fiber volume ratio.

(b) Find the Poisson's Ratio, ν_{12}.

(c) Finder the modulus of elasticity normal to the fiber direction, E_{22}.

(d) What is the Poisson's Ratio, ν_{22}.

(e) Compare these properties with those obtained for the same material system but with $\nu_F = 60\%$ in problem 10.5.

10.16. In a given composite, the coefficient of thermal expansion for the epoxy and the graphite fibers are $+30 \times 10^{-6}$ in/in/°F and -15×10^{-6} in/in/°F respectively. For space application where no thermal distortion can be tolerated what volume fractions of each component are required to make zero expansion and contraction in the fiber direction for an all 0° construction? (Hint: Use the Rule of Mixtures).

10.17. Find the A, B and D matrices for the following composite: 50% volume *Fraction Boron-Epoxy Composite*

$E_{11} = 30.0 \times 10^6$ psi
$E_{22} = 3.0 \times 10^6$ psi
$G_{12} = 1.0 \times 10^6$ psi
$\nu_{12} = 0.22$

Stacking Sequence (each lamina is 0.0125" thick)

θ□45°
θ□0°
θ□90°
θ□-45°
θ□-45°
θ□90°
θ□0°
θ□45°

10.18. Three composite plates are under uniform transverse loading. All the conditions, such as materials, boundary conditions and geometry, etc. are the same except the stacking sequence as shown below. Without using any calculation, indicate which plate will have maximum deflection and will have minimum deflection.

10.19. Consider a Kevlar 49/epoxy composite laminate, whose properties are in Table 10.3 in the text and whose stacking sequence is [0,90,90,0] (i.e. a cross ply laminate). The ply thickness is 0.0055 inches.

(a) Determine the A, B, and D matrix component.
(b) What if any are the couplings in this cross-ply construction that are discussed below Equation (10.66)?
(c) If only in-plane loads are applied, is the plate stiffer in the x direction or y direction, or are they the same?
(d) If only plate bending is considered, is the plate stiffer in the x direction, the y direction, or are they equally stiff?

10.20. Given the following fiber and matrix properties for HM-S/epoxy composite components:

Epoxy	HM-S/Graphite
E_m= 0.5 Msi = 3.45 GPa	E_{f11} = 55.0 Msi = 379.3 GPa
G_m= 0.185 Msi = 1.27 GPa	E_{f22}= 0.9 Msi =6.2 GPa
ρ_m= 0.0440 lb/in^3 = 1.218 gr./cm^3	ν_{f12} = 0.20
	ρ_{f12} = 0.0703 lb/in^3 = 1.946 gr./cm^3

Determine each of the following properties for a unidirectional composite:

E_{11}, E_{12}, E_{33}, G_{12}, G_{13}, υ_{12}, υ_{21}, υ_{13}, υ_{31}, υ_{23} and υ_{32} for the fiber volume fractions of $V_f = 0\%$, 30%, 60%. Which properties increase linearly with volume fraction? Which do not increase linearly with volume fraction?

10.21. Given a cross-ply construction of four lamina of the same composite material system oriented as 0°, 90°, 90°, 0°, each lamina being equally thick, which elements of the [A], [B] and [D] matrices of Equation (10.66) will be equal to zero.

10.22. Given an angle-ply construction of five plies of the same composite material oriented as +θ°/–θ°/+θ°/–θ°/+θ°, each of equal thickness, which elements of the [A], [B] and [D] matrices of Equation (10.66) will be equal to zero.

10.23. Determine the elements of the C_{ij} matrix analogous to the a_{ij} of Equation (10.10) through (10.12) for orthotropic materials.

$$C_{ij}\begin{bmatrix} C_{11} & C_{12} & C_{13} & 0 & 0 & 0 \\ C_{21} & C_{22} & C_{23} & 0 & 0 & 0 \\ C_{31} & C_{32} & C_{33} & 0 & 0 & 0 \\ 0 & 0 & 0 & C_{44} & 0 & 0 \\ 0 & 0 & 0 & 0 & C_{55} & 0 \\ 0 & 0 & 0 & 0 & 0 & C_{56} \end{bmatrix}$$

10.24. A laminate is composed of graphite epoxy (GY70/339) with the following properties:

$E_{11} = 2.89 \times 10^5$ MPa (42×10^6 psi)
$E_{22} = 60.63 \times 10^3$ MPa (0.88×10^6 psi)
$G_{12} = 4.134 \times 10^3$ MPa (0.60×10^6 psi)
$\upsilon_{12} = 0.31$

(a) Determine the elements of the [A], [B], and [D] matrices for a two-ply laminate [+45/-45], where each ply is 0.15mm. (0.006 inches) thick.
(b) What couplings exist as discussed below Equation (10.66) for this laminate?

10.25. For a panel consisting of Boron-Epoxy with the properties

$Q_{11} = 35.32 \times 10^6$ psi
$Q_{22} = 3.532 \times 10^6$ psi
$Q_{12} = 1.06 \times 10^6$ psi
$Q_{66} = 1.50 \times 10^6$ psi

and a stacking sequence of [0°, +45°, -45°, 0°], and a ply thickness of 0.006 inches, determine the elements of the A, B and D matrices. What would the elements be if the ply thickness were 0.0055 inches?

10.26. Determine how the A, B, D matrices are populated for the following two stacking sequences $[0°, \pm 45°, 90°]_{QS}$ and $[0°, \pm 45°, 90°]_S$. The subscript QS mean symmetric Q times where Q = 2, 3,... The composite material is orthotropic and has the properties E_{11}, E_{22}, G_{12}, υ_{12}, with each lamina having thickness h_k. In the [A], [B], and [D] matrices, place an x or an O for each element, where an x shows that the component is non zero, and O shows that the component is zero.

10.27. What type of couplings, as discussed below Equation (10.66) would you expect in the B matrix for (a) and (b) below: (that is, identify the non-zero terms)
(a) 0°/90° laminate
(b) +θ/-θ laminate

10.28. Find the [A], [B] and [D] for the following laminates.

Laminate 1	Laminate 2	Laminate 3	Laminate 4	Thickness
0°	0°	0°	90°	0.1"
90°	0°	90°	0°	0.1"
90°	90°	0°	0°	0.1"
0°	90°	90°	90°	0.1"

Given: $E_{11} = 30 \times 10^6$ psi, $E_{22} = 3 \times 10^6$ psi, $G_{12} = 1 \times 10^6$ psi, $\upsilon = 0.3$.

10.29. In problem 10.28 which laminate is stiffest and which is the least stiff for
(a) In-plane loads in the 0° direction.
(b) In-plane loads in the 90° direction.
(c) Bending in the 0° direction.
(d) Bending in the 90° direction.

10.30. Consider a laminate composed of GY 70/339 graphite/epoxy with the following properties,

$$E_{11} = 2.89 \times 10^5 \text{ MPa } (42 \times 10^6 \text{ psi})$$
$$E_{22} = 6.063 \times 10^3 \text{ MPa } (0.88 \times 10^6 \text{ psi})$$
$$G_{12} = 40134 \times 10^3 \text{ MPa } (0.60 \times 10^6 \text{ psi})$$
$$\upsilon_{12} = 0.31$$

Using the laminate thickness as 0.127mm (0.005 inches) calculate the elements of the [A], [B], and [D] matrices for the following laminates.
(a) $[\pm(45)_2]_s$, $[+45/-45/+45/-45]_s$
(b) $[\pm 45]_s$, $[+45/-45/+45/-45]_s$
(c) $[\pm 45]_{Qs}$, $[+45/-45/+45/-45]$
(d) $[\pm(45)2]_{Qs}$, $[+45/-45/+45/-45]_{Qs}$
Compare the forms of the [A], [B] and [D] matrices between laminate types.

10.31. A composite material has stiffness matrix as follows,

$$[C] = \begin{bmatrix} 100 & -10 & 0 & 0 & 0 & 10 \\ -10 & 50 & 0 & 0 & 0 & 10 \\ 0 & 0 & 50 & 0 & 0 & 0 \\ 0 & 0 & 0 & 20 & 0 & 0 \\ 0 & 0 & 0 & 0 & 20 & 0 \\ 0 & 0 & 0 & 0 & 0 & 20 \end{bmatrix} \times 10^5 \text{ psi}$$

Determine the state of stress if the strains are given by,
$\varepsilon_x = 100$ μin/in, $\varepsilon_y = 50$ μin/in, $\gamma = -100$ μin/in.

10.32. Consider the stress acting on an element of a composite material to be as shown below. The material axes 1,2 are angle θ with respect to the geometry loading axes for the element. Taking the material properties as noted below, find the m-plane displacements u(x,y), v(x,y).

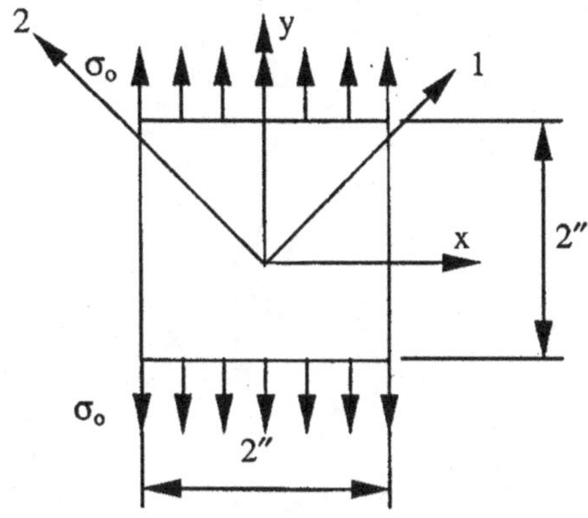

$\sigma_{yy} = \sigma_o = 25$ Ksi
$\theta = 45°$
$E_{11} = 25\times10^6$ psi
$E_{22} = 1\times10^6$ psi
$G_{12} = 0.5\times10^6$ psi
$\nu_{12} = 0.25$

10.33. A 50% boron-epoxy orthotropic material is subjected to combined stress as shown below.

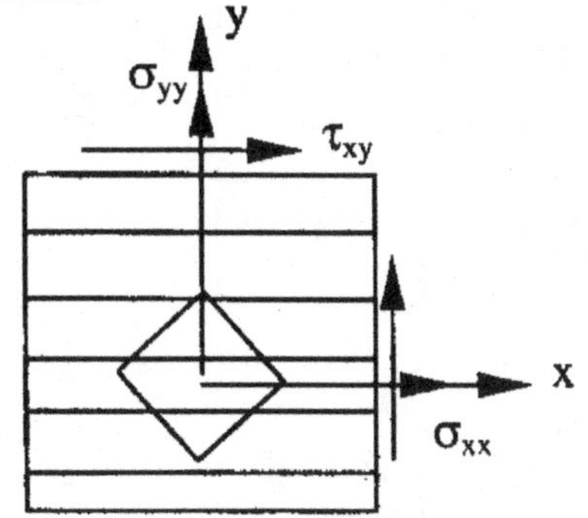

$\sigma_{xx} = 10{,}000$ psi
$\sigma_{yy} = 10{,}000$ psi
$\tau_{xy} = 10{,}000$ psi

- Find the stress on the material element for a 45° rotation about the z-axis in positive sense.

(a) If the strain components in the non-rotated system are given by:

$\varepsilon_x = 260$ µin/in, $\varepsilon_y = 3110$ µin/in, $\gamma = -9090$ µin/in

Find the corresponding strains in the rotated system.

(b) Comment on the corresponding stresses and strains in the rotated system.

10.34. The elastic properties of a unidirectional Kevlar 49/epoxy composite are given in Table 10.3. A cross-ply laminate consists of 4 plies such that the stacking sequence is $[0°, 90°, 90°, 0°]$. Each ply is 0.0055 inches thick.

(a) Determine each element in the $[A]$, $[B]$ and $[D]$ matrices.

(b) What if any coupling exists in this cross-ply configuration?

(c) Concerning in-plane loads only, is the plate stiffer in the x-direction, the y-direction, or are they the same?

(d) Concerning plate bending, is the plate stiffer in the x-direction, the y-direction, or are they equally stiff in bending?

CHAPTER 11

PLATES AND PANELS OF COMPOSITE MATERIALS

11.1 Introduction

In Chapter 10, the constitutive equations for composite materials were developed in detail, describing the relationships between integrated stress resultants (N_x, N_y, N_{xy}), integrated stress couples (M_x, M_y, M_{xy}), in-plane mid-surface strains $(\varepsilon_x^0, \varepsilon_y^0, \varepsilon_{xy}^0)$, and the curvatures $(\kappa_x, \kappa_y, \kappa_{xy})$, as seen in Equation (10.66). These will be utilized with the strain-displacement relations of Equations (10.48) and (10.50) and the equilibrium equations to be developed in Section 11.2 below to develop structural theories for thin plates and panels, the configuration in which composite materials are most generally employed.

11.2 Plate Equilibrium Equations

The integrated stress resultants (N), shear resultants (Q) and stress couples $(\boldsymbol{M})$, with appropriate subscripts, are defined by Equations (2.4) though (2.13), and their positive directions are shown in Figure 2.2, for a rectangular plate, defined as a body of length a in the x-direction, width b in the y-direction, and thickness h in the z-direction, where $h << b, h << a$, i.e. a thin plate.

In mathematically modeling solid materials, including the laminates of Chapter 10, a continuum theory is generally employed. In doing so, a representative material point within the elastic solid or lamina is selected as being macroscopically typical of all material points in the body or lamina. The material point is assumed to be infinitely smaller than any dimension of the structure containing it, but infinitely larger than the size of the molecular lattice spacing of the structured material comprising it. Moreover, the material point is given a convenient shape; and in a Cartesian reference frame that convenient shape is a small cube of dimensions dx, dy, and dz as shown in Figure 1.1.

This cubic material point of dimension dx, dy and dz is termed a control element. The positive values of all stresses acting on each surface of the control element are shown in Figure 1.3, along with how they vary from one surface to another, using the positive sign convention consistent with most scientific literature, and consistent with Figure 1.3. Details of the nomenclature can be found in any text on solid mechanics, including [1.7, 2.7, 2.8, 3.3, 10.1, 10.2]. In addition to the surface stresses acting on the control element shown in Figure 1.3, body force components F_x, F_y and F_z can also act on the body. These body force components such as gravitational, magnetic or centrifugal forces are proportional to the control element volume, i.e., its mass.

A force balance can now be made in the x, y and z directions resulting in three equations of equilibrium. Proceeding exactly as in Section 1.5, one derives Equations (1.5) though (1.8). These three equations comprise the equilibrium equations for a three dimensional elastic body. However, for beam, plate and shell theory, whether involving composite materials or not, one must integrate the stresses across the thickness of the thin walled structures to obtain solutions.

Recalling the definitions of the stress resultants and stress couples defined in (2.4) through (2.13) and (10.54); for a laminated (or sandwich) plate or panel, they are:

$$\begin{bmatrix} N_x \\ N_y \\ N_{xy} \\ Q_x \\ Q_y \end{bmatrix} = \int_{-h/2}^{+h/2} \begin{bmatrix} \sigma_x \\ \sigma_y \\ \sigma_{xy} \\ \sigma_{xz} \\ \sigma_{yz} \end{bmatrix} dz = \sum_{k=1}^{N} \int_{h_{k-1}}^{h_k} \begin{bmatrix} \sigma_x \\ \sigma_y \\ \sigma_{xy} \\ \sigma_{xz} \\ \sigma_{yz} \end{bmatrix}_k dz_k \tag{11.1}$$

$$\begin{bmatrix} M_x \\ M_y \\ M_{xy} \end{bmatrix} = \int_{-h/2}^{h/2} \begin{bmatrix} \sigma_x \\ \sigma_y \\ \sigma_{xy} \end{bmatrix} z\,dz = \sum_{k=1}^{N} \int_{h_{k-1}}^{h_k} \begin{bmatrix} \sigma_x \\ \sigma_y \\ \sigma_{xy} \end{bmatrix}_k z_k\,dz_k \tag{11.2}$$

The first form of each of the above is applicable to a single layer plate, while the second form is necessary for a laminated or sandwich plate due to the stress discontinuities associated with different materials and/or differing orientations in the various plies.

Turning now to (1.5), neglecting the body force term, F_x, for simplicity of this example, integrating term by term across each ply, and summing across the plate provides

$$\sum_{k=1}^{N} \int_{h_{k-1}}^{h_k} \frac{\partial \sigma_{x_k}}{\partial x} dz + \sum_{k=1}^{N} \int_{h_{k-1}}^{h_k} \frac{\partial \sigma_{yx_k}}{\partial y} dz + \sum_{k=1}^{N} \int_{h_{k-1}}^{h_k} \frac{\partial \sigma_{zx_k}}{\partial z} dz = 0 \tag{11.3}$$

In the first two terms integration and differentiation can be interchanged, hence:

$$\frac{\partial}{\partial x}\left[\sum_{k=1}^{N} \int_{h_{k-1}}^{h_k} \sigma_{x_k} dz\right] + \frac{\partial}{\partial y}\left[\sum_{k=1}^{N} \int_{h_{k-1}}^{h_k} \sigma_{yx_k} dz\right] + \sum_{k=1}^{N} \sigma_{zx} \Big]_{h_{k-1}}^{h_k} = 0 \tag{11.4}$$

In the first two terms N_x and N_{yx} appear explicitly as the bracketed quantities. In the third term it is clear that between all plies of a laminated composite plate and between the face and core materials of a sandwich panel the interlaminar shear stresses will cancel each other, and one can define the applied surface shear stresses on the top $(z = h_N)$ and bottom $(z = h_0)$ surfaces as shown below (see Figure 10.1)

$$\sigma_{zx}(h_N) \equiv \tau_{1x} \quad \text{and} \quad \sigma_{zx}(h_o) \equiv \tau_{2x} \tag{11.5}$$

Equation (11.4) can then be written as:

$$\frac{\partial N_x}{\partial x} + \frac{\partial N_{yx}}{\partial y} + \tau_{1x} - \tau_{2x} = 0 \tag{11.6}$$

Similarly, integrating the equilibrium equation in the y-direction provides

$$\frac{\partial N_{xy}}{\partial x} + \frac{\partial N_y}{\partial y} + \tau_{1y} - \tau_{2y} = 0 \tag{11.7}$$

where

$$\sigma_{zy}(h_N) \equiv \tau_{1y} \text{ and } \sigma_{zy}(h_0) \equiv \tau_{2y} \tag{11.8}$$

Likewise equilibrium in the z-direction upon integration and summing provides

$$\frac{\partial Q_x}{\partial x} + \frac{\partial Q_y}{\partial y} + p_1 - p_2 = 0 \tag{11.9}$$

where

$$\sigma_z(h_N) \equiv p_1 \text{ and } \sigma_z(h_0) \equiv p_2 \tag{11.10}$$

It is seen that Equations (11.6), (11.7) and (11.8) are identical to Equations (2.17), (2.18) and (2.16) for a plate made of an isotropic material. The reason is that equilibrium equations are force balances and have nothing to do with the materials comprising the plate.

In addition to the integrated force equilibrium equations above, two equations of moment equilibrium are also needed, one for the x-direction and one for the y-direction. Multiplying equation 1.5 through by zdz, integrating across each ply and summing across all laminae results in the following

$$\sum_{k=1}^{N} \int_{h_{k-1}}^{h_k} \frac{\partial \sigma_{x_k}}{\partial x} z dz + \sum_{k=1}^{N} \int_{h_{k-1}}^{h_k} \frac{\partial \sigma_{yx_k}}{\partial y} z dz + \sum_{k=1}^{N} \int_{h_{k-1}}^{h_k} \frac{\partial \sigma_{zx_k}}{\partial z} z dz = 0$$

Again, in the first two terms integration and summation can be interchanged with differentiation with the result that the first two terms become $(\partial M_x / \partial x) + (\partial M_{xy} / \partial_y)$. However, the third term must be integrated by parts as follows:

$$\sum_{k=1}^{N} \int_{h_{k-1}}^{h_k} \frac{\partial \sigma_{zx_k}}{\partial z} z dz = \sum_{k=1}^{N} \left\{ \left[z\sigma_{zx} \right]_{h_{k-1}}^{h_k} - \int_{h_{k-1}}^{h_k} \sigma_{zx} dz \right\}$$

Here the last term is clearly $-Q_x$. Again in the first term on the right, clearly the moments of all the interlaminar stresses between plies cancel each other out, and the only non-zero terms are the moments of the applied surface shear stresses hence that term becomes

$$h_N \tau_{1x} - h_0 \tau_{2x} = \frac{h}{2}\left[\tau_{1x} + \tau_{2x}\right]$$

Using the former expression, the equation of equilibrium of moments in the x-direction is

$$\frac{\partial M_x}{\partial x} + \frac{\partial M_{xy}}{\partial y} - Q_x + \frac{h}{2}\left[\tau_{1x} + \tau_{2x}\right] = 0 \tag{11.11}$$

Similarly in the y-direction the moment equilibrium equation is

$$\frac{\partial M_{xy}}{\partial x} + \frac{\partial M_y}{\partial y} - Q_y + \frac{h}{2}\left[\tau_{1y} + \tau_{2y}\right] = 0 \tag{11.12}$$

where all the terms are defined above. Again, (11.11) and (11.12) are identical to (2.14) and (2.15). Thus, there are five equilibrium equations for a rectangular plate, regardless of what material or materials are utilized in the plate: (11.6), (11.7), (11.9), (11.11) and (11.12).

11.3 The Bending of Composite Material Laminated Plates: Classical Theory

Consider a plate composed of a laminated composite material that is mid-plane symmetric, i.e. $B_{ij} = 0$, and has no other coupling terms $(\ )_{16} = (\ )_{26} = 0$. Such a plate is called a specially orthotropic plate. Also, assume no surface shear stresses and no hygrothermal effects for simplicity. The plate equilibrium equations for the bending of the plate, due to lateral loads given by Equations (11.9), (11.11) and (11.12) become:

$$\frac{\partial M_x}{\partial x} + \frac{\partial M_{xy}}{\partial y} - Q_x = 0 \tag{11.13}$$

$$\frac{\partial M_{xy}}{\partial x}+\frac{\partial M_y}{\partial y}-Q_y=0 \qquad (11.14)$$

$$\frac{\partial Q_x}{\partial x}+\frac{\partial Q_y}{\partial y}+p(\mathbf{x},\mathbf{y})=0 \qquad (11.15)$$

where $p(\mathbf{x},\mathbf{y})=p_1(x,y)-\mathbf{p_2}(\mathbf{x},\mathbf{y})$. Derivatives of (11.13) and (11.14) can be substituted into (11.15) with the result that:

$$\frac{\partial^2 M_x}{\partial x^2}+2\frac{\partial^2 M_{xy}}{\partial x \partial y}+\frac{\partial^2 M_y}{\partial y^2}=-p(\mathbf{x},\mathbf{y}) \qquad (11.16)$$

Again, it is seen that (11.16) is identical to (2.56), because the above equations are derived from equilibrium considerations alone. From Equation (10.66) and for the case of mid-plane symmetry $(B_{ij}=0)$ and no $(\ )_{16}$ and $(\ )_{26}$ coupling terms, the constitutive relations are:

$$M_x=D_{11}\kappa_x+D_{12}\kappa_y \qquad (11.17)$$

$$M_y=D_{12}\kappa_x+D_{22}\kappa_y \qquad (11.18)$$

$$M_{xy}=2D_{66}\kappa_{xy} \qquad (11.19)$$

where from Equation (10.52)

$$\kappa_x=\frac{\partial\overline{\alpha}}{\partial x},\ \kappa_y=\frac{\partial\overline{\beta}}{\partial y},\ \kappa_{xy}=\frac{1}{2}\left(\frac{\partial\overline{\alpha}}{\partial y}+\frac{\partial\overline{\beta}}{\partial x}\right).$$

It is well known that transverse shear deformation (that is $\varepsilon_{xz}\neq 0, \varepsilon_{yz}\neq 0$) effects are important in plates composed of polymer matrix composite materials in determining maximum deflections, vibration natural frequencies and critical buckling loads. However, it is appropriate to use a simpler stress analysis involving classical theory which neglects transverse shear deformation for preliminary design to determine a "first cut" for stresses, a suitable stacking sequence and an estimate of the required plate thickness.

If, in fact, transverse shear deformation is ignored, then from Section 2.3

$$\bar{\alpha} = -\frac{\partial w}{\partial x},\ \bar{\beta} = -\frac{\partial w}{\partial y},\ \kappa_x = -\frac{\partial^2 w}{\partial x^2},\ \kappa_y = -\frac{\partial^2 w}{\partial y^2},\ \kappa_{xy} = -\frac{\partial^2 w}{\partial x \partial y} \quad (11.20)$$

So, substituting Equation (11.20) into Equations (11.17) through (11.19) results in the following for the case of no transverse shear deformations, i.e., classical plate theory:

$$M_x = -D_{11}\frac{\partial^2 w}{\partial x^2} - D_{12}\frac{\partial^2 w}{\partial y^2} \quad (11.21)$$

$$M_y = -D_{12}\frac{\partial^2 w}{\partial x^2} - D_{22}\frac{\partial^2 w}{\partial y^2} \quad (11.22)$$

$$M_{xy} = -2D_{66}\frac{\partial^2 w}{\partial x \partial y}. \quad (11.23)$$

Substituting derivatives of these three equations in turn into Equation (11.16) results in:

$$D_{11}\frac{\partial^4 w}{\partial x^4} + 2(\mathbf{D_{12} + 2D_{66}})\frac{\partial^4 w}{\partial x^2 \partial y^2} + D_{22}\frac{\partial^4 w}{\partial y^4} = p(\mathbf{x}, \mathbf{y}). \quad (11.24)$$

The above coefficients are usually simplified to:

$$D_{11} \equiv D_1,\ D_{22} \equiv D_2,\ (D_{12} + 2D_{66}) \equiv D_3 \quad (11.25)$$

with the result that (11.24) becomes

$$D_1\frac{\partial^4 w}{\partial x^4} + 2D_3\frac{\partial^4 w}{\partial x^2 \partial y^2} + D_2\frac{\partial^4 w}{\partial y^4} = p(\mathbf{x}, \mathbf{y}). \quad (11.26)$$

This is the governing differential equation for the bending of a plate composed of a composite material, excluding transverse shear deformation, with no coupling terms (that is $B_{ij} = ()_{16} = ()_{26} = 0$), and no hygrothermal terms (that is, $\Delta T = \Delta m = 0$) subjected to a lateral distributed load $p(\mathbf{x}, \mathbf{y})$. Because no 16 nor 26 coupling terms are included, (11.26) refers to a plate that is "specially orthotropic".

As stated previously, neglecting transverse shear deformation and hygrothermal effects can lead to significant errors, as will be shown, but in many cases their neglect results in easier solutions which are useful in preliminary design to "size" the plate initially. Note also that if the plate were isotropic, then in that case $D_{11} = D_{22} = D_1 = D_2 = D_3 = D$, and (11.26) becomes identical to (2.57).

Solutions of Equation (11.26) can be obtained generally in two ways: direct solution of the governing differential equation (11.26), or utilization of an energy principle solution. The latter offers more latitude through attaining satisfactory approximate solutions.

Direct solutions of the governing differential equations for plates of composite materials fall into three categories: Navier solutions, Levy solutions and perturbation solutions. Each has its advantages and disadvantages. However prior to that, boundary conditions need to be discussed.

11.4 Classical Plate Theory Boundary Conditions

In the "classical" (that is, ignoring transverse shear deformation) specially orthotropic plate theory of Section 11.3, two and only two boundary conditions can be satisfied at each edge of the plate because the governing differential equation (11.26) is fourth order in x and fourth order in y. The boundary conditions for a simply supported edge and a clamped edge shown below are identical to those of classical beam theory, where here n is the direction normal to the plate edge and t is the direction parallel or tangent to the edge.

The boundary conditions for the composite material, anisotropic plate are identical to the simply supported, clamped and free edges for an isotropic plate, discussed in detail in Section 2.5.

11.5 Navier Solutions for Rectangular Composite Material Plates

Just as seen earlier for isotropic plates, the Navier approach may be employed. The Navier approach to these solutions for specially orthotropic plates involves separable solutions, as shown below, of the governing differential-plate equation shown in Equation (11.26). Thus

$$
\begin{aligned}
w(x,y) &= \sum_{m=1}^{\infty}\sum_{n=1}^{\infty} A_{mn}\overline{X}_m(x)\overline{Y}_n(y) \\
p(x,y) &= \sum_{m=1}^{\infty}\sum_{n=1}^{\infty} B_{mn}\overline{X}_m(x)\overline{Y}_n(y)
\end{aligned}
\tag{11.27}
$$

In the above, the functions $\overline{X}_m(x)$ and $\overline{Y}_n(y)$ are a uniformly convergent, complete, orthogonal set of functions that satisfy the boundary conditions. Thus, when the complete summation is taken, the exact solution is obtained. From a practical point of view, because of uniform convergence, a finite number of terms are sufficient to provide any desired accuracy.

If one assumes a plate that is simply supported on all four edges, then Equation (11.27) can be written as

$$w(x,y) = \sum_{m=1}^{\infty}\sum_{n=1}^{\infty} A_{mn}\sin\left(\frac{m\pi x}{a}\right)\sin\left(\frac{n\pi y}{b}\right) \tag{11.28}$$

$$p(x,y) = \sum_{m=1}^{\infty}\sum_{n=1}^{\infty} B_{mn}\sin\left(\frac{m\pi x}{a}\right)\sin\left(\frac{n\pi y}{b}\right) \tag{11.29}$$

These half-range sine series (the original Navier problem) satisfy the simply supported boundary conditions on all four edges of the plate shown in Figure 2.1. If the plate has other boundary conditions, then other functions must be used for $\overline{X}_m(x)$ and $\overline{Y}_n(y)$, in which case the approach is labeled the Generalized Navier approach.

To proceed it is first necessary to determine B_{mn} in Equation (11.29) in order to describe the load $p(x,y)$, whether it is continuous of discontinuous. One simply multiplies both the left- and right-hand sides of Equation (11.29) by $\sin\left(\frac{r\pi x}{a}\right)\sin\left(\frac{s\pi y}{b}\right)dxdy$ and integrates both sides from 0 to a in the x-direction and 0 to b in the y-direction, i.e., over the planform area of the plate. It must be remembered that

$$\int_0^a \sin\left(\frac{m\pi x}{a}\right)\sin\left(\frac{r\pi x}{a}\right)dx = \begin{cases} a/2 & \text{if } m = r \\ 0 & \text{if } m \neq r \end{cases} \tag{11.30}$$

Therefore,

$$B_{mn} = \frac{4}{ab}\int_0^a\int_0^b p(x,y)\sin\left(\frac{m\pi x}{a}\right)\sin\left(\frac{n\pi y}{b}\right)dydx \tag{11.31}$$

For example if $p(x,y) = p_0 =$ a constant, a commonly occurring load,

$$B_{mn} = \frac{4p_0}{mn\pi^2}\left[1-(-1)^m\right]\left[1-(-1)^n\right] \tag{11.32}$$

Now the above can be used for an isotropic or orthotropic plate, classical theory or advanced theory (including transverse-shear deformation) and a laminated or single-layer plate. Considering an orthotropic composite panel, using classical plate theory, simply supported on all four edges, Equation (11.26) is used.

Simply substituting Equations (11.28) and (11.29) into Equation (11.26), the equality requires that each term be equated, thus

$$A_{mn} = \frac{B_{mn}}{D_1\left(\frac{m\pi}{a}\right)^4 + 2D_3\left(\frac{m\pi}{a}\right)^2\left(\frac{n\pi}{b}\right)^2 + D_2\left(\frac{n\pi}{b}\right)^4} \tag{11.33}$$

where the flexural stiffness quantities D_1, D_2 and D_3 are defined by Equations (11.25), for a laminated composite panel. For the isotropic case in which $E_{11} = E_{22} = 2G(1+\upsilon)$, Equations (11.25) becomes $D_1 = D_2 = D_3 = D$.

After obtaining the solution for $w(x,y)$ using Equations (11.28), (11.29) and (11.33), one may calculate the magnitude and location of the maximum deflection. By taking the derivative of $w(\boldsymbol{x},\boldsymbol{y})$, and using Equations (11.21) through (11.23), the stress couples M_x, M_y and M_{xy} are determined to find the maximum values and their location. Finally, depending upon whether the panel is a laminate or a single layer, the maximum stresses are determined through calculating the curvatures at the locations of maximum stress couples, using (2.39):

$$\kappa_x = -\frac{\partial^2 w}{\partial x^2}, \quad \kappa_y = -\frac{\partial^2 w}{\partial y^2} \text{ and } \kappa_{xy} = -\frac{\partial^2 w}{\partial x \partial y}$$

Knowing these, one can calculate the stresses in each of the k laminae through the use of (10.53) where in this case of a lateral loading only, there is no in-plane response, i.e. $\varepsilon_{x_0} = \varepsilon_{y_0}$ and ε_{xy_0} are zero, and there are no thermal or moisture effects:

$$\begin{bmatrix} \sigma_x \\ \sigma_y \\ \sigma_{xy} \end{bmatrix}_k = \begin{bmatrix} \overline{Q}_{11} & \overline{Q}_{12} & 0 \\ \overline{Q}_{12} & Q_{22} & 0 \\ 0 & 0 & Q_{66} \end{bmatrix}_k \begin{bmatrix} \kappa_x \\ \kappa_y \\ 2\kappa_{xy} \end{bmatrix} z \tag{11.34}$$

The number of terms necessary to attain a desired accuracy depends upon the particular load $p(\boldsymbol{x},\boldsymbol{y})$, the aspect ratio of the plate (b/a), and the material system of which the plate is fabricated.

11.6 Navier Solution for a Uniformly Loaded Simply Supported Plate – An Example Problem

The case of a uniformly loaded, $p(x, y) = -p_o$ simply supported plate is solved by means of the Navier series solution of Section 11.5, for two composite materials systems: unidirectional and cross-ply, stresses σ_x, σ_y and σ_{xy} are determined for each case at the quarter points and mid-point of the plane . In addition, in this example, the solutions have been examined by utilizing one, three and five terms in the Navier series solution.

In this analysis, it is of course assumed that all plys are perfectly bonded and classical theory is used (that is, ε_{xy} and ε_{yz}, are assumed zero). This results in the in-plane stresses, σ_x, σ_y and σ_{xy} being directly determined, while the transverse shear stresses σ_{xz} and σ_{yz} are determined subsequently.

Using the methods discussed previously it is found that the stresses in each lamina for the case of $p(x, y) = -p_0$ are given by:

$$\begin{Bmatrix} \sigma x \\ \sigma y \\ \sigma xy \end{Bmatrix}_k = +\frac{16 p_0 z}{\pi^4} \sum_{m=1,3,5}^{\infty} \sum_{n=1,3,5}^{\infty} \frac{1}{mnD} \cdot \begin{bmatrix} \left[-\overline{Q}_{11}^k \left(\frac{m}{a}\right)^2 - \overline{Q}_{12}^k \left(\frac{n}{b}\right)^2 \right] \sin\frac{m\pi x}{a} \cdot \sin\frac{n\pi y}{b} \\ \left[-\overline{Q}_{12}^k \left(\frac{m}{a}\right)^2 - \overline{Q}_{22}^k \left(\frac{n}{b}\right)^2 \right] \sin\frac{m\pi x}{a} \cdot \sin\frac{n\pi y}{b} \\ 2\overline{Q}_{66}^k \left(\frac{m}{a}\right)\left(\frac{n}{b}\right) \cos\frac{m\pi x}{a} \cdot \cos\frac{n\pi y}{b} \end{bmatrix}$$

where

$$D = D_{11}\left(\frac{m}{a}\right)^4 + 2D_3\left(\frac{mn}{ab}\right)^2 + D_{22}\left(\frac{n}{b}\right)^4$$

where D_3 is defined by Equation (11.25).

As a numerical example a square plate of $a = b = 12"$ is considered. The total plate thickness is 0.08": eight plies of 0.01" thickness ($h_k = 0.01"$).

The first material system considered is E glass/epoxy, with a fiber volume fraction $V_f = 70\%$, with the following properties:

$E_{11} = 8.8 \times 10^6$ psi $\qquad \nu_{12} = 0.23$

$E_{22} = 3.6 \times 10^6$ psi $\qquad G_{12} = 1.74 \times 10^6$ psi

The stiffnesses $\overline{Q}_{ij}$ are, from (10.43) and (10.44),

0° ply (psi)	90° ply (psi)
$\overline{Q}_{11} = 9.0 \times 10^6$	$\overline{Q}_{11} = 3.68 \times 10^6$
$\overline{Q}_{12} = 0.85 \times 10^6$	$\overline{Q}_{12} = 0.85 \times 10^6$
$\overline{Q}_{22} = 3.68 \times 10^6$	$\overline{Q}_{22} = 9.0 \times 10^6$
$\overline{Q}_{66} = 1.74 \times 10^6$	$\overline{Q}_{66} = 1.74 \times 10^6$

In the following figures, the stresses have been normalized as $\overline{\sigma}_{ij} = \sigma_{ij} / p_0$. In Figure 11.1, the normalized stresses are shown at the plate midpoint ($x = a/2$, $y = b/2$) for a unidirectional ($\theta = 0°$) laminate. In this case the stresses are proportional to the distance from the plate mid-plane. In Figure 11.2, the normalized stresses at plate midpoint are shown for a mid-plane symmetric cross-ply laminate. Here, because the plys alternate between 0° and 90°, the stresses are discontinuous from ply to ply, and in each ply the stresses are larger in the fiber direction than in the 90° direction. In Figure 11.3, these stresses are shown at the quarter point location for the cross-ply plate. Because of symmetry, at the plate center in-plane shear stresses are zero, while at the quarter points a non-zero σ_{xy} exists as shown in Figure 11.3.

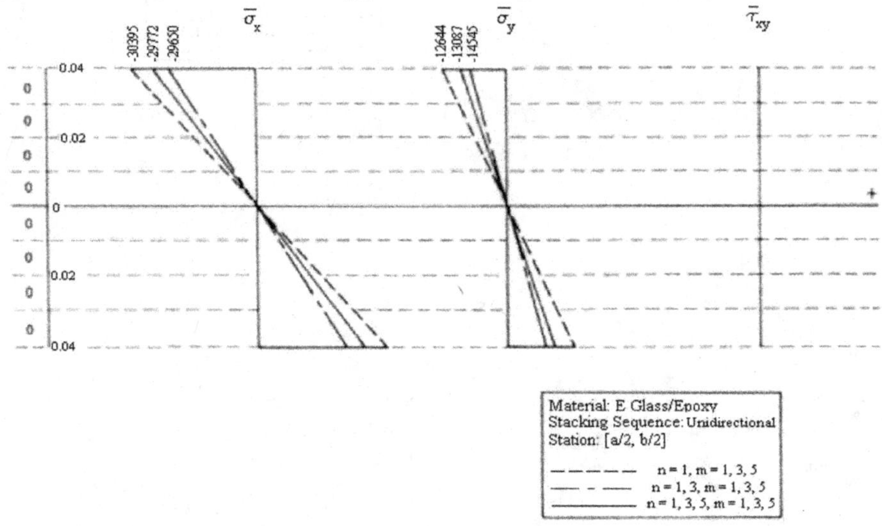

Figure 11.1. Normalized stresses at center of the plate for an E glass/epoxy unidirectional composite.

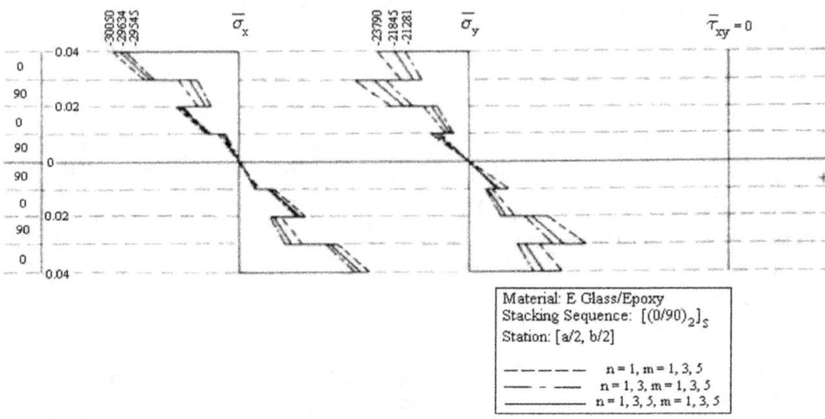

Figure 11.2. Normalized stresses at center of plate for an E glass/epoxy cross-ply composite.

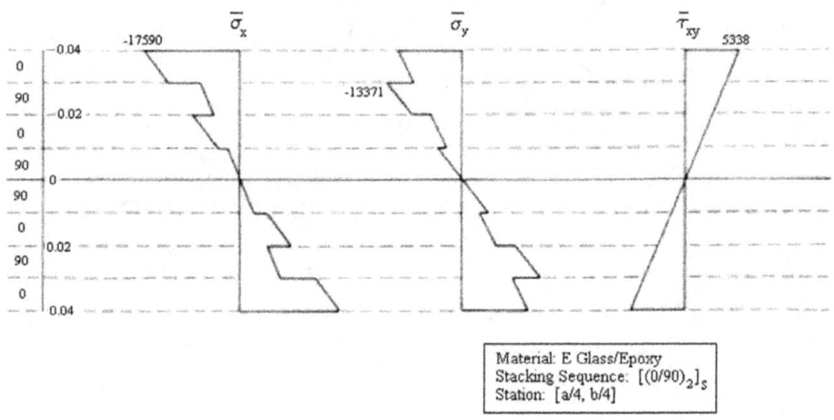

Figure 11.3. Normalized stresses at quarter point of plate for an E glass/epoxy cross-ply composite.

A cross-ply laminate of T300-5208 graphite-epoxy has been used for comparison with the E glass/epoxy laminate. Properties of the graphite-epoxy laminate are as follows for $V_f = 70\%$:

$$E_{11} = 22.2 \times 10^6 \text{ psi} \qquad \nu_{12} = 0.12$$

$$E_{22} = 1.58 \times 10^6 \text{ psi} \qquad G_{12} = 0.81 \times 10^6 \text{ psi}$$

Again normalized stresses have been shown in Figures 11.4 and 11.5 at both plate quarter point and mid-point.

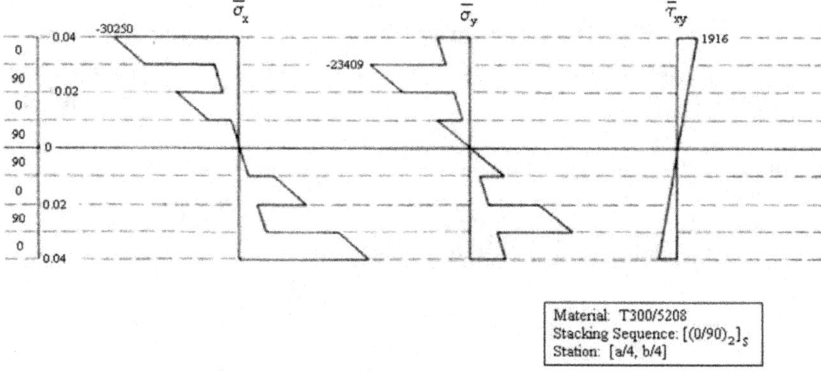

Figure 11.4. Normalized stresses at quarter point of plate of a graphite/epoxy cross-ply composite.

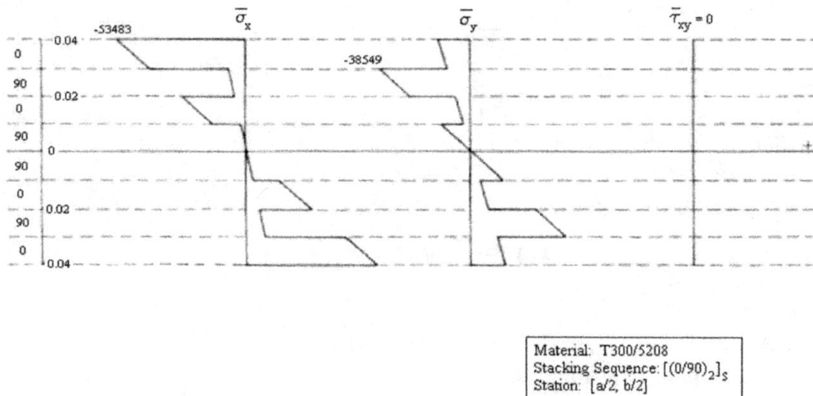

Figure 11.5. Normalized stresses at center of plate of a cross-ply graphite/epoxy composite.

Some conclusions can be drawn from this example set.

1. Solution convergence is rapid within the framework of taking three terms for both m and n for evaluating $\overline{\sigma}_x$, but is not as rapid in calculating $\overline{\sigma}_y$.
2. For the same material there is little difference between the maximum value of the $\overline{\sigma}_x$ stress for both the unidirectional and cross-ply composites at similar plate locations, however, the $\overline{\sigma}_y$ stresses differ significantly.
3. The stress $\overline{\sigma}_y$ at a fixed location for the graphite/epoxy laminate is much smaller relative to the $\overline{\sigma}_x$ value (10%), compared to that in the E glass/epoxy laminate where $\overline{\sigma}_y$ is 33% of the value of $\overline{\sigma}_x$ at the same location.

This example was the work of Wenn-Jinn Liou, a student at the University of Florida.

11.7 Levy Solution for Plates of Composite Materials

The second direct method of solution for the bending of rectangular plates due to lateral loads, is due to Maurice Levy [11.1] who, in 1899, introduced a single infinite-series method of solution for isotropic plate problems, as discussed in Section 3.3. The method can also be used to solve Equation (11.26) for a specially orthotropic composite material plate.

Consider the plate, shown in Figure 2.1, with edges $y = 0$ and $y = \mathrm{b}$ simply supported. The boundary conditions for those edges are

$$\begin{aligned} w(x,0) = w(x,b) = 0 \\ M_y(x,0) = M_y(x,b) = 0 \end{aligned} \tag{11.35}$$

The latter implies that the following equations hold, as shown before in (3.20):

$$\frac{\partial^2 w(x,0)}{\partial y^2} = \frac{\partial^2 w(x,b)}{\partial y^2} = 0 \tag{11.36}$$

Levy assumed the following solution form: a single infinite half-range sine series that satisfies the simply supported boundary conditions on both y edges seen previously in (3.21):

$$w(\boldsymbol{x},\boldsymbol{y}) = \sum_{n=1}^{\infty} \boldsymbol{\phi_n}(\boldsymbol{x}) \sin\left(\frac{n\pi y}{b}\right) \tag{11.37}$$

where $\phi_n(\boldsymbol{x})$ is at this point an unknown function of x. A laterally distributed load $p(\boldsymbol{x},\boldsymbol{y})$ can be expressed as follows:

$$p(x,y) = g(x)h(y) \tag{11.38}$$

where $g(\boldsymbol{x})$ and $h(\boldsymbol{y})$ are specified. Following Section 3.3, the form of Equation (11.37) requires that the $h(\boldsymbol{y})$ portion of the load also be expanded in terms of a half range sine series, such as

$$h(\boldsymbol{y}) = \sum_{n=1}^{\infty} A_n \sin\left(\frac{n\pi y}{b}\right) \tag{11.39}$$

where

$$A_n \quad \frac{2}{b}\int_0^b \boldsymbol{h}(\boldsymbol{y})\mathbf{sin}\left(\frac{n\pi y}{b}\right)dy \tag{11.40}$$

Substituting Equations (11.37) through (11.39) into (11.26) and observing that the equation exists only if it is true term by term, it is seen that, after dividing by D_1 and the trigonometric function:

$$\frac{d^4\phi_n(\boldsymbol{x})}{dx^4}-\frac{2D_3}{D_1}\lambda_n^2\frac{d^2\phi_n(\boldsymbol{x})}{dx^2}+\frac{D_2}{D_1}\lambda_n^4\phi_n(\boldsymbol{x}) \quad \frac{A_n g_n(\boldsymbol{x})}{D_1} \tag{11.41}$$

where $\lambda_n \quad n\pi/b$. Note that Equation (11.41) was derived without specifying any boundary conditions on the x-edges. In fact, the homogeneous solution of Equation (11.41) yields four constants of integration, which are determined through satisfying boundary conditions on those x-edges.

To obtain the homogeneous solution of Equation (11.41) the right-hand side is set equal to zero:

$$\frac{d^4\phi_n(\boldsymbol{x})}{dx^4}(\boldsymbol{x})-\frac{2D_3}{D_1}\lambda_n^2\frac{d^2\phi_n(\boldsymbol{x})}{dx^2}+\frac{D_2}{D_1}\lambda_n^4\phi_n(\boldsymbol{x}) \quad 0 \tag{11.42}$$

After letting $\phi_n(\boldsymbol{x}) \quad e^{sx}$, and dividing the result by e^{sx}, the indicial equation, from (11.42) becomes:

$$s^4-\frac{2D_3}{D_1}\lambda_n^2 s^2+\frac{D_2}{D_1}\lambda_n^4 \quad 0 \tag{11.43}$$

Unlike the case of an isotropic plate where $D_1 \quad D_2 \quad D_3$, see (3.25), such that the roots are easily seen to be $\pm\lambda_n$ and $\pm\lambda_n$ (repeated roots), for this case there are three sets of roots depending upon whether $(D_2/D_1)^{1/2}$ is greater than, equal to or less than D_3/D_1. Hence, for the specially orthotropic composite plate, sandwich or laminate, using the Levy-type solution requires three different forms for the homogeneous solution of $\phi_n(\boldsymbol{x})$ to be put in Equation (11.37) depending on the relative stiffness of the plate in various directions.

For the case, $(D_2/D_1)^{1/2}<(\boldsymbol{D_3}/\boldsymbol{D_1})$

$$\phi_{n_h}(\boldsymbol{x}) \quad C_1\cosh(\lambda_n s_1 x)+C_2\sinh(\lambda_n s_1 x)+C_3\cosh(\lambda_n s_2 x)+C_4\sinh(\lambda_n s_2 x) \tag{11.44}$$

where the roots are

$$s_1 = \sqrt{\left(\frac{D_3}{D_1}\right) + \sqrt{\left(\frac{D_3}{D_1}\right)^2 - \frac{D_2}{D_1}}}, \qquad s_2 = \sqrt{\left(\frac{D_3}{D_1}\right) - \sqrt{\left(\frac{D_3}{D_1}\right)^2 - \frac{D_2}{D_1}}}$$

For the case, $(D_2/D_1)^{1/2} = (D_3/D_1)$

$$\phi_{n_h}(x) = (C_5 + C_6 x)\cosh(\lambda_n s_3 x) + (C_7 + C_8 x)\sinh(\lambda_n s_3 x) \tag{11.45}$$

where the roots are

$$s_3 = \pm\sqrt{\frac{D_3}{D_1}}$$

For the case $(D_2/D_1)^{1/2} > (D_3/D_1)$

$$\begin{aligned} \phi_{n_h}(x) = {} & (C_9 \cos\lambda_n s_5 x + C_{10}\sin\lambda_n s_5 x)\cosh(\lambda_n s_4 x) \\ & + (C_{11}\cos\lambda_n s_5 x + C_{12}\sin\lambda_n s_5 x)\sinh(\lambda_n s_4 x) \end{aligned} \tag{11.46}$$

where the roots are

$$s_4 = \sqrt{\frac{1}{2}\left[\left(\frac{D_2}{D_1}\right)^{1/2} + \frac{D_3}{D_1}\right]}, \qquad s_5 = \sqrt{\frac{1}{2}\left[\left(\frac{D_2}{D_1}\right)^{1/2} - \frac{D_3}{D_1}\right]}$$

Obviously, for a given plate whose materials and orientation have already been specified (the analysis problem) only one of the three cases needs to be solved. However, if one is trying to determine the best material and orientation (the design problem), then more than one case may need to be solved, with the necessity of determining not just four constants, but eight or all twelve to satisfy the edge boundary conditions to determine which construction is best for the design.

Concerning the particular solution, it is noted that if the lateral load, $p(x, y)$, is at most linear in x, hence from Equation (11.38), $g_n(x)$ is at most linear in x, then from Equation (11.41) the particular solution is

$$\phi_{n_p}(x) = \frac{A_n g_n(x)}{\lambda_n^4 D_2} \tag{11.47}$$

Otherwise, one must seek another particular solution. In any case, one must then add the relevant homogeneous ϕ_{n_h} to the particular ϕ_{n_p} to satisfy any set of boundary conditions on the x-edges of the plate. For example, suppose the $x = 0$ edge is simply supported, then from Equation (2.66) the boundary conditions are

$$w(\mathbf{0}, \boldsymbol{y}) = 0, \text{ and } M_x(\mathbf{0}, \boldsymbol{y}) = 0 \rightarrow \frac{\partial^2 w}{\partial x^2}(\mathbf{0}, \boldsymbol{y}) = 0 \tag{11.48}$$

However, when $w(x,y)$ has the form of Equation (11.37) this then implies that:

$$\phi_n(0) = \phi_n''(0) = 0 \tag{11.49}$$

where primes denote differentiation with respect to x.

Similarly, appropriate expressions can be found if the x-edges are clamped or free. Then whatever the relevant form of the boundary conditions on $x = 0$ and $x = a$, the total $\phi_n(\boldsymbol{x}) = \phi_{n_h} + \phi_{n_p}$ and hence $w(x,y)$ is known from Equation (11.37). Then, for a composite-material laminated plate, one must calculate the curvatures, as was done is the previous section for the Navier approach:

$$\kappa_x = -\frac{\partial^2 w}{\partial x^2}, \quad \kappa_y = -\frac{\partial^2 w}{\partial y^2} \quad \text{and} \quad \kappa_{xy} = -\frac{\partial^2 w}{\partial x \partial y}$$

Knowing these, one can calculate the bending stresses in each of the k laminae through the following; which is identical to (11.34):

$$\begin{bmatrix} \sigma_x \\ \sigma_y \\ \sigma_{xy} \end{bmatrix}_k = \begin{bmatrix} \overline{Q}_{11} & \overline{Q}_{12} & 0 \\ \overline{Q}_{12} & \overline{Q}_{22} & 0 \\ 0 & 0 & \overline{Q}_{66} \end{bmatrix}_k \begin{bmatrix} \kappa_x \\ \kappa_y \\ 2\kappa_{xy} \end{bmatrix} z \tag{11.50}$$

The stresses thus derived for each lamina must them be compared with the allowable stresses, determined through some failure criterion to see if structural integrity is retained under a given load (the analysis problem), or if this set of materials and orientation is sufficient for a given load (the design problem).

It is seen that the Levy-type solution is fine for a composite plate with no bending-stretching coupling, i.e., with midplane symmetry, and with two opposite edges simply supported. If two opposite edges are not simply supported then the complexity of the functions necessary to satisfy the boundary conditions on the y = constant edges cause problems.

11.8 Perturbation Solutions for the Bending of a Composite Material Plate With Midplane Symmetry and No Bending-Twisting Coupling

As shown in the previous two sections, the Navier approach is excellent for composite material plates with all four edges simply supported, and the Levy approach is fine for composite material plates with two opposite edges simply supported, regardless of the boundary conditions on the other two edges. But for a composite plate with two opposite edges simply supported, even the Levy approach yields three distinct solutions depending on the relative magnitudes of D_1, D_2 and $D_3 = D_{12} + 2D_{66}$. In addition, there are numerous books and papers available for the solution of isotropic plate problems [2.2-2.5].

Aware of all of the above, and based upon the fact that the solution of the second case of the Levy solution of (11.45) has the same form as that of the isotropic case of (3.26), Vinson showed that the cases of (11.44) and (11.46) can be dealt with as perturbations about the solution of the same plates composed of isotropic materials [11.2, 11.3].

Consider the governing equation for the bending of a composite material plate exhibiting mid-plane symmetry $(B_{ij} = 0)$, no bending-twisting coupling $(D_{16} = D_{26} = 0)$, and no transverse shear deformation (classical theory). Then Equation (11.26) becomes, after dividing both sides by D_1:

$$\frac{\partial^4 w}{\partial x^4} + \frac{2D_3}{D_1}\frac{\partial^4 w}{\partial x^2 \partial y^2} + \frac{D_2}{D_1}\frac{\partial^4 w}{\partial y^4} = \frac{p(\boldsymbol{x},\boldsymbol{y})}{D_1} \tag{11.51}$$

Coordinate stretching is employed by defining the following:

$$\bar{y} = \left(\frac{D_2}{D_1}\right)^{-1/4} \bullet y, \quad \bar{b} = \left(\frac{D_2}{D_1}\right)^{-1/4} \bullet b \tag{11.52}$$

Substituting Equation (11.52) into Equation (11.51) yields

$$\frac{\partial^4 w}{\partial x^4} + 2\left(\frac{D_2}{D_1}\right)^{-1/2} \bullet \left(\frac{D_3}{D_1}\right)\frac{\partial^4 w}{\partial x^2 \partial \bar{y}^2} + \frac{\partial^4 w}{\partial \bar{y}^4} = \frac{p(\boldsymbol{x},\bar{\boldsymbol{y}})}{D_1} \tag{11.53}$$

Next defining a parameter α to be

$$\alpha = 2\left[1 - \left(\frac{D_2}{D_1}\right)^{-1/2}\left(\frac{D_3}{D_1}\right)\right] \tag{11.54}$$

it is seen that substituting Equation (11.54) into Equation (11.53) yields

$$\frac{\partial^4 w}{\partial x^4}+\frac{2\partial^4 w}{\partial x^2 \partial \bar{y}^2}+\frac{\partial^4 w}{\partial \bar{y}^4}-\alpha\frac{\partial^4 w}{\partial x^2 \partial \bar{y}^2}=\frac{p(\boldsymbol{x},\bar{\boldsymbol{y}})}{D_1} \tag{11.55}$$

Note that in this section, α is a perturbation parameter, not to be confused with α, the coefficient of thermal expansion.

If one defines the biharmonic operator, used in all isotropic plate problems (2.58), to be (in the stretched coordinate system)

$$\nabla^4 w=\frac{\partial^4 w}{\partial x^4}+\frac{2\partial^4 w}{\partial x^2 \partial \bar{y}^2}+\frac{\partial^4 w}{\partial \bar{y}^4} \tag{11.56}$$

then Equation (11.55) becomes

$$\nabla^4 w-\alpha\frac{\partial^4 w}{\partial x^2 \partial \bar{y}^2}=\frac{p(\boldsymbol{x},\bar{\boldsymbol{y}})}{D_1} \tag{11.57}$$

Finally, assume the form of the solution for $w(x,\bar{y})$ to be

$$w(\boldsymbol{x},\bar{\boldsymbol{y}})=\sum_{n=0}^{\infty} w_n(\boldsymbol{x},\bar{\boldsymbol{y}})\alpha^n \tag{11.58}$$

which is a perturbation solution employing the "small" parameter α defined in Equation (11.54). Substituting Equation (11.58) into Equation (11.57) and equating all coefficients of α^n to zero, it is easily found that:

$$\nabla^4 w_0=\frac{p(x,\bar{y})}{D_1} \tag{11.59}$$

$$\nabla^4 w_n=-\frac{\partial^4 w_{n-1}(\boldsymbol{x},\bar{\boldsymbol{y}})}{\partial x^2 \partial \bar{y}^2} \qquad n\geq 1 \tag{11.60}$$

It is seen that Equation (11.59) is the governing differential equation for an isotropic plate of stiffness D_1, subjected to the actual lateral load $p(x,\bar{y})$ given by (2.57), with the stretched coordinate $\bar{y}$ defined in Equation (11.52). It is probable that, regardless of boundary conditions on any edge, the solution for w_0 of Equation (11.59) is available in the literature, either exactly or approximately and this was discussed in

Chapter 3. Subsequently w_1, w_2, w_3, and so on are available from solving Equation (11.60), an isotropic plate whose lateral load is $-D_1\left(\partial^4 w_{n-1} / \partial x^2 \partial \overline{y}^{2}\right)$, where w_{n-1} would be determined previously, whose flexural stiffness is D_1 and whose boundary conditions are homogeneous. Where there are non-homogeneous boundary conditions, the non-homogeneous boundary conditions should be taken care of in the w_0 solution, since truncation is anticipated.

This technique is very useful because the "small" perturbation parameter need not be so small; it has been proven that when $|\alpha| < 1$, Equation (11.58) is another form of the exact solution, and $|\alpha| < 1$ covers much of the practical range of composite material properties. Also from a computational point of view, it is seldom necessary to include terms past $n = 1$, in Equation (11.58). This technique can only be used if $|\alpha| < 1$.

The above technique can be very useful. However, even if $|\alpha| \geq 1$, then the composite may fall within another range where for $(D_2 / D_1) << 1$. In that case, the plate behaves as a plate in the x-direction, but because $(D_2 / D_1) << 1$, it behaves as a membrane in the y-direction, with the following simpler governing differential equation,

$$\frac{\partial^4 w}{\partial x^4} + 2\frac{D_3}{D_1}\frac{\partial^4 w}{\partial x^2 \partial y^2} = \frac{p(x,y)}{D_1} \tag{11.61}$$

The solution of (11.61) can be in the form of the following for a plate simply supported on the y edges

$$w(x,y) = \sum_{n=1}^{\infty} \phi_n(x) \sin \lambda_n^* y^* \tag{11.62}$$

where

$$y^* = \left(\frac{2D_3}{D_1}\right)^{-1/2} \bullet y, \quad \lambda_n^* = \frac{n\pi}{b^*}, \quad b^* = \left(\frac{2D_3}{D_1}\right)^{-1/2} \bullet b \tag{11.63}$$

Even if the perturbation technique described by Equations (11.62) and (11.63) is not used it still provides physical insight by showing that if $D_2 << D_1$, then the structure behaves as a plate in the stiffer direction and acts only as a membrane in the weaker direction. Physically, on the x edges, the usual use of two boundary conditions must be used, while in the weaker edges, only one boundary condition on each edge is needed (as with a membrane), that dealing with the lateral displacement set equal to zero.

Finally, if $(\overline{\boldsymbol{b}}/\boldsymbol{a}) = (\boldsymbol{D_2}/\boldsymbol{D_1})^{-1/4}(b/a) > 3$, then the plate behaves purely as a beam in the x-direction, regardless of the boundary conditions on the y-edge, as far as maximum deflection and maximum stresses. Hence, beam solutions have easy application to the solution of many composite plate problems. All of the details on the last two techniques are give in detail in References [11.2] and [11.3].

Incidentally, the techniques described in this section [11.2] were the first use of perturbation techniques involving a material property perturbation, even though geometric perturbations have been utilized for many decades.

11.9 Quasi-Isotropic Composite Panels Subjected to a Lateral Load

When a composite laminate has a stacking sequence in which $D_{11} = D_{22}$, it is referred to as quasi-isotropic. In that case it behaves as an isotropic plate in the determination of lateral deformations, $w(x, y)$, and stress couples, M_x, M_y and M_{xy}. Therefore, for a quasi-isotropic composite plate the methods of Chapter 3 can be employed wherein $D = D_{11} = D_{22}$. When the lateral load is uniform, then Section 3.5 results can be used.

11.10 A Static Analysis of Composite Material Panels Including Transverse Shear Deformation Effects

The previous derivations have involved "classical" plate theory, i.e., they have neglected transverse shear deformation effects. In many composite material laminated plate constructions, transverse-shear deformation effects are important because some of the in-plane plate stiffness quantities are dominated by the placement of in-plane fibers, the plate transverse shear stiffness are dominated by the matrix properties. Therefore, for polymer matrix composite plates transverse shear deformation effects can be significant. Therefore, a more refined theory must be developed. However, because of its simplicity, and the number of solutions available, classical theory is still useful for preliminary design and in analysis to size the structure required in minimum time and effort.

In the simpler classical theory, the neglect of transverse shear deformation effects means that $\varepsilon_{xz} = \varepsilon_{yz} = 0$. To include transverse shear deformation effects, one uses

$$\varepsilon_{xz} = \frac{1}{2}\left(\frac{\partial u}{\partial z} + \frac{\partial w}{\partial x}\right) \tag{11.64}$$

$$\varepsilon_{yz} = \frac{1}{2}\left(\frac{\partial v}{\partial z} + \frac{\partial w}{\partial y}\right) \tag{11.65}$$

Now substituting the admissible forms of the displacement for a plate or panel, Equation (2.1) and (2.2) into Equations (1.20) and (1.21), shows that

$$\varepsilon_{xz} = \frac{1}{2}\left(\overline{\alpha} + \frac{\partial w}{\partial x}\right) \quad (11.66)$$

$$\varepsilon_{yz} = \frac{1}{2}\left(\overline{\beta} + \frac{\partial w}{\partial y}\right) \quad (11.67)$$

No longer are the rotations $\overline{\alpha}$ and $\overline{\beta}$ explicit functions of the derivatives of the lateral deflection w, as shown by Equation (11.20) for classical plate theory. The result is that for this refined theory there are five geometric unknowns, u_0, v_0, w, $\overline{\alpha}$ and $\overline{\beta}$, instead of just the first three in classical theory.

Now one needs to look again at the equilibrium equations, the constitutive equations (stress-strain relations), the strain-displacement relations and the compatibility equations. For the plate, the equilibrium equations are given by Equations (2.14) through (2.18), because they do not change from classical theory. The constitutive equations for a composite material laminated plate and sandwich panel are given by Equations (11.68) through (11.75). The new cogent strain-displacement (kinematic) relations are given above in Equation (11.66) and (11.67). Because the resulting governing equations are in terms of displacements and rotations, any single valued, continuous solution will, by definition, satisfy the compatibility equations.

A plate that is mid-plane symmetric ($B_{ij} = 0$) and has no coupling terms [$(\ )_{16} = (\ )_{26} = (\ )_{45} = 0$]; the constitutive equations for this specially orthotropic plate can be written as follows, where κ, with no subscripts, is a transverse shear coefficient to be discussed later.

$$N_x = A_{11}\varepsilon_x^0 + A_{12}\varepsilon_y^0 \quad (11.68)$$

$$N_y = A_{12}\varepsilon_x^0 + A_{22}\varepsilon_y^0 \quad (11.69)$$

$$N_{xy} = 2A_{66}\varepsilon_{xy}^0 \quad (11.70)$$

$$M_x = D_{11}\kappa_x + D_{12}\kappa_y \quad (11.71)$$

$$M_y = D_{12}\kappa_x + D_{22}\kappa_y \quad (11.72)$$

$$M_{xy} = 2D_{66}\kappa_{xy} \quad (11.73)$$

$$Q_x = 2A_{55}\varepsilon_{xz} = \kappa A_{55}\left(\overline{\alpha} + \frac{\partial w}{\partial x}\right) \quad (11.74)$$

$$Q_y = 2A_{44}\varepsilon_{yz} = \kappa A_{44}\left(\bar{\beta}+\frac{\partial w}{\partial y}\right) \tag{11.75}$$

Because the plate is mid-plane symmetric there is no bending-stretching coupling, hence the in-plane stress resultants (N_x, N_y, N_{xy}) and deflections (u_o, v_o) are uncoupled (separate) from the lateral loads, deflections and rotations. Hence, for a lateral distributed static loading, $p(\boldsymbol{x}, \boldsymbol{y})$, Equations (11.13) through (11.15) and Equations (11.71) through (11.75) are utilized: 8 equations and 8 unknowns.

Substituting Equations (11.71) through (11.75) into Equations (11.13) through (11.15) results in the following set of governing differential equations for a laminated specially orthotropic composite plate subjected to a lateral load, with $\boldsymbol{B_{ij}} = 0$, $(\ )_{16} = (\ )_{26} = (\ )_{45} = 0$, and no applied surface shear stresses (for simplicity)

$$D_{11}\frac{\partial^2\bar{\alpha}}{\partial x^2} + D_{66}\frac{\partial^2\bar{\alpha}}{\partial y^2} + (\boldsymbol{D_{12}} + \boldsymbol{D_{66}})\frac{\partial^2\bar{\beta}}{\partial x\partial y} - \kappa A_{55}\left(\bar{\alpha} + \frac{\partial w}{\partial x}\right) = 0 \tag{11.76}$$

$$(\boldsymbol{D_{12}} + \boldsymbol{D_{66}})\frac{\partial^2\bar{\alpha}}{\partial x\partial y} + D_{66}\frac{\partial^2\bar{\beta}}{\partial x^2} + D_{22}\frac{\partial^2\bar{\beta}}{\partial y^2} - \kappa A_{44}\left(\bar{\beta} + \frac{\partial w}{\partial y}\right) = 0 \tag{11.77}$$

$$\kappa A_{55}\left(\frac{\partial\bar{\alpha}}{\partial x} + \frac{\partial^2 w}{\partial x^2}\right) + \kappa A_{44}\left(\frac{\partial\bar{\beta}}{\partial y} + \frac{\partial^2 w}{\partial y^2}\right) + p(\boldsymbol{x}, \boldsymbol{y}) = 0 \tag{11.78}$$

The inclusion of transverse shear deformation effects results in three coupled partial differential equations with three unknowns, $\bar{\alpha}$, $\bar{\beta}$ and w, contrasted to having one partial differential equation with one unknown, w, in classical plate (panel) theory; see Equation (11.26). Incidentally if one specified that $\bar{\alpha} = -\frac{\partial w}{\partial x}$ and $\bar{\beta} = -\frac{\partial w}{\partial y}$, substituting that into Equations (11.76) through (11.78) reduces the three equations to Equation (11.26), the classical theory composite material plate bending equation. Note that the κ symbol with no subscript in (11.76) through (11.78) is a transverse shear deformation correction factor which is given by Yu, Hodges and Volovoi in [11.4] for an orthotropic material and for an isotropic material can be written as given by Hodges [11.5]

$$\kappa = \frac{5(11-12\nu+34\nu^2-12\nu^3+11\nu^4)}{2(33-40\nu+98\nu^2-40\nu^3+29\nu^4)}.$$

It is interesting to note that leaving out the ν terms in the above equation, $\kappa = 5/6$, the value obtained by Reissner in 1950. Incidentally, there still remains discussion and controversy over the value of the shear correction factor.

The classical plate theory governing partial differential equation is fourth order in both x and y, and therefore requires two and only two boundary conditions on each of the four edges, as discussed in Section 11.4 and 2.5. The refined theory, discussed in this section which includes transverse shear deformation, is really sixth order in both x and y, and therefore requires three boundary conditions on each edge as discussed in Section 11.11 below. See papers by Reddy [11.6], Lo, Christensen and Wu [11.7] DiSciuva [11.8] and Reddy and Phan [11.9].

If the laminated plate is orthotropic but not mid-plane symmetric, i.e., $\boldsymbol{B}_{ij} \neq 0$, the governing equations are more complicated than Equations (11.76) through (11.78) and are given by Whitney [11.10], Vinson [10.1] and are discussed briefly in Section 11.15 below.

11.11 Boundary Conditions for a Plate Including Transverse Shear Deformation

11.11.1 SIMPLY-SUPPORTED EDGE

Again Equation (2.66) holds, but now a third boundary condition is required for the plate bending because (11.76) through (11.78) are sixth order in w with respect to x and y. In addition, since the in-plane and lateral behavior are coupled, a fourth boundary condition enters the picture as well. This has resulted in the use of two different simply supported boundary conditions, both of which are mathematically admissible as natural boundary conditions and are practical structural boundary conditions. By convention the simply supported boundary conditions are given as follows:

$$\begin{aligned}
&\text{S1}(x = \text{constant edge}):\ w = M_x = u_0 = N_{xy} = 0\\
&\text{S1}(y = \text{constant edge}):\ w = M_y = v_0 = N_{yx} = 0\\
&\text{S2}(x = \text{constant edge}):\ w = M_x = N_x = v_0 = 0\\
&\text{S2}(y = \text{constant edge}):\ w = M_y = N_y = u_0 = 0
\end{aligned} \tag{11.79}$$

where u_0 is the mid-surface displacement in the x-direction and v_0 is the mid-surface displacement in the y-direction.

Whether one uses S1 or S2 boundary conditions is determined by the physical aspects of the plate problem being studied.

11.11.2 CLAMPED EDGE

Similarly, for a clamped edge the lateral deflection w and the rotation $\overline{\alpha}$ or $\overline{\beta}$ (for an x = constant edge or a y = constant edge, respectively) are zero (note: the slope is not zero) and the other boundary conditions are analogous to Equation (2.66).

$$
\begin{aligned}
&\text{C1}(x = \text{constant edge}):\ w = \bar{\alpha} = u_0 = N_{xy} = 0 \\
&\text{C1}(y = \text{constant edge}):\ w = \bar{\beta} = v_0 = N_{yx} = 0 \\
&\text{C2}(x = \text{constant edge}):\ w = \bar{\alpha} = N_x = v_0 = 0 \\
&\text{C2}(y = \text{constant edge}):\ w = \bar{\beta} = N_y = u_0 = 0
\end{aligned}
\tag{11.80}
$$

11.11.3 FREE EDGE

The free edge requires three boundary conditions on each edge; therefore, it is no longer necessary to resort to the difficulties of the Kirchhoff boundary conditions for the bending of the plate needed for classical plates which were discussed in (2.67) through (2.69). The bending boundary conditions for the free edge of the plate are:

$$M_n = Q_n = M_{nt} = 0 \tag{11.81}$$

where n and t are directions normal to and tangential with the edge. Again, the in-plane boundary conditions for the free edge are $\boldsymbol{N_n} = \boldsymbol{N_{nt}} = 0$.

11.11.4 OTHER BOUNDARY CONDITIONS

In addition to the above boundary conditions, which are widely used to approximate the actual structural boundary conditions, sometimes it is desirable to consider an edge whose lateral deflection is restrained, whose rotation is restrained or both. The means by which to describe these boundary conditions are given in Section 2.5.

11.12 Composite Plates on An Elastic Foundation or Contacting a Rigid Surface

Consider a composite material plate that is supported on an elastic foundation. In most cases an elastic foundation is modeled as an elastic medium with a constant foundation modulus, i.e., a spring constant per unit planform area, of k in units such as lbs./in./in^2. Therefore, the elastic foundation acts on the plate as a force in the negative direction proportional to the local lateral deflection $w(x,y)$. The force per unit area is $-kw$, because when w is positive the foundation modulus is acting in a negative direction, and vice versa. In order to incorporate the effect of the elastic foundation modeled as above one simply adds another force to the $p(x,y)$ load term. The results are, that for classical theory, Equation (11.26) is modified to be (11.82), and for the refined theory, Equation (11.78) is modified to become Equation (11.83):

$$D_1 \frac{\partial^4 w}{\partial x^4} + 2D_3 \frac{\partial^4 w}{\partial x^2 \partial y^2} + D_2 \frac{\partial^4 w}{\partial y^4} + kw = p(x, y) \tag{11.82}$$

$$\kappa A_{55}\left(\frac{\partial \overline{\alpha}}{\partial x} + \frac{\partial^2 w}{\partial x^2}\right) + \kappa A_{44}\left(\frac{\partial \overline{\beta}}{\partial y} + \frac{\partial^2 w}{\partial y^2}\right) - kw + p(x, y) = 0 \tag{11.83}$$

In addition, if a plate with any boundary conditions, due to the lateral applied load, comes in contact with a rigid smooth surface over part of its area, Hodges [11.5] provides solutions and examples for this important and difficult problem.

For an extended treatment of the modeling of elastic and viscoelastic foundations see Kerr [11.11].

11.13 Solutions for Plates of Composite Materials Including Transverse-Shear Deformation Effects, Simply Supported on All Four Edges

Some solutions are now presented for the equations in Sections 11.10 and 11.11, using the governing differential equations (11.76) through (11.78). In the following κ with no subscript is a transverse shear correction factor, often give as $\pi^2/12$ or 5/6, and discussed above in Section 11.10.

Dobyns [7.5] employed the Navier approach to solve these equations for a composite plate simply supported on all four edges subjected to any lateral load, using the following functions:

$$w(x, y) = \sum_{m=1}^{\infty} \sum_{n=1}^{\infty} C_{mn} \sin\left(\frac{m\pi x}{a}\right) \sin\left(\frac{n\pi y}{b}\right) \tag{11.84}$$

$$\overline{\alpha}(x, y) = \sum_{m=1}^{\infty} \sum_{n=1}^{\infty} A_{mn} \cos\left(\frac{m\pi x}{a}\right) \sin\left(\frac{n\pi y}{b}\right) \tag{11.85}$$

$$\overline{\beta}(x, y) = \sum_{m=1}^{\infty} \sum_{n=1}^{\infty} B_{mn} \sin\left(\frac{m\pi x}{a}\right) \cos\left(\frac{n\pi y}{b}\right) \tag{11.86}$$

$$p(x, y) = \sum_{m=1}^{\infty} \sum_{n=1}^{\infty} q_{mn} \sin\left(\frac{m\pi x}{a}\right) \sin\left(\frac{n\pi y}{b}\right) \tag{11.87}$$

It is seen that Equations (11.84) through (11.87) satisfy the simply supported boundary conditions on all edges given in Equation (11.79).

Substituting these functions into the governing differential equations (11.76) through (11.78) results in the following:

$$\begin{bmatrix} L_{11} & L_{12} & L_{13} \\ L_{12} & L_{22} & L_{23} \\ L_{13} & L_{23} & L_{33} \end{bmatrix} \begin{Bmatrix} A_{mn} \\ B_{mn} \\ C_{mn} \end{Bmatrix} = \begin{Bmatrix} 0 \\ 0 \\ q_{mn} \end{Bmatrix} \tag{11.88}$$

if $\lambda_m \equiv m\pi / a$, $\lambda_n \equiv n\pi / b$ and q_{mn} is the lateral load coefficient of (11.87) above, defined by (11.93) below, then the operators L_{ij} $(i,j = 1,2,3)$ are given by the following:

$$L_{11} = D_{11}\lambda_m^2 + D_{66}\lambda_n^2 + \kappa A_{55} \quad L_{12} = (D_{12} + D_{66})\lambda_m \lambda_n$$

$$L_{13} = \kappa A_{55}\lambda_m, \quad L_{22} = D_{66}\lambda_m^2 + D_{22}\lambda_n^2 + \kappa A_{44}$$

$$L_{23} = \kappa A_{44}\lambda_n, \quad L_{33} = \kappa A_{55}\lambda_m^2 + \kappa A_{44}\lambda_n^2.$$

Solving Equation (11.88), one obtains

$$A_{mn} = \frac{(L_{12}L_{23} - L_{22}L_{13})q_{mn}}{\det} \tag{11.89}$$

$$B_{mn} = \frac{(L_{12}L_{13} - L_{11}L_{23})q_{mn}}{\det} \tag{11.90}$$

$$C_{mn} = \frac{(L_{11}L_{22} - L_{12}^2)q_{mn}}{\det} \tag{11.91}$$

where det is the determinant of the $[L]$ matrix in Equation (11.88).

Having solved the problem to obtain $\overline{\alpha}$, $\overline{\beta}$ and w in (11.84) through (11.86), the curvatures $\kappa_x = (\partial\overline{\alpha} / \partial x)$, $\kappa_y = (\partial\overline{\beta} / \partial y)$ and $\kappa_{xy} = 1/2[(\partial\overline{\alpha} / \partial y) + (\partial\overline{\beta} / \partial y)]$ may be obtained. These then can be substituted back into Equations (11.17) through (11.19) to obtain the stress couples M_x, M_y and M_{xy} to determine the location where they are maximum, necessary in determining where the stresses are maximum.

For a laminated composite plate, to find the bending stresses in each lamina one must use the above equations to find the values for κ_x, κ_y and κ_{xy} in Equation (11.20). Finally, for each lamina the bending stresses can be found using:

$$\begin{bmatrix} \sigma_x \\ \sigma_y \\ \sigma_{xy} \end{bmatrix} = \begin{bmatrix} \overline{Q}_{11} & \overline{Q}_{12} & 0 \\ \overline{Q}_{12} & \overline{Q}_{22} & 0 \\ 0 & 0 & \overline{Q}_{66} \end{bmatrix}_k \begin{bmatrix} \kappa_x \\ \kappa_y \\ 2\kappa_{xy} \end{bmatrix} z \qquad (11.92)$$

The stresses in each lamina in each direction must be compared to the strength of the lamina material using a suitable failure theory. Keep in mind that quite often the failure occurs in the weaker direction in a composite material.

Looking at the load $p(x,y)$ in Equation (11.87), if the lateral load $p(x,y)$ is distributed over the entire lateral surface, then the Euler coefficient, q_{mn} is found to be

$$q_{mn} = \frac{4}{ab}\int_0^a \int_0^b p(x,y)\sin\left(\frac{m\pi x}{a}\right)\sin\left(\frac{n\pi y}{b}\right)dxdy \qquad (11.93)$$

If that load is uniform then,

$$q_{mn} = \frac{4p_0}{mn\pi^2}(1-\cos m\pi)(1-\cos n\pi). \qquad (11.94)$$

For a concentrated load located at $x = \xi$ and $y = \eta$,

$$q_{mn} = \frac{4P}{ab}\sin\left(\frac{m\pi\xi}{a}\right)\sin\left(\frac{n\pi\eta}{b}\right) \qquad (11.95)$$

where P is the total load.

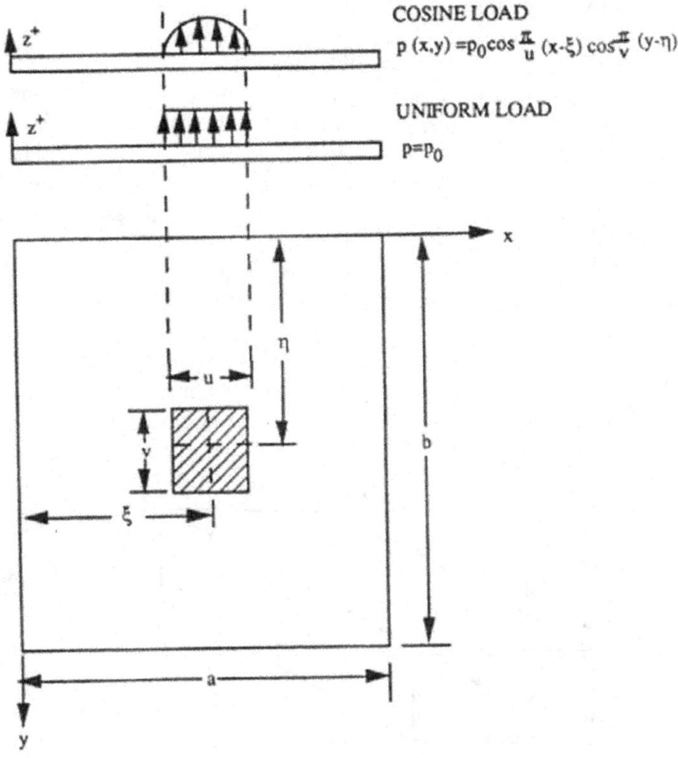

Figure 11.6. Load over a rectangular area. (Reprinted from Reference [11.12]).

For loads over a rectangular area of side lengths u and v whose center is at ξ and η, as shown in Figure 11.6, q_{mn} is given as follows:

$$q_{mn} = \frac{4P\sin\left(\frac{m\pi\eta}{b}\right)\sin\left(\frac{m\pi\xi}{a}\right)\cos\left(\frac{n\pi v}{2b}\right)\cos\left(\frac{m\pi u}{2a}\right)}{abu^2v^2\left(\frac{n}{b}-\frac{1}{v}\right)\left(\frac{n}{b}+\frac{1}{v}\right)\left(\frac{m}{a}-\frac{1}{u}\right)\left(\frac{m}{a}+\frac{1}{u}\right)} \tag{11.96}$$

where P is the total load. Note that when $n/b = 1/v$, $m/a = 1/u$, then $q_{mn} = 0$. Of course, any other lateral load can be characterized by the use of Equation (11.93).

11.14 Some Remarks On Composite Structures

So far in this chapter, plates made of composite materials have been discussed. However, there are complicated constructions, which are made either from composite materials or from isotropic materials, which can be referred to as composite structures. One such structure is a box beam shown below in Figure 11.7, which could be the cross-section of a windmill blade, a water ski, or other representative structural components.

Such a structure will be subjected to tensile or compressive loads in the x-direction, to bending loads about the structural mid-surface, and to torsional loads about the x-axis. In each case one needs to develop the extensional stiffness matrix EA, the flexural stiffness matrix EI and the torsional stiffness matrix GJ, for the rectangular cross-section.

It is probable that in the structural component considered, the top and bottom panels would be identical, as well as each side panel perpendicular to the other one – that will be assumed here, and therefore the subscripts 1 and 2 will be used.

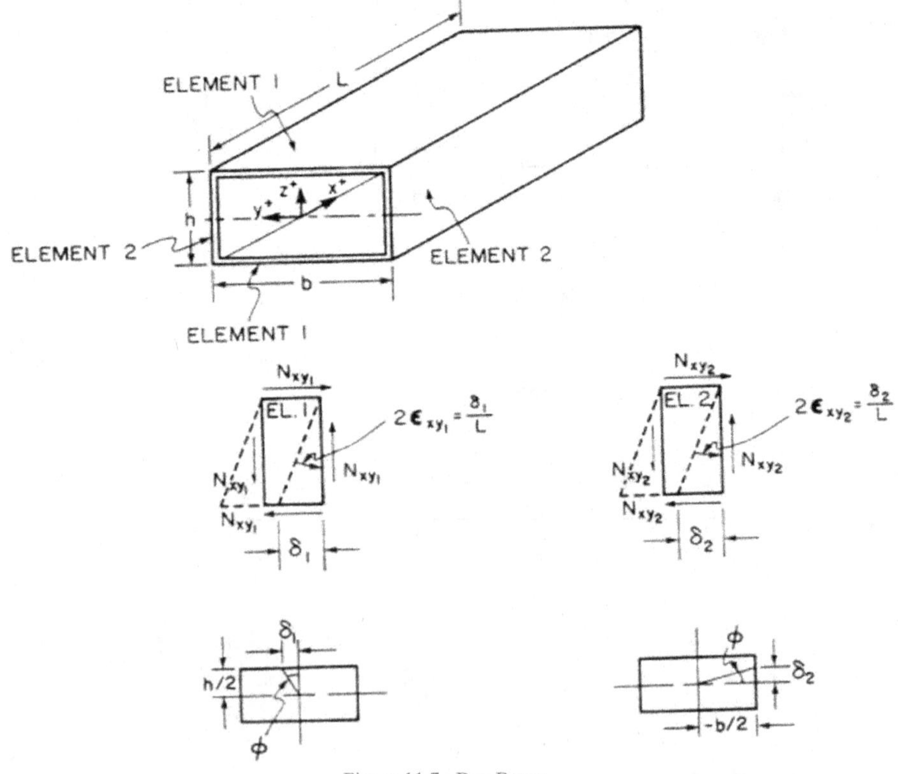

Figure 11.7. Box Beam.

For each of the four panels the extensional relationship in the x-direction involves, for a construction without couplings

$$N_x = A_{11}\varepsilon_x^0$$

where N_x is the force per unit width, and the axial strain, ε_x^0, is identical in each plate element.

Simply adding the contribution of each unit width, the overall load P carried by the overall box beam construction can be written as

$$P = 2N_{x_1}b + 2N_{x_2}h = \left[2(A_{11})_1 b + 2(A_{11})_2 h\right]\varepsilon_x^0$$

Hence, the structural extensional stiffness EA for the rectangular construction of Figure 11.7 is simply

$$EA = 2(A_{11})_1 b + 2(A_{11})_2 h$$

Similarly if the box beam is bent in the *x-z* plane the overall bending moment M will be related to the overall curvature $\boldsymbol{\kappa_x}$, by

$$M = \left[2(D_{11})_1 b + 2(A_{11})_1 b\left(\frac{h}{2}\right)^2 + \frac{2(A_{11})_2 h^3}{12}\right]\kappa_x \tag{11.97}$$

However, if the top and bottom surfaces are thin compared to the overall box height h, then the first term is negligible compared to the other terms, so

$$(EI)_{\text{box beam}} = 2(A_{11})_1 b\left(\frac{h}{2}\right)^2 + \frac{(A_{11})_2 h^3}{6} \tag{11.98}$$

Similar expressions can easily be constructed for the torsional stiffness. Consider the construction of Figure 11.7 subjected to a torsional load T in inch-lbs. about the *x*-axis.

Then it is clear that

$$T = 2\left(N_{xy_1}\right)b\left(\frac{h}{2}\right) + 2\left(N_{xy_2}\right)h\left(\frac{b}{2}\right) \tag{11.99}$$

Now from Equation (10.66), for both elements,

$$N_{xy_i} = 2A_{66_i}\varepsilon_{xy_i}^0 \quad (i = 1,2)$$

If ϕ is the angle of twist caused by the torque T over the length L, then for element 1 and 2

$$\phi = \frac{\delta_1}{(h/2)} \quad \text{and} \quad \phi = \frac{\delta_2}{(b/2)} \tag{11.100}$$

It is also seen that

$$\varepsilon_{xy_1} = \frac{\delta_1}{2L} \quad \text{and} \quad \varepsilon_{xy_2} = \frac{\delta_2}{2L} \tag{11.101}$$

$$T = 2(N_{xy})_1 b\left(\frac{h}{2}\right) + 2(N_{xy})_2 h\left(\frac{b}{2}\right) = 2bh(A_{66})_1 \varepsilon_{xy_1} + 2bh(A_{66})_2 \varepsilon_{xy_2}$$

$$= \frac{bh}{L}(A_{66})_1 \delta_1 + \frac{bh}{L}(A_{66})_2 \delta_2$$

$$T = \frac{bh}{2L}\left[(A_{66})_1 h + (A_{66})_2 b\right]\phi$$

So the GJ, the torsional stiffness of the box beam construction of Figure 11.7 is

$$GJ = \frac{bh}{2}\left[(A_{66})_1 h + (A_{66})_2 b\right] \tag{11.102}$$

The above merely illustrates what one can and must do to develop the basic mechanics of materials global formulation for the extension, bending or twisting of a rectangular section, perhaps composed of very esoteric composite materials but used for a water ski, windmill blade or other shapes for many other purposes.

However, care must be taken to insure that in addition to preventing overall failure of the box beam. As an example, care must be taken to insure that each of the plate structures will not fail at a lower load than the overall structure. This could occur if the plate in compression due to beam bending of the box beam causes the compressive plate to buckle.

11.15 Governing Equations for a Composite Material Plate With Mid-Plane Asymmetry

Consider a rectangular plate in which there are no $(\ )_{16}$ nor $(\ )_{26}$ coupling terms, but which has bending stretching coupling, i.e. $B_{ij} \neq 0$. In that case the equilibrium Equations (11.6) through (11.12), and the strain-displacement Equations (10.50) through (10.52) remain the same.

However, from Equation (10.66) it is seen that the constitutive equations change as shown below.

$$N_x \quad A_{11}\varepsilon_{x_0} + A_{12}\varepsilon_{y_0} + B_{11}\kappa_x + B_{12}\kappa_y$$

$$N_y \quad A_{12}\varepsilon_{x_0} + A_{22}\varepsilon_{y_0} + B_{11}\kappa_x + B_{22}\kappa_y \tag{11.103}$$

$$N_{xy} \quad 2A_{66}\varepsilon_{xy_0} + 2B_{66}\kappa_{xy}$$

Proceeding as before for the mid-plane symmetric rectangular plate of Section 11.3, the resulting three coupled equations using classical plate theory, i.e. no transverse shear deformation, have the following form:

$$\begin{aligned} &A_{11}\left(\frac{\partial^2 u_0}{\partial x^2}\right) + \left[A_{12} + A_{66}\right]\frac{\partial^2 \mathrm{v}_0}{\partial x \partial y} + A_{66}\frac{\partial^2 u_0}{\partial y^2} \\ &- B_{11}\frac{\partial^3 w}{\partial x^3} - \left[B_{12} + B_{66}\right]\frac{\partial^3 w}{\partial x \partial y^2} \quad 0 \end{aligned} \tag{11.104}$$

$$\begin{aligned} &\left[A_{12} + A_{66}\right]\frac{\partial^2 u_0}{\partial x \partial y} + A_{22}\frac{\partial^2 \mathrm{v}_0}{\partial y^2} + A_{66}\frac{\partial^2 \mathrm{v}_0}{\partial x^2} \\ &- \left[B_{12} + B_{66}\right]\frac{\partial^3 w}{\partial x^2 \partial y} - B_{22}\frac{\partial^3 w}{\partial y^3} \quad 0 \end{aligned} \tag{11.105}$$

$$\begin{aligned} &- B_{11}\frac{\partial^3 u_0}{\partial x^3} - \left[B_{12} + 2B_{66}\right]\left(\frac{\partial^3 u_0}{\partial x \partial y^2} + \frac{\partial^3 \mathrm{v}_0}{\partial x^2 \partial y}\right) - B_{22}\frac{\partial^3 \mathrm{v}_0}{\partial y^3} \\ &+ D_1\frac{\partial^4 w}{\partial x^4} + 2D_3\frac{\partial^4 w}{\partial x^2 \partial y^2} + D_2\frac{\partial^4 w}{\partial y^4} \quad p(x,y) \end{aligned} \tag{11.106}$$

Because of the bending-stretching coupling not only are lateral displacements, $w(x,y)$, induced but in-plane displacements, u_0 and v_0, as well; hence, three coupled equations (11.104) through (11.106).

11.16 Governing Equations for a Composite Material Plate With Bending-Twisting Coupling

Looking at Equation (10.66), the moment curvature relations for a rectangular mid-plane symmetric plate with bending-twisting coupling are:

$$
\begin{aligned}
M_x &= D_{11}\kappa_x + D_{12}\kappa_y + 2D_{16}\kappa_{xy} \\
M_y &= D_{12}\kappa_x + D_{22}\kappa_y + 2D_{26}\kappa_{xy} \\
M_{xy} &= D_{16}\kappa_x + D_{26}\kappa_y + 2D_{66}\kappa_{xy}
\end{aligned}
\tag{11.107}
$$

Of course if transverse shear deformation is ignored, i.e., classical theory, then the curvatures are given by (11.20), and the moment curvature relations become:

$$
\begin{aligned}
M_x &= -D_{11}\frac{\partial^2 w}{\partial x^2} - D_{12}\frac{\partial^2 w}{\partial y^2} - 2D_{16}\frac{\partial^2 w}{\partial x \partial y} \\
M_y &= -D_{12}\frac{\partial^2 w}{\partial x^2} - D_{22}\frac{\partial^2 w}{\partial y^2} - 2D_{26}\frac{\partial^2 w}{\partial x \partial y} \\
M_{xy} &= -D_{16}\frac{\partial^2 w}{\partial x^2} - D_{26}\frac{\partial^2 w}{\partial y^2} - 2D_{66}\frac{\partial^2 w}{\partial x \partial y}
\end{aligned}
\tag{11.108}
$$

Substituting these into (11.16), provides the following governing differential equation.

$$
\begin{aligned}
&D_{11}\frac{\partial^4 w}{\partial x^4} + 4D_{16}\frac{\partial^4 w}{\partial x^3 \partial y} + 2D_3\frac{\partial^4 w}{\partial x^2 \partial y^2} + 4D_{26}\frac{\partial^4 w}{\partial x \partial y^3} \\
&+ D_{22}\frac{\partial^4 w}{\partial y^4} = p(x,y)
\end{aligned}
\tag{11.109}
$$

Comparing (11.109) with (11.26), it is seen that due to the presence of the D_{16} and D_{26} bending-twisting coupling terms, odd numbered derivatives appear in the governing differential equation. That precludes the use of both the Navier approach of Section 11.5, and the use of the Levy approach of Section 11.7 in obtaining solutions for plate with bending-twisting coupling. With these complications one may want to obtain solutions using the Theorem of Minimum Potential Energy discussed in Chapter 14 below.

11.17 Concluding Remarks

It appears that there is no end in trying to more adequately describe mathematically the behavior of composite materials utilized in structural components. Unfortunately, the more sophisticated one gets in such descriptions the more difficult the mathematics becomes, as is evidenced in the increasing difficulty observed as one progresses through the sections of Chapter 11.

One additional complication that is important in some composite material structures is that the stiffness (and other properties) are different in tension than they are in compression. This occurs because (1) sometimes the tensile and compressive mechanical properties of both fiber and matrix materials, differ and (2) sometimes it occurs because the matrix material is very weak compared to the fiber (that is $E_m <<< E_f$), such that the fibers buckle in compression under a small load so that for the composite the stiffness in compression differs markedly than the stiffness in tension. Hence, one can idealize a little and say that one has one set of elastic properties in tension and another set of elastic properties in compression. Bert [11.15] has termed this a bimodular material, typical of some composites, certainly typical of aramid (Kevlar) fibers in a rubber matrix that are used in tires, and also typical of certain tissues modeled in biomedical engineering. In this context

$$\begin{bmatrix} A & B \\ B & D \end{bmatrix}_{\text{Tension}} \neq \begin{bmatrix} A & B \\ B & D \end{bmatrix}_{\text{Compression}} \tag{11.110}$$

All of the complications that result are too difficult to treat in this text for those trying to learn the fundamentals of composite material plates and panels.

Lastly, time dependent effects in the stresses, deformation and strains of composite materials are becoming more important design considerations. Viscoelasticity and creep are respected disciplines about which entire books have been written. These effects have been deemed important in some composite material structures. Crossman, Flaggs, Vinson and Wilson have all commented thereupon. Wilson and Vinson [10.26, 10.27] have shown that the effects of viscoelasticity on the buckling resistance of polymer matrix composite material plates is very significant. Similarly, the effect of viscoelasticity on the natural vibration frequencies will also be significant. Many of these effects have been included in a survey article by Reddy [11.16] who has focused primarily on plates composed of composite materials.

11.18 References

11.1. Levy, M. (1899) Sur L'equilibrie Elastique d'une Plaque Rectangulaire, *Compt Rend 129*, pp. 535-539.

11.2. Vinson, J.R. (1961) New Techniques of Solutions for Problems in Orthotropic Plates, *Ph.D. Dissertation*, University of Pennsylvania.

11.3. Vinson, J.R. and Brull, M.A. (1962) New Techniques of Solutions of Problems in Orthotropic Plates, *Transactions of the Fourth United Stated Congress of Applied Mechanics*, Vol. 2, pp. 817-825.

11.4. Yu, W., Hodges, D.H., and Volovoi, V.V. (2002) Asymptotic Construction of Reissner-Like Models for Composite Plates With Accurate Strain Recovery, *International Journal of Solids and Structures*, Vol. 39, No. 20, pp. 5185-5203.

11.5. Hodges, D.H. (2002) Contact Stress from Asymptotic Reissner-Mindlin Plate Theory, *AIAA Journal*, Vol. 41, No. 5, February, pp. 329-331.

11.6. Reddy, J.N. (1984) A Simple High Order Theory for Laminated Composite Plates, *Journal of Applied Mechanics*, ASME, Vol. 51, pp. 745-752.

11.7. Lo, K.H., Christensen, R.M. and Wu, E.M. (1977) A Higher Order Theory for Plate Deformation, Part 2, Laminated Plates, *Journal of Applied Mechanics*, ASME, Vol. 44, pp. 669-676.

11.8. DiSciuva, M. (1984) A Refined Transverse Shear Deformation Theory for Multilayered Anisotropic Plates, *Atti Academia Torino*, Vol. 118, pp. 279-29?.

11.9. Reddy, J.N. and Phan, N.D. (1985) Stability and Vibration of Isotropic, Orthotropic and Laminated Plates Using a Higher-Order Shear Deformation Theory, *Journal of Sound and Vibration*, Vol. 98, pp. 157-17?.

11.10. Whitney, J.M. (1987) *Structural Analysis of Laminated Anisotropic Plates*, Technomic Publishing Co. Inc., Lancaster, Pa.

11.11. Kerr, A.D. (1964) Elastic and Viscoelastic Foundation Models, *Journal of Applied Mechanics*, Vol. 31. pp. 491-498.

11.12. Hyer, M.W. (1981) Some Observations on the Cured Shapes of Thin Unsymmetric Laminates, *Journal of Composite Materials*, Vol. 15, pp. 175-193.

11.13. Hyer, M.W. (1981) Calculations of the Room Temperature Shapes of Unsymmetric Laminates, *Journal of Composite Materials*, Vol. 15, pp. 296-310.

11.14. Hyer, M.W. (1982) The Room Temperature Shapes of Four-Layer Unsymmetric Cross-Ply Laminates, *Journal of Composite Materials*, Vol. 16, pp. 318-340.

11.15. Bert, C.W., Reddy, J.N. Reddy, V.S. and Chao, W.C. (1981) Analysis of Thick Rectangular Plates Laminated of Bimodulus Composite Materials, *AIAA Journal*, Vol. 19, No. 10, October, pp. 1342-1349.

11.16. Reddy, J.N. (1982) Survey of Recent Research in the Analysis of Composite Plates, *Composite Technology Review*, Fall.

11.19 Problems

11.1. The following material properties are given for a unidirectional, 4 ply laminate, $h = 0.020"$

$$A = \begin{bmatrix} 0.84 & 0.00547 & 0 \\ 0.00547 & 0.0176 & 0 \\ 0 & 0 & 0.012 \end{bmatrix} \times 10^6 \text{lb./in.}$$

$$B = 0$$

$$D = \begin{bmatrix} 28.053 & 0.1824 & 0 \\ 0.1824 & 0.5879 & 0 \\ 0 & 0 & 2 \end{bmatrix} \text{lb.} - \text{in.}$$

ρ, the mass density (corresponding to 0.06 lb./in.^{3}) $= 1.554 \times 10^{-4} \text{lb. sec.}^2/\text{in.}^4$

Consider a plate made of the above material with dimensions $a = 20"$, $b = 30"$, $h = 0.020"$. For the first perturbation method of Section 12.8 determine $\bar{b}$ and α. Is α a proper value to use this perturbation technique?

11.2. For a box beam whose dimensions are $b = 4"$, $h = 2"$, $L = 20"$, composed of GY70/339 graphite/epoxy, whose properties are given in Problem 11.9(a), determine the extensional stiffness, EA; the flexural stiffness, EI, and the torsional stiffness, GJ, if the box beam is made of a 4 laminae, unidirectional composite, with a lamina thickness of 0.0055", all fibers being in the length direction.

11.3. Consider a composite material plate of dimensions $0 \le x \le a, 0 \le y \le b$, of thickness h, composed of an E Glass/epoxy, which is modeled as being simply supported on all four edges. It is part of a structural system, which is subjected to a hydraulic load as shown below.

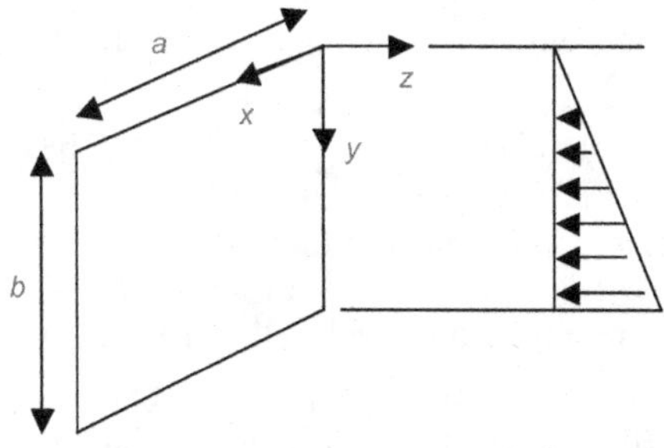

The load is $p(x, y) = \rho y$ where ρ is the weight density of the water.

(a) To utilize the Navier approach determine B_{mn} which is given by

$$B_{mn} = \frac{4}{ab}\int_0^a\int_0^b p(x,y)\sin\left(\frac{m\pi x}{a}\right)\sin\left(\frac{n\pi y}{b}\right)dydx.$$

(b) At what value of x will the maximum deflection occur?
(c) At what value of x will the maximum stress σ_x occur?

11.4. Consider a square plate in which $a = b = 20$", $h = 0,2$", made of a unidirectional Kevlar/epoxy composite, $V_f = 60\%$, whose properties are:

$$E_{11} = 11.02\times10^6\,\text{psi}$$
$$E_{22} = 0.798\times10^6\,\text{psi}$$
$$G_{12} = 0.334\times10^6\,\text{psi}$$
$$\nu_{12} = 0.34$$
$$\rho_w = 0.07\text{lb/in}^3$$

(a) Determine the flexural stiffness matrix $[D]$.
(b) In the first perturbation technique of Section 11.8, calculate $\bar{b}$ and α.
(c) Can this perturbation technique be used for this problem?
(d) What is the total weight of this plate?
(e) If this plate is simply supported on all four edges at what location (i.e., $x = ?$ and $y = ?$) will the maximum deflection occur?
(f) For the plate in (e) above at what location will the maximum bending-stress occur?

11.5. Could the first perturbation solution technique of Section 11.8 be used to obtain solution for the plate of Problem 12.2 subjected to a static lateral load, $p(x,y)$?

11.6. For the panel of Problem 12.9, could the first perturbation method of Section 11.8 be used to solve for deflections and stresses, i.e., is $\alpha < 1$?

11.7. For the plate of Problem 12.5, at what values of x and y will the maximum deflection occur if the plate is subjected to a uniform lateral load $p(x,y) = p_0$ (a constant)?

11.8. Could the perturbation solution technique of Section 11.8 be used to solve problems for the plate of Problem 12.9?

11.9. A square plate, simply supported on all four edges is composed of GY70/339 graphite epoxy. If this square plate is made of four plys with the A and D matrix values shown below, and if the plate is subjected to a uniform lateral load, p_0, which stacking sequence would you choose for a design to have the largest maximum deflection (the most compliant design)? Which stacking sequence has the smallest maximum deflection (the stiffest design)?

GY70/339 graphite epoxy composite

[A] matrix | [D] matrix

a. Unidirectional four ply

$$\begin{bmatrix} 842 & 5.46 & 0 \\ 5.46 & 17.6 & 0 \\ 0 & 0 & 12 \end{bmatrix} \times 10^3 \text{ lb./in.} \qquad \begin{bmatrix} 28.1 & 0.182 & 0 \\ 0.182 & 0.588 & 0 \\ 0 & 0 & 0.4 \end{bmatrix} \text{lb.-in.}$$

b. Crossply $[0^\circ, 90^\circ, 90^\circ, 0^\circ]$ four ply

$$\begin{bmatrix} 430 & 5.46 & 0 \\ 5.46 & 430 & 0 \\ 0 & 0 & 12 \end{bmatrix} \times 10^3 \text{ lb./in.} \qquad \begin{bmatrix} 24.6 & 0.182 & 0 \\ 0.182 & 4.02 & 0 \\ 0 & 0 & 0.4 \end{bmatrix} \text{lb.-in.}$$

c. Angle ply $[\pm 45^\circ]_s$, four ply **(i.e., +45/−45/−45/+45)**

$$\begin{bmatrix} 230 & 206 & 0 \\ 206 & 230 & 0 \\ 0 & 0 & 212 \end{bmatrix} \times 10^3 \text{ lb./in.} \qquad \begin{bmatrix} 7.67 & 6.87 & 5.15 \\ 6.87 & 7.67 & 5.15 \\ 5.15 & 5.15 & 7.07 \end{bmatrix} \text{lb.-in.}$$

11.10. You have been asked to replace an existing aluminum plate structure by a unidirectional Kevlar/epoxy structure using the material properties given in Problem 11.4. The loading on the aluminum plate is all in one direction, both an in-plane tensile load and a bending moment as shown below, and the structure is stiffness critical. Therefore, you must design a unidirectional fiberglass structure to have an extensional stiffness, A_{11} and a flexural stiffness, D_{11}, that equals or exceeds those values for the aluminum structure. The aluminum properties are $E = 10.1 \times 10^6$ psi, $\nu = 0.3$, $\rho = 0.10$ lb./in.^{3} and the aluminum plate is 0.101 inches thick.

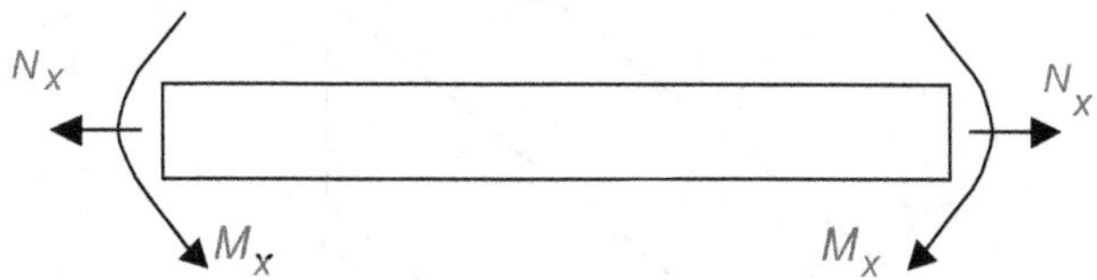

(a) For the existing aluminum structure, what is the extensional stiffness per unit width, $Eh/(1-\nu^2)$?

(b) In the existing structure what is the flexural stiffness per unit width, $Eh^3/12(1-\nu^2)$?

(c) If you replace the aluminum structure with the Kevlar/epoxy structure, what thickness h is required of your composite plate to have A_{11} equal the extensional stiffness of the aluminum structure?

(d) What thickness h is required to your composite plate to have D_{11} equal the flexural stiffness of the aluminum structure?

(e) Which h must your composite design be to achieve the stated design requirement?

(f) Will your composite design be heavier or lighter than the aluminum structure and by what percentage?

11.11. Consider a rectangular panel simply supported on all four edges. The panel measure $a = 25"$, $b = 10"$, where $0 \le x \le a, 0 \le y \le b$. The laminated plate is composed of unidirectional boron/aluminum with the following properties:

E_{11}	32×10^6 psi	ρ	0.0915 lb./in.^{3}
E_{22}	20×10^6 psi	σ_{tu}	250,000 psi
G_{12}	8×10^6 psi	v_f	50%
ply thickness h_k	0.007 in.	ν_{12}	0.35

(a) Determine Q_{11}, Q_{22}, Q_{12} and Q_{66} for a lamina (ply) of this material.

(b) Determine the flexural stiffness D_{11}, D_{22}, D_{12} and D_{66} for a plate made of four ply, unidirectionally oriented (all 0° plys).

(c) If the panel were made of one ply with the fibers oriented at θ $+30^{\circ}$, what is $\overline{Q}_{11}$?

11.12. Consider that four plates identical to the one in Problem 11.12 above are used to fabricate a box beam 60 inches long as shown below.

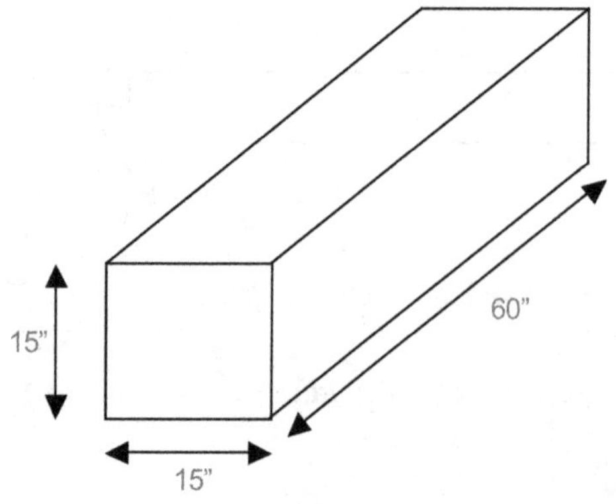

(a) Calculate the axial stiffness, EA, of this box beam.
(b) Calculate the flexural stiffness, EI, of this box beam.
(c) Calculate the torsional stiffness, GJ, of this box beam.

CHAPTER 12

ELASTIC INSTABILITY (BUCKLING) OF COMPOSITE PLATES

12.1 General Considerations

As stated previously, structures usually fail in one of four ways:

- overstressing (strength critical structure)
- over deflection (stiffness critical structure)
- resonant vibration
- buckling.

In monocoque plates, for given plate dimensions, material, boundary conditions, and a given load type (in-plane compression, in-plane shear), only one buckling load will result in actual buckling. This is the lowest eigenvalue of a countable infinity of such eigenvalues. All other eigenvalues exist mathematically, but only the lowest value has physical significance. This differs from natural frequencies in which several eigenvalues can be very important.

For the simplest cases, for columns and isotropic plates, an introduction was given in Chapter 6. While philosophically the simple examples cover the topic of buckling; more complex structures can have several types of buckling instabilities, any one of which can destroy the structure.

There are five major textbooks dealing primarily with elastic stability or buckling. There are authored by Timoshenko and Gere [6.1]*, Bleich [6.2], Brush and Almroth [12.1], Simitses [12.2], and Jones [6.4].

12.2 The Buckling of an Orthotropic Composite Plate Subjected to In-Plane Loads-Classical Theory

From (2.50) through (2.54), it is seen that for a plate there are five equations associated with the in-plane stress resultants N_x, N_y and N_{xy} and the in-plane displacements they cause, namely u_0 and v_0. For the case of a composite material anisotropic plate with mid-plane symmetry ($B_{ij} = 0$) and no thermal or moisture considerations it is seen that the following constitutive equations hold, see (10.66):

$$N_x = A_{11}\varepsilon_x^0 + A_{12}\varepsilon_y^0 + 2A_{16}\varepsilon_{xy}^0 \tag{12.1}$$

---

* Bracketed numbers always refer to references listed at the end of the Chapter indicated by the first number.

$$N_y = A_{12}\varepsilon_x^0 + A_{22}\varepsilon_y^0 + 2A_{26}\varepsilon_{xy}^0 \tag{12.2}$$

$$N_{xy} = A_{16}\varepsilon_x^0 + A_{26}\varepsilon_y^0 + 2A_{66}\varepsilon_{xy}^0. \tag{12.3}$$

Likewise, for the mid-plane symmetric panel, the six governing equations involving $M_x, M_y, M_{xy}, Q_x, Q_y$ and w, are given by (11.9), (11.14), (11.15), and (11.21) through (11.23), the latter three neglecting the D_{16} and D_{26} terms. One can see there is no coupling between in-plane and lateral action for the plate with mid-plane symmetry. Yet it is well known and often observed that in-plane loads do cause lateral deflections through buckling, which is usually disastrous.

The answer to the paradox is that in the above discussion only linear elasticity theory is considered, while the physical event of buckling is a non-linear problem. For brevity, the development of the non-linear theory will not be included herein because it is included in so many other texts, for example, [6.1] and [6.2].

The results of including the terms to predict the advent or inception of buckling for the beam and plate are, modifying (11.26),

$$\begin{aligned} D_1\frac{\partial^4 w}{\partial x^4} + 2D_3\frac{\partial^4 w}{\partial x^2\partial y^2} + D_2\frac{\partial^4 w}{\partial y^4} &= p(x,y) + N_x\frac{\partial^2 w}{\partial x^2} \\ &+ 2N_{xy}\frac{\partial^2 w}{\partial x\partial y} + N_y\frac{\partial^2 w}{\partial y^2} \end{aligned} \tag{12.4}$$

where clearly there is a coupling between the in-plane loads and the lateral deflection.

It should be noted that the buckling loads, like the natural frequencies, are independent of the lateral loads, which will be disregarded in what follows. However, in actual structural analysis, the effect of lateral loads, in combination with the in-plane loads could cause overstressing and failure before the in-plane buckling load is reached. However, the buckling load is still independent of the type or magnitude of the lateral load, as are the natural frequencies. Incidentally, common sense dictates that if one is designing a structure to withstand compressive loads, with the possibility of buckling being the failure mode, one had better design the structure to be mid-plane symmetric, so that $B_{ij} = 0$. Otherwise the bending-stretching coupling would likely cause overstressing before the buckling load is reached.

Looking now at (12.4) for the buckling of the composite plate subjected to an axial load N_x only, and ignoring $p(x,y)$ the equation becomes:

$$D_1\frac{\partial^4 w}{\partial x^4} + 2D_3\frac{\partial^4 w}{\partial x^2\partial y^2} + D_2\frac{\partial^4 w}{\partial y^4} - N_x\frac{\partial^2 w}{\partial x^2} = 0. \tag{12.5}$$

Again, one may assume the buckling mode for a composite orthotropic plate to be that of the Navier solution for the case of the plate simply supported on all four edges:

$$w(x,y) = \sum_{m=1}^{\infty}\sum_{n=1}^{\infty} A_{mn} \sin\frac{m\pi x}{a}\sin\frac{n\pi y}{b}. \tag{12.6}$$

Substituting (12.6) into the homogeneous equation (12.5), it is seen that the equation is satisfied only when N_x has certain values, eigenvalues, namely the critical values, $N_{x\,\mathrm{cr}}$,

$$N_{x\,\mathrm{cr}} = -\frac{\pi^2 a^2}{m^2}\left[D_1\left(\frac{m}{a}\right)^4 + 2D_3\left(\frac{m}{a}\right)^2\left(\frac{n}{b}\right)^2 + D_2\left(\frac{n}{b}\right)^4\right]. \tag{12.7}$$

Again several things are clear: (12.5) is a homogeneous equation, so this is an eigenvalue problem and therefore one cannot determine the value of A_{mn}; and again only the lowest value of $N_{x\,\mathrm{cr}}$ is of any physical importance. However, it is not clear which values of m and n result in the lowest critical buckling load. All values of n appear in the numerator for this case of all edges being simply supported, so $n = 1$ is the necessary value. But m appears several places, and depending upon the value of the flexural stiffnesses D_1, D_2 and D_3, and the length to width ratio, i.e., the aspect ratio, of the plate, a/b, it is not clear which value of m will provide the lowest value of $N_{x\,\mathrm{cr}}$. However, for a given plate this is easily determined computationally.

What about the buckling loads of composite material plates with boundary conditions other than simply supported? In those cases, quite often the Minimum Potential Energy Theorem is used in which trial functions for the lateral deflection are selected as follows. It is seen that all combinations of beam vibrational mode shapes are applicable for plates with various boundary conditions. These have been developed by Warburton [8.3] and all derivatives and integrals of those functions catalogued conveniently by Young and Felgar [3.1] for easy use.

The buckling loads calculated in this section do not include transverse shear deformation effects, and are therefore only approximate – but they are useful for preliminary design, because of their relative simplicity. If transverse shear deformation were included, the buckling loads are lower than those calculated with classical theory. Therefore the buckling loads calculated, neglecting transverse shear deformation, are not conservative.

12.3 Buckling of a Composite Plate on an Elastic Foundation

Referring to the previous discussion regarding plates on an elastic foundation in Section 11.12, the governing differential equation for the buckling of a specially orthotropic composite plate on an elastic foundation can be written as follows. In previous sections, the simplest model for an elastic foundation was used. In what follows, a more sophisticated and accurate foundation model is used. Also, the buckling of an isotropic plate on an elastic foundation using the simplest foundation model was discussed in Section 6.6.

$$D_1 \frac{\partial^4 w}{\partial x^4} + 2D_3 \frac{\partial^4 w}{\partial x^2 \partial y^2} + D_2 \frac{\partial^4 w}{\partial y^4} = N_x \frac{\partial^2 w}{\partial x^2} - p_i \tag{12.8}$$

Following the research of Paliwal and Ghosh [12.3] a Kerr foundation [12.4] is used, which involves two spring layers and a shear layer employing the constants k_1, k_2 and G, where k_1 and k_2 are the foundation moduli of the upper and lower spring layers respectively, while G is the shear modulus of the shear layer.

The lateral deflection is given by

$$w(x,y) = w_1(x,y) + w_2(x,y)\,. \tag{12.9}$$

The contact pressure p_1 and p_2 under the plate are

$$p_1(x,y) = k_1 w_1 = k_1(w - w_2) \tag{12.10}$$

$$p_2(x,y) = k_2 w \tag{12.11}$$

The governing differential equation for the shear layer is:

$$k_2 w_2 - G\nabla^2 w_2 = p_1 \tag{12.12}$$

Eliminating w_2 from (12.12) and (12.10), and substituting the value of p_i from (12.8) one obtains:

$$\begin{aligned} &-\left[1+\frac{k_2}{k_1}\right]\left[D_1 \frac{\partial^4 w}{\partial x^4} + 2D_3 \frac{\partial^4 w}{\partial x^2 \partial y^2} + D_2 \frac{\partial^4 w}{\partial y^4} + p\frac{\partial^2 w}{\partial x^2}\right] \\ &+\frac{G}{k_1}\left\{D_1\left(\frac{\partial^6 w}{\partial x^6} + \frac{\partial^6 w}{\partial x^4 \partial y^2}\right) + 2D_3\left(\frac{\partial^6 w}{\partial x^4 \partial y^2} + \frac{\partial^6 w}{\partial x^2 \partial y^4}\right)\right. \\ &\left.+D_2\left(\frac{\partial^6 w}{\partial x^2 \partial y^4} + \frac{\partial^6 w}{\partial y^6}\right) + p\left(\frac{\partial^4 w}{\partial x^4} + \frac{\partial^4 w}{\partial x^2 \partial y^2}\right)\right\} \\ &= k_2 w - G\left(\frac{\partial^2 w}{\partial x^2} + \frac{\partial^2 w}{\partial y^2}\right) \end{aligned} \tag{12.13}$$

In the composite plate one may assume (12.6) for the lateral deflection $w(x,y)$. Substituting (12.6) into (12.13) and letting the plate aspect ratio $a/b = c$, results in the following for a specially orthotropic plate simply supported on all four edges, remembering that for this plate also $n = 1$ only:

$$\overline{N}_x\left[\left(1+\frac{\lambda_2}{\lambda_1}\right)+\frac{\mu}{\lambda_1}\left\{1+\left(\frac{m}{c}\right)^2\right\}\right]$$
$$=\lambda_2\left(\frac{c}{m}\right)^2+\mu\left\{1+\left(\frac{c}{m}\right)^2\right\}+\left\{\sqrt{\frac{D_1}{D_2}}\left(\frac{m}{c}\right)^2\right.$$
$$\left.+\frac{2D_3}{\sqrt{D_1D_2}}+\sqrt{\frac{D_2}{D_1}}\left(\frac{c}{m}\right)^2\right\}\left[1+\frac{\lambda_2}{\lambda_1}+\frac{\mu}{\lambda_1}\left\{1+\left(\frac{m}{c}\right)^2\right\}\right] \tag{12.14}$$

In Equations (12.14) through (12.16) $\overline{N}_x = N_{x\,\mathrm{cr}}b^2/\pi^2(D_1/D_2)^{1/2}$.

Equation (12.14) is the solution for the Kerr foundation. Paliwal and Ghosh give the solutions for the Winkler and Pasternak foundation, which are given below, for a plate simply supported on all four edges:

$$\overline{N}_x = \sqrt{\frac{D_1}{D_2}}\left(\frac{m}{c}\right)^2+\frac{2D_3}{\sqrt{D_1D_2}}+\sqrt{\frac{D_2}{D_1}}\left(\frac{c}{m}\right)^2+\lambda\left(\frac{c}{m}\right)^2 \tag{12.15}$$

$$\overline{N}_x = \sqrt{\frac{D_1}{D_2}}\left(\frac{m}{c}\right)^2+\frac{2D_3}{\sqrt{D_1D_2}}+\sqrt{\frac{D_2}{D_1}}\left(\frac{c}{m}\right)^2$$
$$+\lambda\left(\frac{c}{m}\right)^2+\mu\left\{1+\left(\frac{c}{m}\right)^2\right\} \tag{12.16}$$

where λ is the non-dimensional foundation modulus of the spring layer, μ is the non-dimensional shear modulus, and k is the foundation modulus,

$$\lambda = kb^4/\pi^4\sqrt{D_1D_2}$$

$$\mu = Gb^4/\pi^2\sqrt{D_1D_2}$$

Similarly, Paliwal and Ghosh studied the buckling of a composite plate subjected to in-plane compressive loads in two directions. In that case the governing equation is given by

$$D_1\frac{\partial^4 w}{\partial x^4}+2D_3\frac{\partial^4 w}{\partial x^2\partial y^2}+D_2\frac{\partial^4 w}{\partial y^4} = N_x\frac{\partial^2 w}{\partial x^2}+N_y\frac{\partial^2 w}{\partial y^2}-p_1. \tag{12.17}$$

The final buckling load is found from the following equation, where $\alpha = N_y/N_x$:

$$
\begin{aligned}
&\overline{N}_x\left[\left(1+\frac{\lambda_2}{\lambda_1}\right)\left\{\left(\frac{m}{c}\right)^2+\alpha n^2\right\}\right.\\
&\qquad\left.+\frac{\mu}{\lambda_1}\left\{\left(\frac{m}{c}\right)^2+(\alpha+1)\left(\frac{m}{c}\right)^2 n^2+\alpha n^4\right\}\right]\\
&\quad\lambda_2+\mu\left\{\left(\frac{m}{c}\right)^2+n^2\right\}+\left(1+\frac{\lambda_2}{\lambda_1}\right)\\
&\times\left\{\sqrt{\frac{D_1}{D_2}}\left(\frac{m}{c}\right)^4+\frac{2D_3}{\sqrt{D_1D_2}}\left(\frac{m}{c}\right)^2 n^2+\sqrt{\frac{D_2}{D_1}}n^4\right\}\\
&+\frac{\mu}{\lambda_1}\left[\sqrt{\frac{D_1}{D_2}}\left\{\left(\frac{m}{c}\right)^6+\left(\frac{m}{c}\right)^4 n^2\right\}+\frac{2D_3}{\sqrt{D_1D_2}}\left\{\left(\frac{m}{c}\right)^4 n^2\right.\right.\\
&\left.\left.+\left(\frac{m}{c}\right)^2 n^4\right\}+\sqrt{\frac{D_2}{D_1}}\left\{\left(\frac{m}{c}\right)^2 n^4+n^6\right\}\right]
\end{aligned}
\qquad (12.18)
$$

Postbuckling behavior of composite plates is beyond the scope of this basic text. However, the work of Minguet, Dugundji and Lagace [12.5] provides an introduction to this topic. Also, a number of NASA reports by Nemeth [12.6] treat the buckling of composite plates subjected to thermal and mechanical loads.

12.4 References

12.1. Brush, D.O. and Almroth, B.O. (1975) *Buckling of Bars, Plates and Shells*, McGraw-Hill Book Co., Inc., New York.

12.2. Simitses, G.J. (1976) *An Introduction to the Elastic Stability of Structures*, Prentice-Hall, Inc., Englewood Cliffs, New Jersey.

12.3. Paliwal, D.N. and Ghosh, S.K. (2000) Stability of Orthotropic Plates on a Kerr Foundation, Vol. 38, No. 10, October, pp. 1994-1997.

12.4. Kerr, A.D. (1964) Elastic and Viscoelastic Models, *Journal of Applied Mechanics*, Vol. 31, September, pp. 491-498.

12.5. Minguet, P., Dugundji, J. and Lagace, P.A. (1989) Postbuckling Behavior of Laminated Plates Using a Direct Energy-Minimization Technique, *AIAA Journal*, Vol. 27, No. 12, pp. 1785-1792.

12.6. Nemeth, M.P. (2003) Buckling Behavior of Long Anisotropic Plates Subjected to Fully Restrained Thermal Expansion, NASA/TP-2003-212131, February.

12.5 Problems

12.1. Find the critical buckling load, $N_{x_{cr}}$ in lbs./in. for a plate simply supported on all four edges made of a material whose flexural stiffness properties are given as follows and whose thickness is 1 inch.

$$[\boldsymbol{D}] = \begin{bmatrix} 1.63 & 0.028 & 0 \\ 0.028 & 0.160 & 0 \\ 0 & 0 & 0.037 \end{bmatrix} \times 10^6 \text{lb.-in.}$$

(a) If $a = 30$ inches and $b = 20$ inches.
(b) If $a = 50$ inches and $b = 12$ inches.

12.2. Consider a plate measuring 16" x 16" in planform of $[\mathbf{0°, 90°, 90°, 0°}]$, of total thickness 0.022". The $[D]$ matrix for this construction is

$$D = \begin{bmatrix} 8.70 & 0.242 & 0 \\ 0.242 & 1.854 & 0 \\ 0 & 0 & 0.296 \end{bmatrix} \text{lb.} - \text{in.}$$

If the plate is subjected to an in-plane compressive load in the $\theta = 0^{\circ}$ direction, what is the critical buckling load per inch of the edge distance, $N_{x_{cr}}$, using classical plate theory?

12.3. In designing a test facility to demonstrate the buckling of the plate of Problem 12.2, what load cell capacity (force capability) is needed to attain the loads necessary to buckle the plate?

12.4. Determine the critical buckling load, $N_{x_{cr}}$, for the same panel as in Problem 12.9.

12.5. Consider a plate of dimensions a = 18" and b = 12", composed of a laminated composite material whose lamina properties are:

$E_1 = 18.5 \times 10^6$ psi $\qquad \nu_{12} = 0.30$

$E_2 = 1.64 \times 10^6$ psi $\qquad G_{12} = 0.87 \times 10^6$ psi

The stacking sequence of the plate is $[0°, 90°, 90°, 0°]$ in which each lamina is 0.006" = h_k. The plate is simply supported on each edge.

(a) What are A_{11}, B_{11} and D_{11} for this plate?
(b) At what values of x and y will the maximum deflection occur if the plate is subjected to a uniform lateral load $p(x, y) = p_0$ (a constant)?
(c) At which values of x and y would maximum ply stresses occur?

(d) Calculate the critical buckling load per unit width, $N_{x_{cr}}$, if the plate is subjected to a uniform compressive load in the x direction.

12.6. Consider a Kevlar 49/epoxy composite, whose properties are given in Table 10.3 of the text, and whose weight density is $\rho_w = 0.06$ **lbs./in.**^{3}. A plate whose stacking sequence is $[0°, 90°, 90°, 0°]$ is fabricated wherein each ply is 0.0055" thick. The plate is $20'' \times 16''$ in planform dimensions, and is simply supported on all four edges.

(a) Determine $A_{11}, A_{12}, A_{22}, A_{66}, D_{11}, D_{12}, D_{22}$ and D_{66}.

(b) Could the first perturbation solution technique of Section 11.8 be used to solve problems for this plate if it were subjected to a lateral load $p(x, y)$?

(c) If the plate is subjected to an in-plane compressive load in the x- direction only, what is the critical buckling load per inch of edge distance, $N_{x_{cr}}$, using classical plate theory?

12.7. For a plate simply supported on all four edges that is 6 inches wide and 15 inches long made up of the unidirectional four ply graphite epoxy described in (a) of Problem 11.10, what is the critical buckling load, $N_{x_{cr}}$, if the compressive load is applied parallel to the longer direction of the plate?

12.8. Given a Kevlar/epoxy rectangular plate, with the unidirectional material properties given in Table 10.3, for a plate of dimensions $16'' \times 12''$, and a thickness of $0.1''$, as shown below, simply supported on all four edges. The fibers are all aligned in the longer direction.

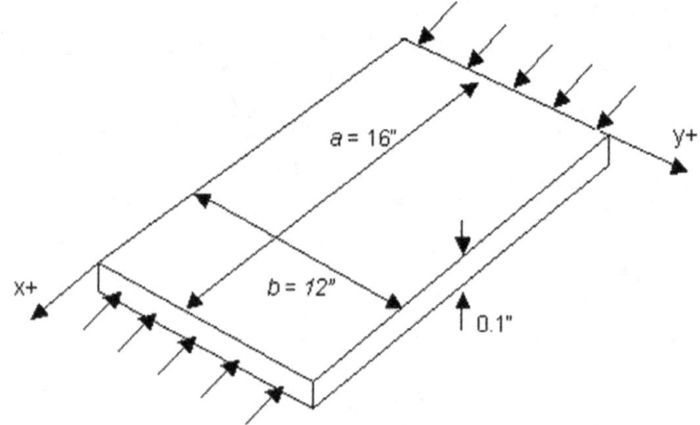

(a) What is the critical load per unit inch, $N_{x_{cr}}$, to cause plate buckling of the plate?

(b) What is the stress in the load direction at buckling?

(c) Will the plate be overstressed before it could buckle?

(d) What is the total weight of this plate?

(e) What thickness would this plate have to be to have the buckling stress equal to the compressive strength of the composite material? Assume here that the compressive strength is equal to the tensile strength (That is not always true!).

12.9. A panel simply supported on all four edges, measuring $a = 30"$, $b = 10"$, composed of T300-5208 graphite epoxy, composed of laminae with the following properties:

$Q_{11} = 22.36 \times 10^6$ psi $\quad \rho_w = 0.05$ lb./in.^{3}

$Q_{22} = 1.591 \times 10^6$ psi

$Q_{12} = 0.4773 \times 10^6$ psi $\quad h_k = 0.0055"$

$Q_{66} = 0.81 \times 10^6$ psi

In the October 1986 issue of the AIAA Journal, M.P. Nemeth discuses the conditions in which one can ignore D_{16} and D_{26} in determining the buckling load for a composite plate. He defines:

$$\gamma = \frac{D_{16}}{\left(D_{11}{}^3 D_{22}\right)^{1/4}} \text{ and } \delta = \frac{D_{26}}{\left(D_{11} D_{22}{}^3\right)^{1/4}}$$

If both of these ratios are less than 0.18, one can use Equation (12.7) to determine the buckling load within 2% of the correct value for a plate simply supported on all four edges. If either of the ratios is greater than 0.18 one must replace the left hand side of Equation (12.4) with the left hand side of Equation (12.109), which negates the use of the Navier and Levy methods being used, thus complicating the solution.

For a four ply panel with stacking sequence of $[+45^o, -45^o, -45^o, +45^o]$, determine γ and δ to see if the simpler solution can be used.

CHAPTER 13

LINEAR AND NONLINEAR VIBRATION OF COMPOSITE PLATES

13.1 Dynamic Effects on Panels of Composite Materials

Seldom in real life is a structure subjected only to static loads. More often products and structures are subjected to vehicular, impact, crash, earthquake, handling, or fabricating dynamic loads. In the linear-elastic range, dynamic effects can be divided into two categories: natural vibrations and forced vibrations, and the latter can be further subdivided into one-time events (an impact) or recurring loads (such as cyclic loading). These will be discussed in turn.

Physically every elastic continuous body has an infinity of natural frequencies, only a few of which are of practical significance. When a structure is excited cyclically at a natural frequency, it takes little input energy for the amplitude to grow until one of four things happens:

(1) The amplitude of vibration grows until the ultimate strength of a brittle material is exceeded and the material and structure fails.
(2) Portions of the structure exceed the yield strength, plastically deform and the dynamic response behavior changes drastically.
(3) The amplitude grows until nonlinear effects become significant, and there is no natural frequency.
(4) Due to damping or other mechanisms the amplitude is limited, but as the natural vibration continues, fatigue failures may occur.

Physically, when a structure is undergoing a natural vibration the sum of the potential energy and kinetic energy remains constant if no damping is present. This can be termed a conservative system. However, the energy is compartmentalized, i.e., if a structure is truly vibrating in one mode of natural vibration it will not change by commencing to vibrate in some other mode of natural vibration at some other natural frequency. In a complex structure if two components have vibrational natural frequencies that are identical, then when one component is excited, the other component will also be excited. It is for this very reason that duplicative natural frequencies are to be avoided. Also in complex structures of course the structural natural frequencies can be coupled involving all components.

Mathematically, natural vibration problems are called eigenvalue problems. They are represented by homogeneous equations, for which nontrivial solutions only occur at certain characteristic (*eigen*, in German) values of a parameter, from which the natural frequencies are determined. In a vibration at a natural frequency, the displacement field comprises a normal mode for that natural frequency. At any two different natural frequencies the corresponding normal modes are mathematical functions that are orthogonal to each other (hence the compartmentalization of the energy). The normal modes comprise the solutions to the homogeneous governing differential equations, and non-trivial solutions of the equations occur only at the eigenvalues for those equations and boundary conditions.

If there is a forcing function, then the particular solution for the specific forcing function (which can be cyclical or a one time dynamic impact load) is added onto the homogeneous solution, which involves the natural frequencies and mode shapes. Physically, any dynamic load excites each and every one of the normal modes and corresponding natural frequencies. Usually, only a relatively few are large enough to be of concern. The largest amplitude of response will be in those mode shapes whose natural frequencies are closest to the oscillatory component of the forcing function.

When there are no natural frequencies close to the oscillatory portion of the dynamic load, then the structure will respond at each time, t, in deflection and stresses that correspond only to the magnitude and spatial distribution of the load at that time t. Such a condition results in solving the worst-case static problem in which the largest load at some time, t, is applied. This is termed a quasistatic case. However, if the dynamic load oscillatory component is close to one of more natural frequencies, then the structural response can be much larger than the value obtained from a quasistatic calculation, and that increase can be represented by a dynamic load factor.

In what follows natural frequencies are treated first, then forced linear vibrations and finally nonlinear large-amplitude vibrations are discussed.

13.2 Natural Flexural Vibrations of Rectangular Plates: Classical Theory

Consider a rectangular composite material plate that is mid-plane symmetric such that $B_{ij} = 0$. If this plate is quasi-isotropic, i.e., $D_1 = D_2 = D_3 = D$, then the governing differential equation is given by Equation (2.57) for the classical theory, i.e., no transverse-shear deformation, and repeated here as

$$D\left[\frac{\partial^4 w}{\partial x^4}+2\frac{\partial^4 w}{\partial x^2 \partial y^2}+\frac{\partial^4 w}{\partial y^4}\right] = p(x,y) \tag{13.1}$$

For dynamic loads, using d'Alembert's Principle, the equation is written as

$$D\left[\frac{\partial^4 w}{\partial x^4}+2\frac{\partial^4 w}{\partial x^2 \partial y^2}+\frac{\partial^4 w}{\partial y^4}\right] = p(x,y,t)-\rho_m h\frac{\partial^2 w}{\partial t^2} \tag{13.2}$$

where the last term is the mass per unit planform area times the acceleration. So the natural vibrations for a quasi-isotropic composite plate parallels exactly the discussion in Section 7.3.

If the composite plate is specially orthotropic and mid-plane symmetric, then for the natural vibration problem the governing differential equation is written as follows:

$$D_1\frac{\partial^4 w}{\partial x^4}+2D_3\frac{\partial^4 w}{\partial x^2\partial y^2}+D_2\frac{\partial^4 w}{\partial y^4}=-\rho_m h\frac{\partial^2 w}{\partial t^2} \tag{13.3}$$

Again, if all four edges are simply supported then the mode shapes are given by Equation (7.11) with the result that the natural circular frequency in radians per second is given by

$$\omega_{mn}=\frac{\pi^2}{\sqrt{\rho_m h}}\left[D_1\left(\frac{m}{a}\right)^4+2D_3\left(\frac{m}{a}\right)^2\left(\frac{n}{b}\right)^2+D_2\left(\frac{n}{b}\right)^4\right]^{1/2} \tag{13.4}$$

The natural frequency in Hz is given by Equation (7.13). As in the case of isotropic plate vibrations m and n are integers relating to the mode shapes. Also it is seen that if the plate is isotropic, $D_1=D_2=D_3=D$, (13.4) becomes (7.12).

Keep in mind that for Equation (13.4) and other equations for the frequencies for natural vibrations for thin walled structures, to accurately describe the motion, the maximum deflection must be limited to some fraction of the plate thickness since the theory is linear. Above that level of motion, nonlinear effects become increasingly significant.

One major reference for the free vibration of rectangular isotropic and composite plates is authored by Leissa [13.1].

13.3 Natural Flexural Vibrations of Composite Material Plates Including Transverse Shear Deformation Effects

The governing partial differential equations for a composite plate or panel that is specially orthotropic and mid-plane symmetric subjected to a lateral static load $p(x,y)$ are given in Equations (11.24) through (11.26). If one now wishes to find the natural frequencies of this composite plate, that has mid-surface symmetry ($B_{ij}=0$), no other couplings $(\ )_{16}=(\ )_{26}=(\ )_{45}=0$, but includes transverse shear deformation, $\varepsilon_{xz}\neq 0$, $\varepsilon_{yz}\neq 0$, then one sets $p(x,y)=0$ in Equation (11.26), but adds $-\rho_m(\partial^2 w/\partial t^2)$ to the right-hand side where ρ_m is the average mass density of the plate material. So, $\rho_m h$ in (13.7) below is the mass density per unit planform area. In addition, because $\overline{\alpha}$ and $\overline{\beta}$ are both dependent variables that are independent of w, there will be an oscillatory motion of the lineal element across the plate thickness about the mid-surface of the plate. This results in the last term on the left-hand side of Equations (13.5) and (13.6) becoming $I(\partial^2\overline{\alpha}/\partial t^2)$ and $I(\partial^2\overline{\beta}/\partial t^2)$ respectively, as shown below:

$$D_{11}\frac{\partial^2\overline{\alpha}}{\partial x^2}+D_{66}\frac{\partial^2\overline{\alpha}}{\partial y^2}+(D_{12}+D_{66})\frac{\partial^2\overline{\beta}}{\partial x\partial y}-\kappa A_{55}\left(\overline{\alpha}+\frac{\partial w}{\partial x}\right)-I\frac{\partial^2\overline{\alpha}}{\partial t^2}=0 \tag{13.5}$$

$$(D_{12}+D_{66})\frac{\partial^2\overline{\alpha}}{\partial x\partial y}+D_{66}\frac{\partial^2\overline{\beta}}{\partial x^2}+D_{22}\frac{\partial^2\overline{\beta}}{\partial y^2}-\kappa A_{44}\left(\overline{\beta}+\frac{\partial w}{\partial y}\right)-I\frac{\partial^2\overline{\beta}}{\partial t^2}=0 \tag{13.6}$$

$$\kappa A_{55}\left(\frac{\partial\overline{\alpha}}{\partial x}+\frac{\partial^2 w}{\partial x^2}\right)+\kappa A_{44}\left(\frac{\partial\overline{\beta}}{\partial y}+\frac{\partial^2 w}{\partial y^2}\right)-\rho_m h\frac{\partial^2 w}{\partial t^2}=0 \tag{13.7}$$

where $\rho_m h$, the plate mass density per unit planform area, in (13.4) and (13.7) above, is given by

$$\rho_m h=\sum_{k=1}^{N}\rho_k(h_k-h_{k-1}) \tag{13.8}$$

where ρ_k is the *mass* density of the *k*th lamina material, and here *I* is

$$I=\sum_{k=1}^{N}\rho_m z_k^2 dz_k. \tag{13.9}$$

In Equations (13.5) through (13.7) the κ's, without subscripts, are transverse-shear deformation correction parameters, as discussed earlier in Section 11.10.

Similar to the Navier procedure used in previous analyses and following Dobyns [7.5] for the simply supported plate, looking at (11.84) through (11.86), let

$$w(x,y,t)=\sum_{m=1}^{\infty}\sum_{n=1}^{\infty}C'_{mn}\sin\left(\frac{m\pi x}{a}\right)\sin\left(\frac{n\pi y}{b}\right)e^{i\omega t} \tag{13.10}$$

$$\overline{\alpha}(x,y,t)=\sum_{m=1}^{\infty}\sum_{n=1}^{\infty}A'_{mn}\cos\left(\frac{m\pi x}{a}\right)\sin\left(\frac{n\pi y}{b}\right)e^{i\omega t} \tag{13.11}$$

$$\overline{\beta}(x,y,t)=\sum_{m=1}^{\infty}\sum_{n=1}^{\infty}B'_{mn}\sin\left(\frac{m\pi x}{a}\right)\cos\left(\frac{n\pi y}{b}\right)e^{i\omega t} \tag{13.12}$$

Substituting these equations into the dynamic governing equations above results in a set of homogeneous equations that can be solved for the natural frequencies of vibration

$$\begin{bmatrix} L'_{11} & L_{12} & L_{13} \\ L_{12} & L'_{22} & L_{23} \\ L_{13} & L_{23} & L'_{33} \end{bmatrix}\begin{Bmatrix} A'_{mn} \\ B'_{mn} \\ C'_{mn} \end{Bmatrix}=\begin{Bmatrix} 0 \\ 0 \\ 0 \end{Bmatrix} \tag{13.13}$$

where the unprimed L quantities were defined below Equation (11.88) and

$$L'_{11} = L_{11} - \frac{\rho_m h^3}{12}\omega_{mn}^2$$

$$L'_{22} = L_{22} - \frac{\rho_m h^3}{12}\omega_{mn}^2 \qquad (13.14)$$

$$L'_{33} = L_{33} - \rho_m h \omega_{mn}^2$$

Three eigenvalues (natural circular frequencies) result from solving Equation (13.13) for each value of m and n. However, two of the frequencies are significantly higher than the other because they are associated with the rotatory inertia terms, which are the last terms on the left-hand sides of Equations (13.5) and (13.6) and are very seldom important in structural responses. If they are neglected then $L'_{11} = L_{11}$ and $L'_{22} = L_{22}$ above, and the square of the remaining natural frequency can be easily found to be

$$\omega_{mn}^2 = [QL_{33} + 2L_{12}L_{23}L_{13} - L_{22}L_{13}^2 - L_{11}L_{23}^2] / \rho_m h Q \qquad (13.15)$$

where, here, $Q = L_{11}L_{22} - L_{12}^2$. Also,

$$A'_{mn} = \frac{L_{12}L_{23} - L_{22}L_{13}}{Q} C'_{mn}$$

$$B'_{mn} = \frac{L_{12}L_{13} - L_{11}L_{23}}{Q} C'_{mn}$$

If transverse-shear deformation effects were neglected, Equation (13.15) would reduce to Equation (13.4). Then, if the plate is isotropic (13.15) becomes identical to (7.12).

In composite material structures, matrix cracks can occur. The effect of these cracks upon the natural vibrations of composite panels, particularly sandwich panels with composite facings has been treated by Birman and Simitses.

For non-linear large amplitude vibrations of composite material plates, see Wu and Vinson [13.2, 13.3].

13.4 Forced Vibration Response of a Composite Material Plate Subjected to a Dynamic Lateral Load

Dobyns [7.5] then goes on to develop the solutions for the simply supported laminated composite plate subjected to a dynamic lateral load $p(x,y,t)$, neglecting the rotatory inertia terms discussed above, utilizing a convolution integral $P(t)$ as seen below in (13.19). Incidentally, the convolution integral is also known as the superposition integral and the Duhamel integral.

The solutions to Equations (13.5) through (13.7), modified to include a dynamic distributed lateral load $p(x,y,t)$ and neglecting the rotatory inertia terms are given by

$$w(x,y,t) = \frac{1}{\rho_m h}\sum_{m=1}^{\infty}\sum_{n=1}^{\infty}\left(\frac{q_{mn}}{\omega_{mn}}\right)\sin\left(\frac{m\pi x}{a}\right)\sin\left(\frac{n\pi y}{b}\right)P(t) \tag{13.16}$$

$$\overline{\alpha}(x,y,t) = \frac{1}{\rho_m h}\sum_{m=1}^{\infty}\sum_{n=1}^{\infty}\left(\frac{q_{mn}}{\omega_{mn}}\right)\frac{(L_{12}L_{23}-L_{22}L_{13})}{Q}\cos\left(\frac{m\pi x}{a}\right)\sin\left(\frac{n\pi y}{b}\right)P(t) \tag{13.17}$$

$$\overline{\beta}(x,y,t) = \frac{1}{\rho_m h}\sum_{m=1}^{\infty}\sum_{n=1}^{\infty}\left(\frac{q_{mn}}{\omega_{mn}}\right)\frac{(L_{12}L_{13}-L_{11}L_{23})}{Q}\sin\left(\frac{m\pi x}{a}\right)\cos\left(\frac{n\pi y}{b}\right)P(t) \tag{13.18}$$

where

$$P(t) = \int_0^t F(\tau)\sin[\omega_{mn}(t-\tau)]d\tau \tag{13.19}$$

and q_{mn} is the coefficient of the lateral-load function expanded in series form [see Equations (11.29) through (11.32) where there B_{mn} is used.

So for a given lateral distributed load $p(x,y,t)$, if a solution of the form given by Equations (13.16) through (13.18) is applicable, then the curvatures κ_x, κ_y and κ_{xy} for the plate can be found from Equation (10.52), and the stresses in each lamina are found from Equation (10.53). The function $P(t)$ has been solved analytically for several representative forcing functions shown in Figure 13.2.

For the <u>sine pulse</u>, the forcing function $F(t)$ and the convolution integral $P(t)$ are

$$F(t) = F_0\sin(\pi t/t_1) \qquad 0 \le t \le t_1$$

$$F(t) = 0 \qquad t > t_1$$

$$P(t) = \int_0^t F(t)\sin[\omega_{mn}(t-\tau)]d\tau$$

$$= \frac{F_0 t_1[\pi \sin \omega_{mn} t - \omega_{mn} t_1 \sin(\pi\, t/t_1)]}{\pi^2 - t_1^2 \omega_{mn}^2} \quad \text{for } 0 \le t \le t_1 \qquad (13.20)$$

$$P(t) = \frac{F_0 \pi\, t_1[\sin \omega_{mn} t - \sin \omega_{mn}(t-t_1)]}{\pi^2 - t_1^2 \omega_{mn}^2} \quad \text{for } t \ge t_1 \qquad (13.21)$$

For the stepped pulse the forcing function $F(t)$ and the convolution integral $P(t)$ are given

$$F(t) = F_0 \qquad 0 \le t \le t_1$$

$$F(t) = 0 \qquad t > t_1$$

$$P(t) = \int_0^t F(\tau)\sin[\omega_{mn}(t-\tau)]d\tau = \frac{F_0}{\omega_{mn}}[1-\cos(\omega_{mn} t)] \quad \text{for } 0 \le t \le t_1 \qquad (13.22)$$

$$P(t) = \frac{F_0}{\omega_{mn}}\{\cos[\omega_{mn}(t-t_1)] - \cos(\omega_{mn} t\} \quad \text{for } t > t_1 \qquad (13.23)$$

For a triangular pulse:

$$F(t) = F_0(1-t/t_1) \qquad 0 \le t \le t_1$$

$$F(t) = 0 \qquad t > t_1$$

$$P(t) = \int_o^t F(\tau)\sin[\omega_{mn}(t-\tau)]d\tau$$

$$= \frac{F_0}{\omega_{mn}}\left[1-\cos(\omega_{mn} t) + \frac{1}{\omega_{mn} t_1}\sin(\omega_{mn} t) - \frac{t}{t_1}\right] \quad \text{for } 0 \le t \le t_1 \qquad (13.24)$$

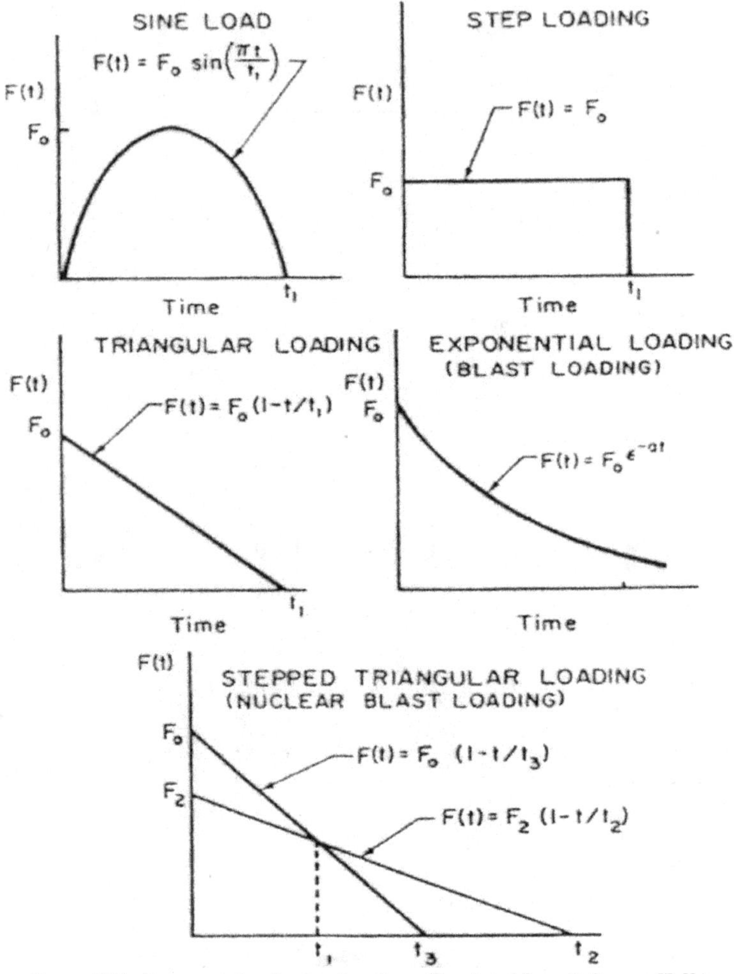

Figure 13.2. Representative forcing functions. (Reprinted from Reference [7.5].)

$$P(t) = F_0\left\{-\frac{1}{\omega_{mn}}\cos(\omega_{mn}t)+\frac{2}{\omega_{mn}^2 t_1}\cos\omega_{mn}\left(t-\frac{t_1}{2}\right)\sin\omega_{mn}\left(\frac{t_1}{2}\right)\right\} \quad \text{for } t > t_1 \tag{13.25}$$

The stepped triangular pulse of Figure 13.2 simulates a nuclear-blast loading [13.1] where the pressure pulse consists of a long-duration phase of several seconds due to the overpressure and a short-duration phase of a few milliseconds due to the shock wave reflection. The short-duration phase has twice the pressure of the long-duration phase.

$$F(t) = F_0(1 - t/t_3) \qquad 0 \le t \le t_1$$

$$F(t) = F_2(1 - t/t_2) \qquad t \le t \le t_2$$

$$F(t) = 0 \qquad t > t_2$$

$$P(t) = \int_0^t F(\tau)\sin[\omega_{mn}(t-\tau)]d\tau$$

$$= \frac{F_0}{\omega_{mn}}\left[1 - \cos(\omega_{mn}t) + \frac{1}{\omega_{mn}t_3}\sin(\omega_{mn}t) - \frac{t}{t_3}\right] \text{ for } 0 < t \le t_1 \qquad (13.26)$$

$$\begin{aligned} P(t) = F_0 \Bigg\{ & \frac{1}{\omega_{mn}}\left(1 - \frac{t}{t_3}\right)\cos[\omega_{mn}(t-t_1)] - \frac{1}{\omega_{mn}}\cos(\omega_{mn}t) \\ & - \frac{1}{\omega_{mn}^2 t_3}\sin[\omega_{mn}(t-t_1)] + \frac{1}{\omega_{mn}^2 t_3}\sin(\omega_{mn}t)\Bigg\} \\ & + F_2\Bigg\{\frac{1}{\omega_{mn}}\left(1 - \frac{t}{t_2}\right) - \frac{1}{\omega_{mn}}\left(1 - \frac{t}{t_2}\right)\cos[\omega_{mn}(t-t_1)] \\ & + \frac{1}{\omega_{mn}^2 t_2}\sin[\omega_{mn}(t-t_1)]\Bigg\} \qquad \text{for } t_1 \le t \le t_2 \end{aligned} \qquad (13.27)$$

$$\begin{aligned} P(t) = F_0 \Bigg[& \frac{1}{\omega_{mn}}\left(1 - \frac{t_1}{t_3}\right)\cos[\omega_{mn}(t-t_1)] - \frac{1}{\omega_{mn}}\cos(\omega_{mn}t) \\ & + \frac{1}{\omega_{mn}^2 t_3}\{\sin(\omega_{mn}t) - \sin[\omega_{mn}(t-t_1)]\}\Bigg] \\ & + F_2\Bigg[\frac{1}{\omega_{mn}}\left(\frac{t_1}{t_2} - 1\right)\cos[\omega_{mn}(t-t_1)] \\ & - \frac{1}{\omega_{mn}^2 t_2}\{\sin[\omega_{mn}(t-t_2)] - \sin[\omega_{mn}(t-t_1)]\}\Bigg] \qquad \text{for } t > t_2 \end{aligned} \qquad (13.28)$$

Lastly, the exponential pulse of Figure 13.2 may be used to simulate a high explosive (non-nuclear) blast loading when the decay parameter α is empirically determined to fit the pressure pulse of the actual blast. The equations are

$$F(t) \quad F_0 e^{-at} \qquad 0 \leq t \leq \infty$$

$$P(t) \quad \int_0^t F(\tau)\sin[\omega_{mn}(t-\tau)d\tau$$

$$\frac{F_0[\omega_{mn}e^{-at} + \alpha\sin(\omega_{mn}t) - \omega_{mn}\cos(\omega_{mn}t)]}{\alpha^2 + \omega_{mn}^2} \quad \text{for } t > 0 \qquad (13.29)$$

It should be noted that although the forcing-function equations given above are used herein to investigate the dynamic response of a composite material plate, these equations are useful for many other purposes.

Dobyns [7.5] concludes that the equations presented in this section allow one to analyze a composite material panel subjected to dynamic loads with only a little more effort than is required for the same panel subjected to static loads. He stated that one does not have to rely upon approximate design curves or arbitrary dynamic-magnification load factors.

With the information presented to this point, the necessary equations for the study of a composite plate without the various couplings, but that include transverse shear deformation, have been developed. The plate may be subjected to various static loads and a variety of dynamic loads. These loads and solutions can be used singly-or superimposed-to describe a complex dynamic input. (Those same load functions of this Section, since they are functions of time only, can be used for beams, shells and many other structural configurations.) With the solutions for $\overline{\alpha}(x,y)$, $\overline{\beta}(x,y)$ and w, maximum deflections and stresses can be determined for deflection stiffness-critical and strength-critical structures.

For further reading on impact load effects on composite structures, see the text by Abrate [13.4], and research by Bert [13.5].

13.5 Vibration Damping

Damping of composite structures is clearly beyond the scope of this basic textbook. However, it would be a mistake not to mention that composite structures incorporate significant damping through the intrinsic properties of composite materials compared to metallic structures.

For the study of vibration damping, the text by Nashif, Jones, and Henderson [7.4] is excellent. Also, the text by Inman [13.6] concerning vibrations, vibrations damping, control, measurement and stability provides much needed information useful to the study of composite material structural vibrations.

13.6 References

13.1. Leissa, A.W. (1973) The Free Vibration of Rectangular Plates, *Journal of Sound and Vibration*, Vol. 31, No. 3, pp. 257-293.
13.2. Wu, C-I and Vinson, J.R. (1969) On the Nonlinear Oscillations of Plates Composed of Composite Materials, *Journal of Composite Materials*, Vol. 3, pp. 548-561, July.
13.3. Wu, C-I and Vinson, J.R. (1971) Non-linear Oscillations of Laminated Specially Orthotropic Plates with Clamped and Simply Supported Edges, *Journal of the Acoustical Society of America*, Vol. 49, No. ? (Part 2), pp. 1561-1567, May.
13.4. Abrate, S. (1998) *Impact on Composite Structures*, Cambridge University Press.
13.5. Bert, C.W. (1991) Research on Dynamic Behavior of Composite and Sandwich Plates – V: Part 1, *Shock and Vibration Digest*, Vol. 23, pp. 3-14.
13.6. Inman, D.J. (1989) *Vibration with Control Measurement and Stability*, Prentice Hall, Englewood Cliffs, New Jersey.

13.7 Problems

13.1. Find the fundamental natural frequency in Hz (cps) for each of the plates of Problem 12.1, if the mass density of the material is

$$\rho_m = 1.8 \times 10^{-4} \frac{\text{lb.} - \text{sec.}^2}{\text{in.}^4}$$

13.2. Consider the plate of problem 11.1. If it is simply supported on all four edges, what is its fundamental natural frequency in cycles per seconds neglecting transverse shear deformation?

13.3. What is the fundamental natural frequency of the plate of Problem 12.2 in Hz (i.e. cycles per second), using classical plate theory? The weight density of the composite is 0.06 lb./in.^{3}.

13.4. (a) The plate of Problems 12.2, 12.3 and 13.3 will be used in an environment in which it will be exposed to a sinusoidal frequency of 6 Hz. Is it likely there will be a vibration problem requiring detailed study? Why?
(b) What about 12 Hz? Why?

13.5. For a plate or panel, what are the four ways in which it may fail or become subjected to a condition which may terminate its usefulness?

13.6. Determine the fundamental natural frequency in Hz (cycles per second) for the panel of Problem 12.9 made of four plys, unidirectionally oriented **(all 0°** plys).

13.7. What is the fundamental natural vibration frequency in Hz for the plate of Problem 12.6. Assume a weight density for the composite to be $\rho_w = 0.06$ lb./in.^{3}.

13.8. Suppose the plate of Problem 13.7 were designed to be subjected to a continuing harmonic forcing function at:

(a) 38 to 48 Hz
(b) 10 Hz.

Would there be a problem structurally with this due to dynamic effects? Why?

13.9. What is the fundamental natural frequency of the plate of Problem 12.7 in Hz., using classical plate theory?

13.10. If the fundamental natural frequency were calculated including the effects of transverse shear deformation, would that frequency be higher, lower or equal to the frequency calculated in Problem 13.9 above?

13.11. A square plate, simply supported on all four edges is composed of GY70/339 graphite epoxy. If this square plate is made of four plys with the A and D matrix values shown below, which stacking sequence would you choose for a design to have the highest fundamental natural frequency? Which stacking sequence has the lowest fundamental natural frequency?

GY70/339 graphite epoxy composite

[A] matrix [D] matrix

a. Unidirectional four ply

$$\begin{bmatrix} 842 & 5.46 & 0 \\ 5.46 & 17.6 & 0 \\ 0 & 0 & 12 \end{bmatrix} \times 10^3 \text{ lb./in.} \qquad \begin{bmatrix} 28.1 & 0.182 & 0 \\ 0.182 & 0.588 & 0 \\ 0 & 0 & 0.4 \end{bmatrix} \text{ lb.-in.}$$

b. Crossply $[0^\circ, 90^\circ, 90^\circ, 0^\circ]$ four ply

$$\begin{bmatrix} 430 & 5.46 & 0 \\ 5.46 & 430 & 0 \\ 0 & 0 & 12 \end{bmatrix} \times 10^3 \text{ lb./in.} \qquad \begin{bmatrix} 24.6 & 0.182 & 0 \\ 0.182 & 4.02 & 0 \\ 0 & 0 & 0.4 \end{bmatrix} \text{ lb.-in.}$$

c. Angle ply $[\pm 45^\circ]_S$, four ply

$$\begin{bmatrix} 230 & 206 & 0 \\ 206 & 230 & 0 \\ 0 & 0 & 212 \end{bmatrix} \times 10^3 \text{ lb./in.} \qquad \begin{bmatrix} 7.67 & 6.87 & 5.15 \\ 6.87 & 7.67 & 5.15 \\ 5.15 & 5.15 & 7.07 \end{bmatrix} \text{ lb.-in.}$$

13.12. Consider a square plate with length and width of 12 inches, and thickness of $h = 0.020''$, composed of graphite/epoxy whose stiffness matrix properties are given in Problem 13.11a. Calculate the natural frequency f_{23} in cycles per second (i.e., $m = 2, n = 3,\ f_{mn} = \omega_{mn} / 2\pi$).

13.13. Does a natural frequency of vibration of a plate clamped on all four edges, subjected to a lateral distributed load $p(x, y) = p_0$, where p_0 is a constant, depend on the value of the load p_0?

13.14. Consider a rectangular panel simply supported on all four edges. The panel measure $a = 25''$, $b = 10''$, where $0 \le x \le a, 0 \le y \le b$. The laminated plate is composed of a unidirectional boron/aluminum metal matrix composite with the following properties; where the fibers are oriented in the long direction.

E_{11}	32×10^6 psi	ρ	0.0915 lb./in.^{3}
E_{22}	20×10^6 psi	σ_{lu}	250,000 psi
G_{12}	8×10^6 psi	v_f	50%
ply thickness h_k	0.007 in.	ν_{12}	0.35

Determine the fundamental natural frequency in Hertz (cycles per second) for the panel if the laminate is made of four laminae, all oriented at 0°.

13.15. Consider a rectangular composite plate whose stiffness matrices are given in Problem 11.1. The plate is 15 inches wide, 60 inches long, simply supported on all four edges, is 0.020 inches thick, whose weight density is 0.06 lbs./in.^{3}, and the fibers are in the longer direction.

(a) If an in-plane compressive load, $\boldsymbol{N_x}$, is applied in the direction parallel to the longer dimension, what is the critical buckling load, $N_{x_{cr}}$, using classical plate theory?
(b) Using classical plate theory what is the fundamental natural frequency in Hz.?
(c) If transverse shear deformation effects were included in the above calculations would the buckling load and fundamental natural frequency be higher, the same, or lower?

CHAPTER 14

ENERGY METHODS FOR COMPOSITE MATERIAL STRUCTURES

14.1 Introduction

Many composite material structures not only involve anisotropy, multilayer considerations and transverse shear deformation, but also have hygrothermal effects. Because of these complications, plus any caused by complicated loads, obtaining approximate solutions through the use of energy methods may be the best way to proceed, if not the only way, rather than expending much time and effort in the hope of obtaining an analytical solution.

As stated in Chapter 8, dealing with isotropic plates, with energy methods, one can always obtain a good approximate solution, no matter what the complications caused by the structural configuration, the loads or the boundary conditions.

14.2 A Rectangular Composite Material Plate Subjected to Lateral and Hygrothermal Loads

A detailed study by Sloan [14.1] shows clearly what is involved in analyzing composite panels to accurately account for the effects of anisotropy, transverse shear deformation, thermal and hygrothermal effects. The results are also shown in the text by Vinson and Sierakowski [1.7].

The stress-strain equations to be considered are given by Equation (10.43), wherein the $\overline{Q}_{ij}$ are given by Equation (10.44). The strain-displacement equations are given by Equation (10.48), the form of the panel displacements by Equation (10.49), from which Equation (10.50) results. By neglecting ε_z and σ_z the constitutive relations for the laminate reduce to Equation (10.66). Sloan [14.1] has shown that even for this problem, the ε_z and σ_z can be ignored.

Employing the Theorem of Minimum Potential Energy, Equation (8.1), for the plate under discussion it is seen that summing the strain-energy-density functions for each lamina across the N laminae that comprise the plate gives the total potential energy as

$$V = \frac{1}{2}\sum_{k=1}^{N}\iint_A \int_{h_{k-1}}^{h_k} \{\sigma_x[\varepsilon_x - \alpha_x \Delta T - \beta_x \Delta m]$$

$$+\sigma_y[\varepsilon_y - \alpha_y \Delta T - \beta_y \Delta m] + \sigma_{xz}[2\varepsilon_{xz}]$$

$$+\sigma_{yz}[2\varepsilon_{yz}] + \sigma_{xy}\left[2\left(\varepsilon_{xy} - \alpha_{xy}\Delta T - \beta_{xy}\Delta m\right)\right]\} \mathrm{d}z\mathrm{d}A \tag{14.1}$$

$$-\iint_A p(x,y)w(x,y)\mathrm{d}A$$

Here A refers to the planform area of the plate whose dimensions are $0 \leq x \leq a, 0 \leq y \leq b$ and $h/2 \leq z \leq -h/2$. It is noted that the strains used in the strain-energy relations are the isothermal strains, hence one notes the differences between total strain and the thermal and hygrothermal strains in Equations (4.2) and (10.30).

Now, substituting the constitutive Equations (10.66) and the strain-displacement relations (10.50) into Equation (14.1) results in the following:

$$
\begin{aligned}
V = \iint_A \Bigg\{ & \frac{A_{11}}{2}\left(\frac{\partial u_0}{\partial x}\right)^2 + B_{11}\frac{\partial u_0}{\partial x}\frac{\partial \overline{\alpha}}{\partial x} + \frac{D_{11}}{2}\left(\frac{\partial \overline{\alpha}}{\partial x}\right)^2 + A_{12}\frac{\partial u_0}{\partial x}\frac{\partial v_0}{\partial y} \\
& + B_{12}\left[\frac{\partial u_0}{\partial x}\frac{\partial \overline{\beta}}{\partial y} + \frac{\partial v_0}{\partial y}\frac{\partial \overline{\alpha}}{\partial x}\right] + D_{12}\frac{\partial \overline{\beta}}{\partial y}\frac{\partial \overline{\alpha}}{\partial x} + A_{16}\left[\frac{\partial u_0}{\partial x}\frac{\partial u_0}{\partial y} + \frac{\partial u_0}{\partial x}\frac{\partial v_0}{\partial x}\right] \\
& + D_{16}\left[\frac{\partial \overline{\alpha}}{\partial x}\frac{\partial \overline{\alpha}}{\partial y} + \frac{\partial \overline{\alpha}}{\partial x}\frac{\partial \overline{\beta}}{\partial x}\right] + B_{16}\left[\frac{\partial u_0}{\partial x}\frac{\partial \overline{\alpha}}{\partial y} + \frac{\partial u_0}{\partial x}\frac{\partial \overline{\beta}}{\partial x} + \frac{\partial u_0}{\partial y}\frac{\partial \overline{\alpha}}{\partial x} + \frac{\partial v_0}{\partial x}\frac{\partial \overline{\alpha}}{\partial x}\right] \\
& + \frac{A_{22}}{2}\left(\frac{\partial v_0}{\partial y}\right)^2 + B_{22}\frac{\partial v_0}{\partial y}\frac{\partial \overline{\beta}}{\partial y} + \frac{D_{22}}{2}\left(\frac{\partial \overline{\beta}}{\partial y}\right)^2 + A_{26}\left[\frac{\partial v_0}{\partial y}\frac{\partial u_0}{\partial y} + \frac{\partial v_0}{\partial y}\frac{\partial v_0}{\partial x}\right] \\
& + B_{26}\left[\frac{\partial v_0}{\partial y}\frac{\partial \overline{\alpha}}{\partial y} + \frac{\partial v_0}{\partial y}\frac{\partial \overline{\beta}}{\partial x} + \frac{\partial u_0}{\partial y}\frac{\partial \overline{\beta}}{\partial y} + \frac{\partial v_0}{\partial x}\frac{\partial \overline{\beta}}{\partial y}\right] + D_{26}\left[\frac{\partial \overline{\alpha}}{\partial y}\frac{\partial \overline{\beta}}{\partial y} + \frac{\partial \overline{\beta}}{\partial x}\frac{\partial \overline{\beta}}{\partial y}\right] \\
& + A_{45}\left[\overline{\alpha}\overline{\beta} + \overline{\alpha}\frac{\partial w}{\partial y} + \overline{\beta}\frac{\partial w}{\partial x} + \frac{\partial w}{\partial x}\frac{\partial w}{\partial y}\right] + A_{55}\left[\frac{\overline{\alpha}^2}{2} + \overline{\alpha}\frac{\partial w}{\partial x} + \frac{1}{2}\left(\frac{\partial w}{\partial x}\right)^2\right] \\
& + A_{44}\left[\frac{\overline{\beta}^2}{2} + \overline{\beta}\frac{\partial w}{\partial y} + \frac{1}{2}\left(\frac{\partial w}{\partial y}\right)^2\right] + A_{66}\left[\frac{1}{2}\left(\frac{\partial u_0}{\partial y}\right)^2 + \frac{\partial u_0}{\partial y}\frac{\partial v_0}{\partial x} + \frac{1}{2}\left(\frac{\partial v_0}{\partial x}\right)^2\right] \\
& + B_{66}\left[\frac{\partial u_0}{\partial y}\frac{\partial \overline{\alpha}}{\partial y} + \frac{\partial u_0}{\partial y}\frac{\partial \overline{\beta}}{\partial x} + \frac{\partial v_0}{\partial x}\frac{\partial \overline{\alpha}}{\partial y} + \frac{\partial v_0}{\partial x}\frac{\partial \overline{\beta}}{\partial x}\right] \\
& + D_{66}\left[\frac{1}{2}\left(\frac{\partial \overline{\alpha}}{\partial y}\right)^2 + \frac{\partial \overline{\alpha}}{\partial y}\frac{\partial \overline{\beta}}{\partial x} + \frac{1}{2}\left(\frac{\partial \overline{\beta}}{\partial x}\right)^2\right] - \frac{\partial u_0}{\partial x}\left(N_{1j}^T + N_{1j}^m\right) \\
& - \frac{\partial v_0}{\partial y}\left(N_{2j}^T + N_{2j}^m\right) - \left(\frac{\partial u_0}{\partial y} + \frac{\partial v_0}{\partial x}\right)\left(N_{6j}^T + N_{6j}^m\right) - \frac{\partial \overline{\alpha}}{\partial x}\left(M_{1j}^T + M_{1j}^m\right) \\
& - \frac{\partial \overline{\beta}}{\partial y}\left(M_{2j}^T + M_{2j}^m\right) - \left(\frac{\partial \overline{\alpha}}{\partial y} + \frac{\partial \overline{\beta}}{\partial x}\right)\left(M_{6j}^T + M_{6j}^m\right) \\
& + T^* + M^* + \overline{MT}^* - p(x,y)w(x,y)\Bigg\}\, \mathrm{d}A
\end{aligned}
\tag{14.2}
$$

In the above, all quantities (except displacements) are defined by Equations (10.58) through (10.61), Equations (10.63) through (10.65) and the following. Note that the α_i and β_i (unbarred) below are the coefficients of thermal and hygrothermal expansion respectively:

$$
T^* = \frac{1}{2}\sum_{k=1}^{N}\int_{h_{k-1}}^{h_k}\left[\overline{Q}_{ij}\right]_k\left[\alpha_i\right]_k\left[\alpha_j\right]_k(\Delta T(z,t))^2\,\mathrm{d}z
$$

$$M^* = \frac{1}{2}\sum_{k=1}^{N}\int_{h_{k-1}}^{h_k}\left[\overline{Q}_{ij}\right]_k\left[\beta_i\right]_k\left[\beta_j\right]_k(\Delta m(z,t))^2\, dz$$

$$\overline{MT}^* = \sum_{k=1}^{N}\int_{h_{k-1}}^{h_k}\left[\overline{Q}_{ij}\right]_k\left[\alpha_i\right]\left[\beta_j\right]\Delta T(z,t)\Delta m(z,t)\, dz$$

As written Equation (14.2) provides the expression to use in the analysis of composite material rectangular plates of constant thickness, wherein one uses the appropriate values of the [*A*], [*B*] and [*D*] stiffness matrix quantities given by (10.58), (10.59), and (10.63).

Equation (14.2) is the most general formulation and it is seen that without the surface load term there are 30 terms to represent the strain energy in the composite material panel. Referring to Equation (10.66) and the ensuing discussion it is seen that if the laminate has no stretching-shearing coupling $(A_{16} = A_{26} = 0)$ then two terms would be dropped; if no twisting-stretching coupling $(B_{16} = B_{26} = 0)$ two more would be dropped; likewise two more are dropped if there were no bending-twisting coupling $(D_{16} = D_{26} = 0)$. If the laminate were symmetric about the mid-plane, a very common construction, then five B_{ij} terms would be canceled out, because there would be no bending-stretching coupling.

One can now proceed to solve any problem involving composite material plates subjected to any set of loads, thermal and moisture conditions through employing (14.2). For trial functions one can use the appropriate functions of Section 14.6 below in each of the *x* and *y* directions. Now to proceed with a number of examples to illustrate the use. One could start by redoing the examples of Sections 8.3, 8.4, 8.5 and 8.7 for a composite material plate, but they will not be repeated herein. Rather, the following are presented.

14.3 In-Plane Shear Strength Determination of Composite Materials in Composite Panels

In this illustrative problem using the Theorem of Minimum Potential Energy, consider a simple test procedure to determine the in-plane shear strength of laminated composite materials, as well as other orthotropic and isotropic advanced material systems; see the recent publication by Vinson [14.2]. The test apparatus shown in Figure 14.1 is simple, inexpensive, and the flat rectangular plate test specimen is not restricted in size or aspect ratio. In addition to its use for laminated composite materials, the test can also be used for foam core sandwich panels. In sandwich panels the tests can be used to determine the in-plane shear strengths of the faces, the core and/or the adhesive bond between face and core. The shear stresses developed vary linearly in the thickness direction and are constant over the entire planform area.

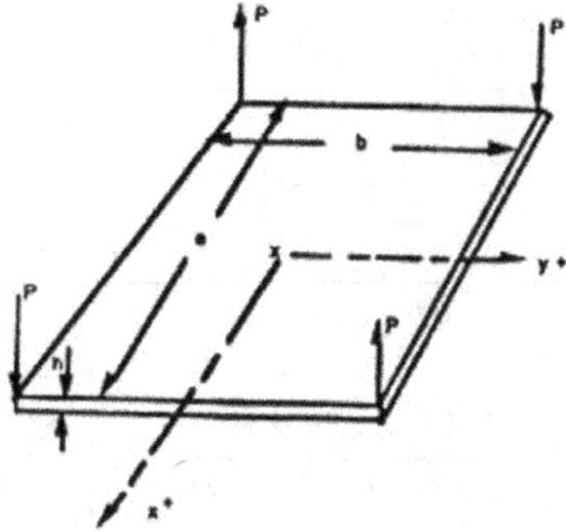

Figure 14.1. Test method and geometry for the determination of in-plane shear strength. (Reprinted from Reference [14.2].)

Consider a panel of the material to be tested to be rectangular in planform, with dimensions a in the x-direction and b in the y-direction. The panel is of constant thickness h and if anisotropic, the material principal axes 1-2, should be aligned with the structural axes x-y. The rectangular panel is placed in a test machine such that the loads P are applied at each of the four corners as shown in Figure 14.1. The loads P are recorded as applied until the test specimen fails. As a result, the in-plane shear strength of the material is easily calculated from the equations developed below.

The Theorem of Minimum Potential Energy is utilized to obtain solutions. It is assumed that the material system is specially orthotropic, i.e., the material axes (1-2) are aligned with the structural axes (x-y). It is further assumed that the test specimen is mid-plane symmetric, i.e., no bending-stretching coupling, $B_{ij} = 0$, no stretching-shearing coupling $(A_{16} = A_{26} = 0)$, and no bending-twisting coupling $(D_{16} = D_{26} = 0)$. The following methods of analysis can be altered to include laminates or sandwich panels asymmetric to the mid-surface, i.e., $B_{ij} \neq 0$.

The potential energy V for the structure and loading of Figure 14.1 is given by Equation (14.2), which for this case is

$$
\begin{aligned}
V = \int_{-a/2}^{a/2}\int_{-b/2}^{b/2} & \left\{ \frac{A_{11}}{2}\left(\frac{\partial u_0}{\partial x}\right)^2 + \frac{D_{11}}{2}\left(\frac{\partial \overline{\alpha}}{\partial x}\right)^2 \right. \\
& + A_{12}\frac{\partial u_0}{\partial x}\frac{\partial \mathrm{v}_0}{\partial y} + D_{12}\frac{\partial \overline{\beta}}{\partial y}\frac{\partial \overline{\alpha}}{\partial x} + \frac{A_{22}}{2}\left(\frac{\partial \mathrm{v}_0}{\partial y}\right)^2 + \frac{D_{22}}{2}\left(\frac{\partial \overline{\beta}}{\partial y}\right)^2 \\
& + A_{66}\left[\frac{1}{2}\left(\frac{\partial u_0}{\partial y}\right)^2 + \frac{\partial u_0}{\partial y}\frac{\partial \mathrm{v}_0}{\partial x} + \frac{1}{2}\left(\frac{\partial \mathrm{v}_0}{\partial x}\right)^2\right] \\
& \left. + D_{66}\left[\frac{1}{2}\left(\frac{\partial \overline{\alpha}}{\partial y}\right)^2 + \frac{\partial \overline{\alpha}}{\partial y}\frac{\partial \overline{\beta}}{\partial y} + \frac{1}{2}\left(\frac{\partial \overline{\beta}}{\partial x}\right)^2\right] \right\} \mathrm{d}y\mathrm{d}x \\
& - Pw\left(\tfrac{a}{2},\tfrac{b}{2}\right) - Pw\left(-\tfrac{a}{2},-\tfrac{b}{2}\right) \\
& + Pw\left(\tfrac{a}{2},-\tfrac{b}{2}\right) + Pw\left(-\tfrac{a}{2},\tfrac{b}{2}\right)
\end{aligned}
\tag{14.3}
$$

where u_0 and v_0 are the mid-plane in-plane displacements in the x and y-directions, respectively, and w is the lateral displacement. All displacements are positive in the positive coordinate direction. The means to portray the concentrated loads at the corners of the plate is clearly seen in the last two lines of Equation (14.3).

Neglecting transverse-shear deformation effects, then the strain-displacement relations are given by (2.27) and (2.28). The in-plane and flexural stiffness matrix quantities are given by (10.58) and (10.63). Note that these stiffness quantities are valid for both composite laminate and sandwich construction.

To insure complete generality, the following forms for the displacements are assumed as trial functions, where the numbered coefficients are constants to be determined by boundary conditions and the variational operation:

$$
\begin{aligned}
w(x,y) = {} & C_1 + C_2 x + C_3 y + C_4 x^2 + C_5 xy + C_6 y^2 \\
& + C_7 x^3 + C_8 x^2 y + C_9 xy^2 + C_{10} y^3 + C_{11} x^4 \\
& + C_{12} x^3 y + C_{13} x^2 y^2 + C_{14} xy^3 + C_{15} y^4 \\
u_0(x,y) = {} & A_1 + A_2 x + A_3 y + A_4 x^2 + A_5 xy + A_6 y^2 \\
\mathrm{v}_0(x,y) = {} & B_1 + B_2 x + B_3 y + B_4 x^2 + B_5 xy + B_6 y^2
\end{aligned}
\tag{14.4}
$$

When, as an analyst, one has trouble deciding on a suitable trial function for the deflection, include as many terms in the polynomial such that the highest power in the polynomial is equal to the highest order in the differential equations.

The following physical conditions are used to simplify the above assumed functions:

$$w(0,0) = 0; \qquad u_0(0,0) = 0;$$

$$v_0(0,0) = 0; \quad \frac{\partial w}{\partial x}(0,0) = 0; \quad \frac{\partial w}{\partial y}(0,0) = 0;$$

$$N_x = N_{xy} = 0 \text{ on } x = \pm(a/2) \text{ edges}; \quad M_x = M_{xy} = 0 \text{ on } x = \pm(a/2) \text{ edges};$$

$$N_y = N_{xy} = 0 \text{ on } y = \pm(b/2) \text{ edges}; \quad M_y = M_{xy} = 0 \text{ on } y = \pm(b/2) \text{ edges};$$

$$w(a/2, b/2) = w(-a/2, -b/2) = -w(a/2, -b/2) = -w(-a/2, b/2)$$

The result of satisfying all of the above is that the trial functions given by (14.4) are reduced to the following:

$$w(x,y) = C_5 xy$$

$$u_0(x,y) = A_2 x + A_3 y \qquad v_0(x,y) = B_2 x + B_3 y \tag{14.5}$$

Substituting Equation (14.5) into (14.3), and setting the variation equal to zero ($\delta V = 0$) results in the following relationships:

$$C_5 = \frac{P}{4D_{66}}, \quad A_3 = -B_2, \quad A_2 = B_3 = 0$$

Therefore:

$$w(x,y) = \frac{P}{4D_{66}} xy, \quad u_0(x,y) = A_3 y, \quad v_0(x,y) = -Ax \tag{14.6}$$

From Equation (14.6) it is seen that no curvature exists in the loaded panel, and that if the panel is of monocoque construction or a laminate in which each lamina is oriented the same as all other laminae, then $D_{66} = \frac{G_{12}h^3}{12}$, where G_{12} is the in-plane shear modulus of the material, and h is the panel thickness.

For a laminated composite plate the in-plane shear stress for the kth lamina is given by

$$(\sigma_{xy})_k = \left[2\bar{Q}_{66}\right]_k \left[\varepsilon_{xy}^0 + \kappa_{xy} z\right] = -\frac{\left[2\bar{Q}_{66}\right]_k Pz}{4D_{66}} \tag{14.7}$$

since for this test, from Equation (14.6), $\varepsilon_{xy}^0 = 0$ and $\kappa_{xy} = -C_5 = -P/4D_{66}$.

In Equation (14.7), for a specially orthotropic material $\bar{Q}_{66} = G_{12}$, the in-plane shear modulus of the material, and D_{66} is given by

$$D_{66} = \frac{1}{3}\sum_{k=1}^{N}\left(\bar{Q}_{66}\right)_k\left(h_k^3 - h_{k-1}^3\right)$$

Equation (14.7) provides an easy way to calculate the shear strength of the failed material simply by measuring the load P at failure, and the location z, i.e., the distance from the midsurface of the panel, of the initial failure site. Likewise, if one is only interested in overall panel in-plane shear strength, then knowing the load P at panel failure, and using $z = h/2$, provides the "panel" in-plane shear strength.

For instance, if the plate is an isotropic single-layer material then

$$\sigma_{xy}(\pm h/2) = \mu\frac{3P}{h^2} \text{ or } \left|\sigma_{xy}\right| = \frac{3P}{h^2} \tag{14.8}$$

Equation (14.8) is also applicable if the plate is composed of an orthotropic single-layer material. In either case the shear modulus is not needed to calculate the in-plane shear strength.

14.4 Cantilevered Anisotropic Composite Plate Subjected to a Uniform Lateral Load

In this example, a fin on a flight vehicle is studied to determine what angle θ of the fiber orientation would produce the greatest angle of twist, if the fins were of a unidirectional composite and the applied lateral load were of a uniform pressure p_0. In other words how does one maximize bending-twisting coupling.

Starting with Equation (14.2), classical theory is assumed, the potential energy expression is given by (14.9) below, where $\bar{\alpha} = -\partial w/\partial x$ and $\bar{\beta} = -\partial w/\partial y$. The planform view of the fin is shown in Figure 14.2, where the positive direction of θ is shown in Figure 10.10.

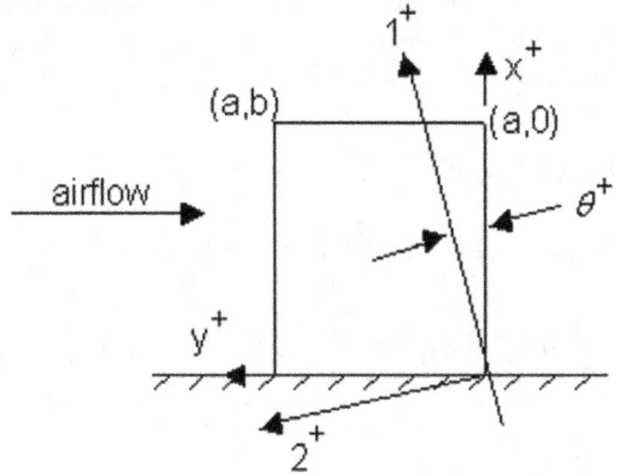

Figure 14.2. Planform view of the fin.

So for classical theory

$$V = \int_0^a \int_0^b \left\{ \frac{D_{11}}{2}\left(\frac{\partial^2 w}{\partial x^2}\right)^2 + D_{12}\left(\frac{\partial^2 w}{\partial x^2}\right)\left(\frac{\partial^2 w}{\partial y^2}\right) + \frac{D_{22}}{2}\left(\frac{\partial^2 w}{\partial y^2}\right)^2 \right.$$
$$+ 2D_{16}\left[\left(\frac{\partial^2 w}{\partial x^2}\right)\left(\frac{\partial^2 w}{\partial x \partial y}\right)\right] + 2D_{26}\left[\left(\frac{\partial^2 w}{\partial x \partial y}\right)\left(\frac{\partial^2 w}{\partial y^2}\right)\right] \tag{14.9}$$
$$\left. + 2D_{66}\left(\frac{\partial^2 w}{\partial x \partial y}\right)^2 - p(x,y)\,w(x,y) \right\} \mathrm{d}x\,\mathrm{d}y$$

Assuming that the stacking sequence is unidirectional, then from (10.58) and (10.63)

$$A_{ij} = \overline{Q}_{ij} h$$
$$D_{ij} = \frac{\overline{Q}_{ij} h^3}{12} \tag{14.10}$$

where h is the total laminate thickness. Assuming $p(x,y) = p_0$ a constant, and that $u_0 = v_0 = 0$, the trial function for the lateral deflection can be assumed to be

$$w(x,y) = (x^4 - 4ax^3 + 6a^2x^2)(A + By) \tag{14.11}$$

where the function of x is from the solution of a cantilever beam subjected to a uniform lateral load, and the function of y allows for an angle of twist (or an angle of attack).

Substituting (14.11) into (14.9) results in

$$
\begin{aligned}
V = \int_0^a \int_0^b & 72D_{11}(x^4 - 4ax^3 + 6a^2x^2 - 4a^3x + a^4)(A^2 + 2ABy + B^2y^2) \\
& + D_{12}(0) + 32D_{66}(x^6 - 6ax^5 + 15a^2x^4 - 18a^3x^3 + 9a^4x^2)B^2 \\
& + \mathrm{D}_{22}(0) \\
& + 96D_{16}(AB + B^2y)(x^5 - 5ax^4 + 10a^2x^3 - 9a^3x^2 + 3a^4x) \\
& + D_{26}(0) \\
& - p_0(x^4 - 4ax^3 + 6a^2x^2)(A + By)\}\,\mathrm{d}y\,\mathrm{d}x
\end{aligned}
\tag{14.12}
$$

Next one performs the integration across the planform area,

$$
\begin{aligned}
V = & \frac{72}{5}D_{11}a^5\left(A^2b + ABb^2 + \frac{B^2b^3}{3}\right) + \frac{144}{7}D_{66}B^2ba^7 \\
& + 16D_{16}a^6bB\left(A + \frac{Bb}{2}\right) - \frac{6}{5}p_0a^5b\left[A + \frac{Bb}{2}\right]
\end{aligned}
\tag{14.13}
$$

One now takes the variation of A and B and sets the result equal to zero.

$$
\begin{aligned}
\delta V = 0 = & \frac{72}{5}D_{11}a^5b\left[2A\delta A + Bb\delta A + Ab\delta B + b^2\frac{2}{3}B\delta B\right] \\
& + \frac{144}{7}D_{66}ba^7\,2B\delta B \\
& + 16D_{16}a^6b\left[\mathbf{B\delta A + A\delta B + bB\delta B}\right] \\
& - \frac{6}{5}p_0a^5b\left[\delta A + \frac{b}{2}\delta B\right] = 0
\end{aligned}
\tag{14.14}
$$

Finally collecting terms

$$\delta V = 0 = \delta A\left[\frac{144}{5}D_{11}a^5bA + \frac{72}{5}D_{11}a^5bBb + 16D_{16}a^6bB - \frac{6}{5}p_0a^5b\right]$$
$$+\,\delta B\left[\frac{72}{5}D_{11}a^5b^2A + \frac{72}{5}D_{11}a^5b^3\left(\frac{2}{3}\right)B\right.$$
$$+\frac{144}{7}D_{66}ba^7 2B + 16D_{16}a^6bA + 16D_{16}a^6b^2B \qquad (14.15)$$
$$\left.-\frac{6}{5}\frac{p_0a^5b^2}{2}\right] = 0$$

one finds the two governing differential equations

$$\frac{144}{5}D_{11}a^5bA + \frac{72}{5}D_{11}a^5b^2B + 16D_{16}a^6bB - \frac{6}{5}p_0a^5b = 0 \qquad (14.16)$$

and

$$\frac{72}{5}D_{11}a^5b^2A + \frac{144}{15}D_{11}a^5b^3B + \frac{288}{7}D_{66}ba^7B$$
$$+16D_{16}a^6bA + 16D_{16}a^6b^2B - \frac{6}{5}\frac{p_0a^5b^2}{2} = 0 \qquad (14.17)$$

From these two equations both A and B can be solved. It is found that A is given by

$$A = -\frac{bB}{2} - \frac{5}{9}\left(\frac{D_{16}}{D_{11}}\right)aB + \frac{p_0}{24D_{11}} \qquad (14.18)$$

The constant B can be written as

$$B = -\frac{105}{378}\frac{(D_{16}/D_{11})(a/b)p_0}{bD_{11}\{\ \}} \qquad (14.19)$$

where $\{\ \} = 756D_{11}^2b^2\left[1 + \frac{3240}{189}\left(\frac{D_{66}}{D_{11}}\right)\left(\frac{a}{b}\right)^2 - \frac{700}{189}\left(\frac{D_{16}}{D_{11}}\right)^2\left(\frac{a}{b}\right)^2\right]$.

From (14.11) it is seen that the two tip deflections are given by:

$$w(a,b) = 3a^4(A + Bb)$$
$$w(a,0) = 3a^4A \qquad (14.20)$$

So the twist or the angle of attack of the plate at the outer ($x = a$) edge is

$$\frac{w(a,b)-w(a,0)}{b} = 3a^4 B = 3a^4\left(-\frac{105}{378}\right)\frac{(D_{16}/D_{11})(a/b)p_0}{bD_{11}\left\{1+\frac{3240}{189}\left(\frac{D_{66}}{D_{11}}\right)\left(\frac{a}{b}\right)^2-\frac{700}{189}\left(\frac{D_{16}}{D_{11}}\right)^2\left(\frac{a}{b}\right)^2\right\}} \tag{14.21}$$

From laminate mechanics where $m = \cos\theta$ and $n = \sin\theta$,

$$\frac{D_{16}}{D_{11}} = \frac{-mn^3Q_{22}+m^3nQ_{11}-mn(m^2-n^2)(Q_{12}+2Q_{66})}{Q_{11}m^4+2(Q_{12}+2Q_{66})m^2n^2+Q_{22}n^4}$$

$$\frac{D_{12}}{D_{11}} = \frac{(Q_{11}+Q_{22}-4Q_{66})m^2n^2+Q_{12}(m^4+n^4)}{Q_{11}m^4+2(Q_{12}+2Q_{66})m^2n^2+Q_{22}n^4}$$

$$\frac{D_{66}}{D_{11}} = \frac{(Q_{11}+Q_{22}-2Q_{12})m^2n^2+Q_{66}(m^2-n^2)^2}{Q_{11}m^4+2(Q_{12}+2Q_{66})m^2n^2+Q_{22}n^4}$$

A further approximation is that since E_{22}/E_{11}, G_{12}/E_{11} and ν_{21} are very small, simply ignore them. Therefore, using these assumptions the tip angle of twist can be written as

$$\frac{w(a,b)-w(a,0)}{b} \approx \frac{\frac{(12)3a^4}{E_{11}bh^3}\left\{-\frac{105}{376}\left(\frac{a}{b}\right)p_0\left(\frac{n}{m}\right)\right\}}{1+\frac{3240}{189}\left(\frac{a}{b}\right)^2\left(\frac{n}{m}\right)^2-\frac{700}{189}\left(\frac{a}{b}\right)^2\left(\frac{n}{m}\right)^2} \tag{14.22}$$

In (14.22), calling the tip angle of twist the Factor of Merit (FM), $\frac{n}{m} = \tan\theta$ and $\left(\frac{n}{m}\right)\left(\frac{a}{b}\right) = \phi$, (14.22) is as follows:

$$\text{FM} = \frac{36a^4p_0}{(378)bh^3E_{11}}\left\{\frac{-105\phi}{\left(1+\frac{2540}{189}\phi^2\right)}\right\} \tag{14.23}$$

To find the optimum value of ϕ,

$$\frac{\partial \mathrm{FM}}{\partial \phi} = 0 = \frac{36a^4 p_0}{(378)bh^3 E_{11}} \frac{\left(1+\frac{2540}{189}\phi^2\right)(-105)+105(\phi)\left(\frac{2(2540)}{189}\phi\right)}{\left(1+\frac{2540}{189}\phi^2\right)^2} \tag{14.24}$$

Finally it is found that

$$\tan\theta = \left(\frac{b}{a}\right)(0.2728) \tag{14.25}$$

For an example, if $a = 62.5\text{mm}$ and $b = 80\text{mm}$, then $\tan\theta = \left(\frac{80}{62.5}\right)(0.2728) = 0.349$

$$\theta = -19.25^{\circ} \tag{14.26}$$

So, the optimum fiber angle for this problem, with this configuration, and these dimensions is 19.25° in the swept back direction, see Figure 14.2. Also, from (14.23),

$$FM = \frac{0.1137 p_0 a^4}{bD_{11}} \tag{14.27}$$

All of this illustrates that regardless of the problem complexity one may always find a solution. One can argue about the assumptions made: with all that were made it is seen that the maximum bending-twisting coupling occurs when the fiber angle is $\theta = 19.25^{\circ}$. Now having found this, one may return to the original problem and negate each assumption and proceed to obtain more complicated solutions. If the optimum angle θ does not change significantly, then the original assumptions were good assumptions.

14.5 Use of the Theorem of Minimum Potential Energy to Determine Buckling Loads in Composite Plates

As a practical example, consider a specially orthotropic composite material plate shown below in Figure 14.3, where each x = constant edge is simply supported, and the y edges are clamped ($y = 0$) and free ($y = b$). This is typical of a flange on a ladder side rail and many other structural applications involving open cross-sections. Consider this to be

Case I:

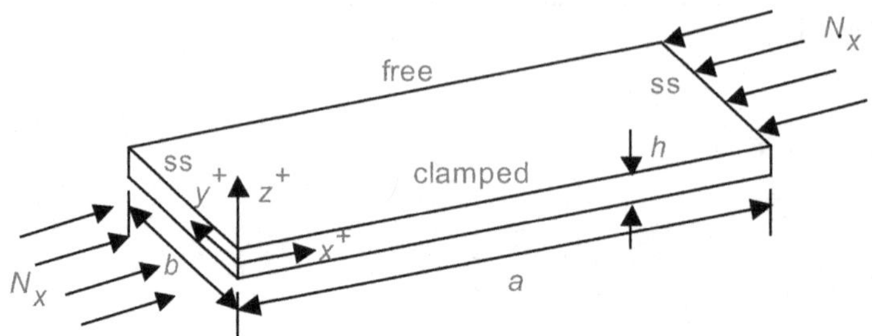

Figure 14.3. Buckling of a Flange Plate

The plate is subjected to a uniform compressive in-plane load of N_x (lbs./in. of width) in the x direction as shown in the Figure 14.3. The Theorem of Minimum Potential Energy is used to determine the critical buckling load, $N_{x_{cr}}$.

From Equation (14.2), and using classical theory i.e. $\overline{\alpha} = -\partial w/\partial x$, $\overline{\beta} = -\partial w/\partial y$, the potential energy expression is:

$$V = \int_0^a \int_0^b \left\{ \frac{D_{11}}{2}\left(\frac{\partial^2 w}{\partial x^2}\right)^2 + D_{12}\left(\frac{\partial^2 w}{\partial x^2}\right)\left(\frac{\partial^2 w}{\partial y^2}\right) + \frac{D_{22}}{2}\left(\frac{\partial^2 w}{\partial y^2}\right)^2 + 2D_{66}\left(\frac{\partial^2 w}{\partial x \partial y}\right)^2 \right\} dydx - \int_0^a \int_0^b \frac{N_x}{2}\left(\frac{\partial w}{\partial x}\right)^2 dxdy \tag{14.28}$$

In this expression the effect of the in-plane load for buckling is given by the latter integral, where the effect of an in-plane load on the lateral deflection is seen.

To use the Theorem of Minimum Potential Energy, let the deflection be a separable solution of the x and y variables

$$w(x,y) = f(x)g(y)$$

Since the edges are simply supported at $x = 0$ and $x = a$, let

$$f(x) = A_n \sin\frac{n\pi x}{a}$$

which satisfies the boundary conditions, and let

$$g(y) \square (y^4 - 4y^3 b + 6b^2 y^2)$$

because this is proportional to the exact deflection function for a cantilevered beam subjected to a uniform lateral load and satisfies the clamped-free boundary conditions. Therefore,

$$w(x,y) = A_n \sin\frac{n\pi x}{a}\left(y^4 - 4y^3 b + 6b^2 y^2\right) \tag{14.29}$$

Substituting (14.29) and its derivatives into (14.28) gives

$$\begin{aligned} V = \int_0^a\int_0^b \Bigg\{ & \left\{\frac{D_{11}}{2}\left[A_n^2 \frac{n^4\pi^4}{a^4}\sin^2\frac{n\pi x}{a}\left(y^4 - 4by^3 + 6b^2y^2\right)\right]\right. \\ & + D_{12}\left\{\left[-A_n\frac{n^2\pi^2}{a^2}\sin\frac{n\pi x}{a}\left(y^4 - 4by^3 + 6b^2y^2\right)\right] x\right. \\ & \left.\left[+A_n \sin\frac{n\pi x}{a}\left(12y^2 - 24by + 12b^2\right)\right]\right\} \\ & + \frac{D_{22}}{2}\left[A_n^2\sin^2\frac{n\pi x}{a}\left(12y^2 - 24by + 12b^2\right)^2\right] \\ & \left. + 2D_{66}\left[A_n^2\frac{n^2\pi^2}{a^2}\cos^2\frac{n\pi x}{a}\left(4y^3 - 12by^2 + 12by^2\right)\right]\right\}\Bigg\} dydx \\ & - \frac{N_x}{2}\int_0^a\int_0^b A_n^2\frac{n^2\pi^2}{a^2}\cos^2\frac{n\pi x}{a}\left(y^4 - 4by^3 + 6b^2y^2\right) dydx \end{aligned} \tag{14.30}$$

Performing the spatial integration the final expression for the potential energy is:

$$\begin{aligned} V = \left(A_n^2\right)\Bigg\{ & \frac{D_{11}\pi^4 n^4 b^9}{2a^4}\left(\frac{104}{45}\right) - \frac{12}{7}D_{12}\frac{n^2\pi^2}{a^2}b^7 + \frac{72}{5}D_{22}b^5 \\ & + \frac{144}{7}D_{66}n^2\pi^2\frac{b^7}{a^2} - N_x n^2\pi^2\frac{b^9}{a^2}\left(\frac{52}{45}\right)\Bigg\} \end{aligned} \tag{14.31}$$

Taking variations of (14.31) for this eigenvalue problem one sees that the bracket must be zero.

$$\delta V = 0 = 2A_n \delta A_n\{\ \} = 0, \text{ so } \{\ \} = 0$$

From the above the critical buckling load $N_{x_{cr}}$ is seen to be

$$N_x = \frac{\pi^2 D_{11}}{b^2}\left[n^2\left(\frac{b}{a}\right)^2 - \frac{(3)(45)}{(7)(13)\pi^2}\frac{D_{12}}{D_{11}} + \frac{(18)(9)}{13\pi^4 n^2}\left(\frac{D_{22}}{D_{11}}\right)\left(\frac{a}{b}\right)^2 + \frac{(36)(45)}{(7)(13)\pi^2}\frac{D_{66}}{D_{11}}\right] \tag{14.32}$$

This is the solution for this case in general, written so that the relative influence for various stiffness terms can be assessed.

First, it is important to determine the wave number n yielding the lowest critical buckling load, because this is the only one of physical significance.

$$\frac{\partial N_x}{\partial n} = 0 = \frac{\pi^2 D_{11}}{b^2}\left[2n\left(\frac{b}{a}\right)^2 - \frac{2(18)(9)}{(13)(\pi^4)}\left(\frac{D_{22}}{D_{11}}\right)\left(\frac{a}{b}\right)^2 \frac{1}{n^3}\right] = 0.$$

Therefore,

$$n = \left[\frac{(18)(9)}{(13)(\pi^4)}\left(\frac{D_{22}}{D_{11}}\right)\right]^{1/4}\left(\frac{a}{b}\right) \tag{14.33}$$

Now, substituting (14.33) into (14.32) results in

$$N_{x_{cr}} = \frac{\pi^2 D_{11}}{b^2}\left\{\left(\frac{18}{13}\right)^{1/2}\frac{3}{\pi^2}\left(\frac{D_{22}}{D_{11}}\right)^{1/2} - \frac{(3)(45)}{(7)(13)\pi^2}\left(\frac{D_{12}}{D_{11}}\right) + \left(\frac{18}{13}\right)^{1/2}\frac{3}{\pi^2}\left(\frac{D_{22}}{D_{11}}\right)^{1/2} + \frac{(36)(45)}{(7)(13)\pi^2}\left(\frac{D_{66}}{D_{11}}\right)\right\} \tag{14.34}$$

Note the lowest critical buckling load for this case given in (14.34) is independent of the integer n, and the aspect ratio (a/b) !. It is also interesting to note that the first and the third terms above are identical. This means that in (14.34) the $\frac{D_{11}}{2}\left(\frac{\partial^2 w}{\partial x^2}\right)^2$ term and the $\frac{D_{22}}{2}\left(\frac{\partial^2 w}{\partial y^2}\right)^2$ term contribute the same amount of strain energy regardless of the aspect ratio. This in turn means that if D_{22} is reduced from one material to another the curvature in the y direction increases such that the strain energy remains the same, and of course the reverse is true.

Another set of boundary conditions are exemplified by:

Case II: $x =$ constant edges clamped, $y =$ constant edges clamped-free

Here the following deflection function can apply

$$w(x,y) = A_n\left[1-\cos\frac{2n\pi x}{2}\right](y^4-4y^3b+6b^2y^2) \tag{14.35}$$

In this case, following the above,

$$N_x = \frac{\pi^2 D_{11}}{b^2}\left\{\frac{n^2}{4}\left(\frac{b}{a}\right)^2 - \frac{135}{(13)(7)\pi^2}\left(\frac{D_{12}}{D_{11}}\right) + \frac{27}{4n^2\pi^4}\left(\frac{a}{b}\right)^2 + \frac{(45)(72)}{13\pi^2}\left(\frac{D_{66}}{D_{11}}\right)\right\} \tag{14.36}$$

Setting the derivative of N_x with respect to n equals zero in the above gives:

$$n = \left[\frac{(27)}{\pi^4}\left(\frac{D_{22}}{D_{11}}\right)\right]^{1/4}\left(\frac{a}{b}\right) \tag{14.37}$$

Therefore, after substituting (14.37) into (14.36) the minimum critical buckling load per unit width is:

$$N_{x\,cr} = \frac{\pi^2 D_{11}}{b^2}\left\{\frac{(27)^{1/2}}{\pi^2}\left(\frac{D_{22}}{D_{11}}\right)^{1/2} - \frac{135}{(13)(7)\pi^2}\left(\frac{D_{12}}{D_{11}}\right) + \frac{(27)^{1/2}}{\pi^2}\left(\frac{D_{22}}{D_{11}}\right)^{1/2} + \frac{(45)(36)}{(13)(7)\pi^2}\left(\frac{D_{66}}{D_{11}}\right)\right\} \tag{14.38}$$

Again the conclusions reached for Case II are identical to those of Case I.

Case III: Let the ends be simply supported and the y edges simple-free

In this case, one can use the following:

$$w(x,y) = A_n y \sin\frac{n\pi x}{a} \tag{14.39}$$

From the analysis analogous to the above, one finds that for the lowest buckling load, $n = 1$, and

$$N_{x\,cr} = \frac{\pi^2 D_{11}}{b^2}\left[\left(\frac{b}{a}\right)^2 + \frac{12}{\pi^2}\left(\frac{D_{66}}{D_{11}}\right)\right] \tag{14.40}$$

Therefore since $n = 1$ is the lowest buckling load, this is overall buckling, no crippling.

14.6 Trial Functions for Various Boundary Conditions for Composite Material Rectangular Plates

To satisfy various boundary conditions for rectangular composite material plates, the lateral deflection $w(x,y)$ may take many forms as long as they are single valued and continuous.

For all combinations of clamped and simply supported boundary conditions Wu and Vinson [13.2, 13.3] provide functions that can be used which involve characteristic beam functions that formulated by Warburton [8.3] and completely characterized by Young and Felgar [3.1].

Another example are the functions used by Causbie and Lagace [14.3] for the study of composite material rectangular plates subjected to in-plane compressive buckling loads in the x-direction. They assumed the following:

1. Simply supported on all edges

$$w = \sum_{m=1}^{6}\sum_{n=1}^{2} A_{mn}\sin\frac{m\pi x}{a}\sin\frac{n\pi y}{b} \tag{14.43}$$

2. SS on loaded edges, SS on one side, free on the other side

$$w = \sum_{m=1}^{3}\sum_{n=1}^{3}\sum_{p=1}^{3}\sin\left(\frac{m\pi x}{a}\right)\left[A_{mn}\sinh\left(\frac{ny}{b}\right) + B_{mp}\sin\left(\frac{p\pi y}{b}\right)\right] \tag{14.44}$$

3. Clamped on loaded ends; simply supported on the sides

$$w = \sum_{n=1}^{6}\sum_{m=1}^{2} A_{mn}\left\{\cos\left[\frac{(m+1)\pi x}{a}\right] - \cos\left[\frac{(m-1)\pi x}{a}\right]\right\}\sin\left(\frac{n\pi y}{b}\right) \tag{14.45}$$

4. Clamped on loaded ends; simply supported on one side; free on the other side

$$w = \sum_{n=1}^{5}\sum_{m=1}^{3}\sum_{p=1}^{3}\left\{\cos\left[\frac{(m+1)\pi x}{a}\right]-\cos\left[\frac{(m-1)\pi x}{a}\right]\right\}x + \left[A_{mn}\sinh\left(\frac{ny}{b}\right)+B_{mp}\sin\left(\frac{p\pi y}{b}\right)\right] \tag{14.46}$$

14.7 Elastic Stability of a Composite Panel Including Transverse Shear Deformation and Hygrothermal Effects

In this section, a general buckling theory is formulated that accounts for the hygrothermal effects as well as transverse shear deformation and all of the couplings discussed in earlier sections as performed by Flaggs [14.4]. Again the Theorem of Minimum Potential Energy is employed, so the strain energy of the plate is identical to that previously used, namely Equation (8.36) for isotropic plates and (14.2) for an anisotropic plate, whether monocoque or sandwich. However, in the absence of a lateral load $p(x, y)$, the last term of Equation (14.2) is absent, but in its place are the effects of in-plane stress resultants N_x, N_{xy}, and N_y, that can cause an elastic instability. Equations (8.36) and (14.2) become:

$$V = \int_R W\mathrm{d}R\,[\text{See (8.36) or (14.2)}] - \iint\left\{N_x\left[\frac{\partial u_0}{\partial x}+\frac{1}{2}\left(\frac{\partial w}{\partial x}\right)^2\right] + N_{xy}\left[\frac{\partial u_0}{\partial y}+\frac{\partial v_0}{\partial x}+\frac{\partial w}{\partial x}\frac{\partial w}{\partial y}\right]+N_y\left[\frac{\partial v_0}{\partial y}+\frac{1}{2}\left(\frac{\partial w}{\partial y}\right)^2\right]\right\}\mathrm{d}A \tag{14.47}$$

These terms are treated in detail in References [6.1] and [6.2], and will not be developed in detail here, but they are standard for isotropic panels as well as composite laminates.

The buckling loads, N_x, N_{xy}, or N_y, are determined by finding the value of the load at which bifurcation occurs, that is, loads at which the plate can be in equilibrium in both a strain configuration (i.e., $w = 0$) and in a slightly deformed $(w \neq 0)$ configuration. This is accomplished through setting the variation of the potential energy V in Equation (14.47) equal to zero, as in Equation (8.2), as shown in Equations (8.44) through (8.47). This operation results in an eigenvalue problem that can be solved for nontrivial solutions that are discrete values of the applied loads. The lowest critical load is the actual physical buckling load. Unlike solving for several natural frequencies, all of which could be important, only the lowest buckling load has any physical meaning.

In solving this, u_0, v_0, w, $\overline{\alpha}$, and $\overline{\beta}$ could be considered as unknowns to be found through the solution. In the following, simply to illustrate an alternative approach, for the rotations $\overline{\alpha}$ and $\overline{\beta}$, the rotations are solved for in terms of the lateral deflection, w. To do this, consider the laminate to be a beam, hence modifying Equation (14.2) to only have an x-dependence. Solving that problem results in three equations:

$$A_{11}\frac{d^2u_0}{dx^2}+B_{11}\frac{d^2\overline{\alpha}}{dx^2}=0 \tag{14.48}$$

$$A_{55}\frac{d\overline{\alpha}}{dx}-A_{55}\frac{d^2w}{dx^2}-N_x\frac{d^2w}{dx^2}=0 \tag{14.49}$$

$$-A_{55}\overline{\alpha}-A_{55}\frac{dw}{dx}+B_{11}\frac{d^2u_0}{dx^2}+D_{11}\frac{d^2\overline{\alpha}}{dx^2}=0 \tag{14.50}$$

These Equations can be solved for $\overline{\alpha}$, the result being as follows. Even though $\overline{\alpha}$ is determined for a beam, a y-dependence is permitted since a plate is being considered.

$$\overline{\alpha}(x,y)=-\frac{\partial w}{\partial x}-\left(\frac{D_{11}A_{11}-B_{11}^2}{A_{11}}\right)\left(\frac{A_{55}-N_x}{A_{55}^2}\right)\frac{\partial^3 w}{\partial x^3} \tag{14.51}$$

In Equation (14.51) note that the first bracketed term is precisely the reduced flexural stiffness term discussed previously. Similarly, assuming a beam in the y-direction, it is found that

$$\overline{\beta}(x,y)=-\frac{\partial w}{\partial y}-\left(\frac{D_{22}A_{22}-B_{22}^2}{A_{22}}\right)\left(\frac{A_{44}-N_y}{A_{44}^2}\right)\frac{\partial^3 w}{\partial y^3} \tag{14.52}$$

Equations (14.51) and (14.52) can now be substituted into Equation (14.47) so that the potential energy expression contains only u_0, v_0, and w as unknown functions.

From Equations (11.79) and (11.80) the simply supported case $S1$ and the clamped case $C1$ are chosen as a good set of examples to investigate various effects. The forms of displacements chosen are given in the paper [14.4] by Flaggs and Vinson.

However, they chose a form for w that differs from the usual assumptions. It must be remembered that an admissible function for the displacement must satisfy at least the geometric boundary conditions (i.e. those involving the lateral displacement and its first derivative) but that function is not unique, hence in this case the following is chosen:

$$\begin{aligned} w(x,y)=\sum_{m=1}^{\infty}\sum_{n=1}^{\infty}W_{mn}&\left\{\cos\left[\frac{(m-1)\pi x}{a}\right]-\cos\left[\frac{(m+1)\pi x}{a}\right]\right\}\\ \times&\left\{\cos\left[\frac{(n-1)\pi y}{b}\right]-\cos\left[\frac{(n+1)\pi y}{b}\right]\right\}\end{aligned} \tag{19.53}$$

The displacements can now be substituted into the Potential Energy expression, Equation (14.47). Then taking the variation with respect to the unknown amplitudes

W_{mn}, Γ_{mn}, and Λ_{mn} results in the eigenvalue problem below, where one can define $N_y = \lambda N_x$ where here λ is the ratio of the applied N_y to the applied N_x.

$$\begin{bmatrix} E_{11} & E_{12} & E_{13} \\ E_{21} & E_{22} & E_{23} \\ E_{31} & E_{32} & E_{33} \end{bmatrix} \begin{Bmatrix} \Gamma_{mn} \\ \Lambda_{mn} \\ W_{mn} \end{Bmatrix} = N_x \begin{bmatrix} 0 & 0 & F_{13} \\ 0 & 0 & F_{23} \\ F_{31} & F_{32} & F_{33} \end{bmatrix} \begin{Bmatrix} \Gamma_{mn} \\ \Lambda_{mn} \\ W_{mn} \end{Bmatrix} + N_x^2 \begin{bmatrix} 0 & 0 & 0 \\ 0 & 0 & 0 \\ 0 & 0 & G_{33} \end{bmatrix} \begin{Bmatrix} \Gamma_{mn} \\ \Lambda_{mn} \\ W_{mn} \end{Bmatrix} \tag{14.54}$$

In (14.54), the E_{ij}, F_{ij}, and G_{ij} quantities are lengthy expressions that can be found through the variational process of setting $\delta V = 0$, in (14.47), and also given in [14.4].

Equation (14.54) can be simplified by uncoupling the third set of simultaneous linear algebraic equations above, by substituting for the Γ_{mn} and Λ_{mn} equations in terms of W_{mn}, resulting in

$$[E]\{\boldsymbol{W_{mn}}\} = N_x[F]\{W_{mn}\} + (N_x)^2[G]\{W_{mn}\} \tag{14.55}$$

For a symmetrically laminated $[0, 45, -45, 90]_{4s}$ T300/5208 graphite-epoxy composite plate with simply supported and clamped boundary conditions, the buckling loads using the generally laminated plate theory are calculated for both steady state and transient hygrothermal conditions. Figure 14.4 and 14.5 show the effects on the applied buckling load, N_x, of different steady-state hygrothermal environments for clamped and simply-supported boundary conditions. The effects of temperature and moisture are both quite clear. Note, along the abscissa, one sees the combination of moisture and temperature that combine to produce buckling with no applied axially compressive mechanical load. It is straightforward to develop analogous plots for various sandwich panels.

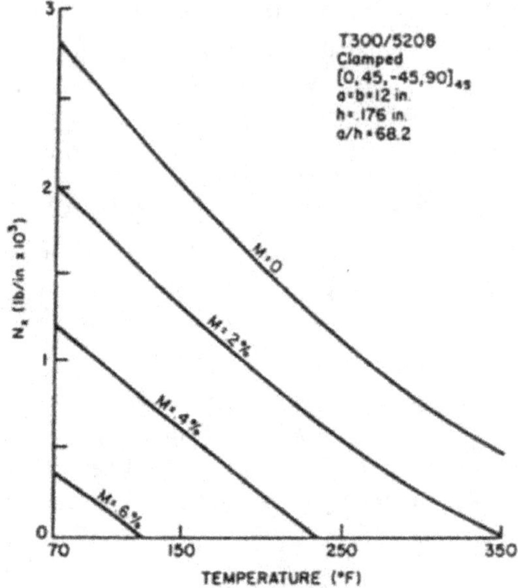

Figure 14.4. Buckling load per unit width as a function of temperature and moisture.

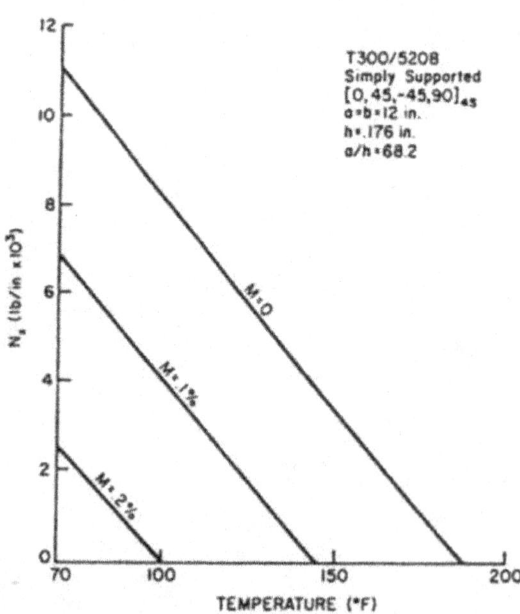

Figure 14.5. Buckling load per unit width as a function of temperature and moisture.

14.8 References

14.1. Sloan, J.G. (1979) The Behavior of Rectangular Composite Material Plates Under Lateral and Hygrothermal Loads, *MMAE Thesis*, University of Delaware, (also, AFOSR-TR-78-1477, July 1978).

14.2. Vinson, J.R. (1997) On Determining the In-Plane Shear Strength of Laminated Composite Materials, *Applied Mechanics in the Americas*, Vol. 4, eds. L.A. Godoy, M. Rysz and L.E. Suarez, pp. 179-182.

14.3. Causbie, S.M. and Lagace, P.A. (1988) Buckling and Final Failure of Graphite/PEEK Stiffened Sections, AIAA Journal, Vol. 26, No. 9, September, pp. 1100-1106.

14.4. Flaggs, D.L. (1978) Elastic Stability of Generally Laminated Composite Plates Including Hygrothermal Effects, MMAE Thesis, University of Delaware.

14.9 Problems

14.1. Consider a composite beam of flexural stiffness bD_{11} clamped on each end and subjected to a uniform lateral load $q(x) = q_0$ lb/in. Using the Theorem of Minimum Potential Energy (MPE) and an assumed deflection function of:

$$w(x) = A\left[1-\cos\frac{2\pi x}{L}\right]$$

where A is the amplitude and L the beam length, find, using classical beam theory:

a. The magnitude and location of the maximum deflection.
b. The magnitude and location of the maximum stresses.
c. Does this form of the deflection function satisfy all of the boundary conditions necessary to use MPE for this problem?

14.2. Consider an orthotropic composite panel shown in Figure 10.11, which has the following boundary conditions:

$x = 0$ clamped
$x = a$ simply supported
$y = 0$ clamped
$y = b$ free

Select a suitable function for the lateral deflection $w(x,y)$ with which to utilize the Theorem of Minimum Potential Energy for this composite plate to determine the deflection of the panel when it is subjected to a laterally distributed load $p(x,y)$.

14.3. Consider a simply supported composite beam subjected to the load

$$q(x) = \bar{q} \sin \frac{\pi x}{L}$$

where $\bar{q}$ is a constant. Using the Theorem of Minimum Potential Energy, and letting the lateral deflection be

$$w(x) = W \sin \frac{\pi x}{L}$$

a. Determine W.
b. What and where is the maximum stress?

14.4. Consider a simply supported composite beam subjected to the load

$$q(x) = \bar{q}\left[1 - \sin \frac{\pi x}{L}\right]$$

where $\bar{q}$ is a constant. Using the Theorem of Minimum Potential Energy, and lettering the lateral deflection be

$$w(x) = W \sin \frac{\pi x}{L}$$

a. Determine W.
b. What and where is the maximum stress?

14.5. The Theorem of Minimum Potential Energy is to be used to analyze an orthotropic composite rectangular composite plate. The plate is midplane symmetric, has no moisture or thermal loading, does include transverse shear deformation effects, has no in-plane displacements, and is subjected to a lateral distributed load $p(x,y)$ only. What is the explicit expression for the potential energy V to use in solving this problem?

14.6. The exact solution for a simply supported beam subjected to a uniform lateral load per unit length of q_0 is:

$$w(x) = \frac{q_0 x}{24EI}\left(x^3 - 2Lx^2 + L^3\right)$$

Using the Theorem of Minimum Potential Energy and a deflection function of

$$w(x) = \sum_{n=1}^{m} A_n \sin\left(\frac{n\pi x}{L}\right)$$

a. Determine A_n.
b. Where and what is the maximum deflection?
c. Where and what is the maximum face stress?
d. Compare these results with the exact values. What are the percentage differences?

CHAPTER 15

GOVERNING EQUATIONS FOR PLATES AND PANELS OF SANDWICH CONSTRUCTION

15.1 Constitutive Equations for a Sandwich Plate

Consider a cross-section of a sandwich structure shown in Figure 15.1 below. The two face thicknesses are designated as t_f, and the core depth is labeled h_c. In this initial example, the faces are identical in thickness and material, whether isotropic or an anisotropic composite.

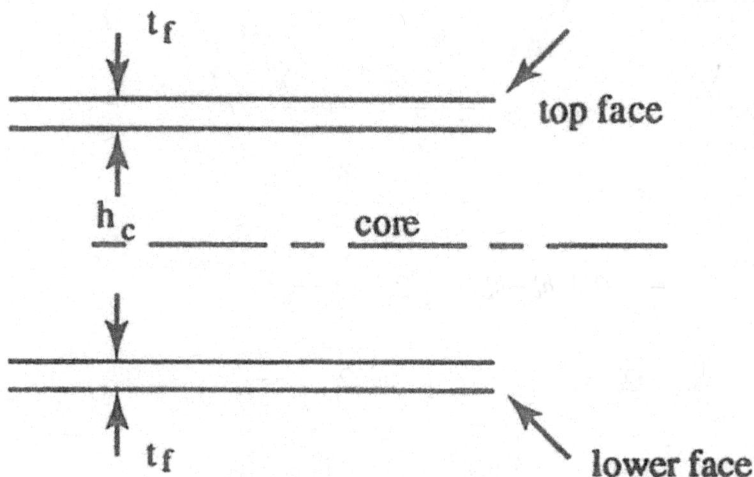

Figure 15.1. Cross section of a symmetric sandwich structure.

For a sandwich plate or panel, the equilibrium equations and the strain-displacement relations remain the same as they are for a monocoque isotropic or composite laminate plate or panel. See (2.14)-(2.18), (1.16)-(1.21), and (10.47)-(10.52). Only the constitutive equations differ from the monocoque structures.

To illustrate how the same methods used to determine the stiffness matrix quantities for a laminated structure given in Figure 10.12 can be used to obtain the stiffness quantities for a sandwich plate, simply define lamina 1 as the lower face, lamina 2 as the core and lamina 3 as the upper face.

Therefore, in this example if the materials are isotropic $\overline{Q}_{ij}$ Q_{ij}, and for face quantities use subscript *f*, and for core quantities use subscript *c*. From (10.58)

$$
\begin{aligned}
A_{ij} &= \sum_{k=1}^{3}\left[\overline{Q}_{ij}\right]_k\left[h_k - h_{k-1}\right] = \sum_{k=1}^{3}\left[Q_{ij}\right]_k\left[h_k - h_{k-1}\right] \\
&= \left[Q_{ij}\right]_f\left[-\frac{h_c}{2} - \left(-\frac{h_c}{2} - t_f\right)\right] + \left[Q_{ij}\right]_c\left[\left(\frac{h_c}{2}\right) - \left(-\frac{h_c}{2}\right)\right] \\
&\quad + \left[Q_{ij}\right]_f\left[\left(\frac{h_c}{2} + t_f\right) - \frac{h_c}{2}\right] = \left[Q_{ij}\right]_f\left[t_f\right] + \left[Q_{ij}\right]_c h_c + \left[Q_{ij}\right]_f\left[t_f\right] \\
&= \left[Q_{ij}\right]_f\left(2t_f\right) + \left[Q_{ij}\right]_c h_c .
\end{aligned}
\tag{15.1}
$$

For i and $j = 1$ or 2, to calculate the in-plane stiffness terms, $\left[Q_{ij}\right]_f = \frac{E_f}{(1-\nu_f^2)}$ and $\left[Q_{ij}\right]_c = \frac{E_c}{(1-\nu_c^2)}$.

$$
A_{11} = A_{22} = \frac{E_f}{(1-\nu_f^2)}(2t_f) + \frac{E_c}{(1-\nu_c^2)}h_c .
\tag{15.2}
$$

Similarly, from (10.63)

$$
\begin{aligned}
D_{ij} &= \frac{1}{3}\sum_{k=1}^{3}\left[\overline{Q}_{ij}\right]_k\left[h_k^3 - h_{k-1}^3\right] = \frac{1}{3}\left[Q_{ij}\right]_f\left[\left(-\frac{h_c}{2}\right)^3 - \left(-\frac{h_c}{2} - t_f\right)^3\right] \\
&\quad + \frac{1}{3}\left[Q_{ij}\right]_c\left[\left(\frac{h_c}{2}\right)^3 - \left(-\frac{h_c}{2}\right)^3\right] + \frac{1}{3}\left[Q_{ij}\right]_f\left[\left(\frac{h_c}{2} + t_f\right)^3 - \left(\frac{h_c}{2}\right)^3\right]
\end{aligned}
\tag{15.3}
$$

For i and $j = 1$ or 2, the resulting flexural stiffness quantities are seen to be

$$
\begin{aligned}
D_1 = D_2 &= \frac{1}{3}\frac{2E_f}{(1-\nu_f^2)}\left[\left(\frac{3}{4}\right)\frac{h_c^2}{2}t_f + \frac{3}{2}h_c t_f^2 + t_f^3\right] + \frac{E_c}{(1-\nu_c^2)}\frac{h_c^3}{4} \\
&= \frac{1}{2}\frac{E_f h_c^2 t_f}{(1-\nu_f^2)}\left[1 + \frac{1}{6}\frac{E_c}{E_f}\frac{(1-\nu_f^2)}{(1-\nu_c^2)}\frac{h_c}{t_f}\right].
\end{aligned}
\tag{15.4}
$$

Now for many sandwich structures both $\frac{E_c}{E_f} << 1$, and $\frac{t_f}{h_c} << 1$, such that if $\frac{E_c h_c}{E_f t_f} << 1$ the second term in the bracket of (15.4) is very small compared to unity. In that case

$$D_1 = D_2 = \frac{1}{2}\frac{E_f h_c^2 t_f}{(1-\nu_f^2)}, \tag{15.5}$$

which is the widely used expression for the flexural stiffness for an isotropic sandwich structure.

Likewise all A, B and D quantities can be derived from the laminate analysis, for a sandwich structure, isotropic or anisotropic, mid-plane symmetric or asymmetric. Subsequently, quite often the resulting expressions can be simplified as shown above. For example, consider a mid-surface symmetric rectangular sandwich panel, wherein both faces are made of a unidirectional continuous fiber composite with the material 1 axis coincident with the panel's x-axis etc., and the core is assumed to contribute nothing to the stiffness matrix. In that case

$$A_{ij} = (\overline{Q}_{ij})_f 2t_f = (Q_{ij})_f 2t_f \qquad (i, j = 1, 2, 6)$$

$$B_{ij} = 0$$

$$D_{ij} = \frac{(\overline{Q}_{ij})_f t_f h_c^2}{2} = \frac{(Q_{ij})_f t_f h_c^2}{2} \qquad (i, j = 1, 2, 6)$$

or

$$A_{ij} = \begin{bmatrix} \dfrac{2E_{11}t_f}{(1-\nu_{12}\nu_{21})} & \dfrac{2\nu_{21}E_{11}t_f}{(1-\nu_{12}\nu_{21})} & 0 \\ \dfrac{2\nu_{21}E_{11}t_f}{(1-\nu_{12}\nu_{21})} & \dfrac{2E_{22}t_f}{(1-\nu_{12}\nu_{21})} & 0 \\ 0 & 0 & 2G_{12}t_f \end{bmatrix} \tag{15.6}$$

$$D_{ij} = \begin{bmatrix} \dfrac{E_{11}h_c^2 t_f}{2(1-\nu_{12}\nu_{21})} & \dfrac{\nu_{21}E_{11}h_c^2 t_f}{2(1-\nu_{12}\nu_{21})} & 0 \\ \dfrac{\nu_{21}E_{11}h_c^2 t_f}{2(1-\nu_{12}\nu_{21})} & \dfrac{E_{22}h_c^2 t_f}{2(1-\nu_{12}\nu_{21})} & 0 \\ 0 & 0 & \dfrac{G_{12}h_c^2 t_f}{2} \end{bmatrix}. \tag{15.7}$$

It should be noted that for this sandwich

$$(D_{ij}) = (A_{ij})h_c^2/4.$$

For the transverse shear quantities A_{44} and A_{55}, following the same procedures for the isotropic sandwich

$$A_{44} = A_{55} = G_c h_c + 2t_f G_f \tag{15.8}$$

However, in some cases the loads are such that the core material is compressed significantly, and then the question arises as to how does the core behave. According to Sikarskie and Mercado [15.1], most of the common core materials behave linearly in shear initially, and then behave nonlinearly: elastic-perfectly plastic for PVC foams, and bilinear for end grain balsa, as examples in the extremes. Sikarskie and Mercado then analyze sandwich beams under four point bending and sandwich plates showing the growth of damage and behavior with the nonlinear core materials.

Many modern sandwich structures involve foam cores that are compressible, and under some loadings the upper and lower faces undergo differing deformation patterns. This occurs particularly under lateral localized loads, and can lead to premature failure of the structure. Frostig [15.2] and numerous colleagues including Baruch [15.3], Patel [15.4], Shenhar [15.5], and Thomsen [15.6] have studied these problems extensively regarding buckling, vibrations, and delamination. Also, Frostig [15.2] emphasizes that a stiffener edge support always causes stress concentrations that affect the faces as well as the skin-core interfaces with any type of loading. Computational models for sandwich panels and shells are also discussed by Noor, Burton and Bert [15.7].

15.2 Governing Equations for Sandwich Plates and Panels

Using the constitutive equations of the previous Section, all of the governing equations derived in Chapter 10 apply to sandwich panels. The only change is to use the sandwich stiffness matrices of Section 15.1 for the A_{ij}, B_{ij} and D_{ij} stiffness matrices.

Therefore, for a sandwich panel that is specially orthotropic (i.e. no $(\ )_{16}$ and $(\ )_{26}$ terms), mid-plane symmetric (i.e. no B_{ij} terms), if classical plate theory is used, the governing equation for a sandwich panels is given by Equation (11.26). If transverse shear deformation effects are included then (11.76) through (11.78) apply. Thus all of the material included in Chapter 11 applies if the sandwich stiffness quantities of Section 15.1 are used. This result is that this is a short chapter.

15.3 Minimum Potential Energy Theorem for Sandwich Plates

Analogous to the above the overall expression for the potential energy for a sandwich panel is given by (14.2), remembering to use the stiffness matrix properties of Section 15.1 above for the sandwich panel. Again, because of this, Chapter 15 is very short.

15.4 Solutions to Problems Involving Sandwich Panels

Any solution for the lateral and in-plane displacements for anisotropic or isotropic plates, with or without transverse shear deformation effects can be used for the solution to the analogous sandwich plate problem if the proper stiffness matrix quantities are used.

However, care must be taken to subsequently describe the stresses in the faces and core. Again, the very accurate expressions for the stresses of (11.34) can be used. If it is assumed that all of the in-plane and bending stresses are face stresses, and that the core only resists transverse shear loads, then the stresses of (11.34) are reduced to the following where ($i = x, y$)

$$\sigma_{f_i} = \frac{N_i}{2t_f} + \frac{M_i}{t_f h_c} \tag{15.9}$$

and the transverse shear stresses, σ_{iz}, in the core and the faces are found by using Equations (10.46) for σ_{xz_k} and σ_{yz_k} and (10.48).

The solutions to many problems involving sandwich plates and panels are given in textbooks by Plantema [15.8], Allen [15.9], Zenkert [15.10, 15.11] and Vinson [10.1].

15.5 References

15.1. Sikarski, D.L. and Mercado, L.L. (1997) On the Response of a Sandwich Panel with a Bilinear Core.

15.2. Frostig, Y. (1997) Bending of Curved Sandwich Panels with Transversely Flexible Cores – Closed-Form High-Order Theory, *ASME*, Dallas.

15.3. Frostig, Y. and Baruch, M. (1996) Localized Load Effects in High-Order Bending of Sandwich Panels with Transversely Flexible Core, *Journal of the ASCE*, Engineering Mechanics Division, Vol. 122, No. 11.

15.4. Patel, D. and Frostig, Y. (1995) High Order Bending of Piecewise Uniform Sandwich Beams with a Tapered Transition Zone and a Transversely Flexible Core, *Composite Structures*, Vol. 31, pp. 151-162.

15.5. Frostig, Y. and Shenhar, I. (1995) High-Order Bending of Sandwich Beams with a Transversely Flexible Core and Unsymmetrical Laminated Composite Skins, *Composites Engineering*, Vol. 5, No. 4, pp. 405-414.

15.6. Thomsen, O.T. and Frostig, Y. (1997) Localized Bending Effects in Sandwich Panels: Photoelastic Investigation Versus High-Order Sandwich Theory Results, *Composites Engineering*, in progress.

15.7. Noor, A.K., Burton, W.S. and Bert, C.W. (1996) Computational Models for Sandwich Panels and Shells, *Applied Mechanics Review*, ASME, Vol. 49, pp. 155-199.

15.8. Plantema, F.J. (1966) *Sandwich Construction: The Bending and Buckling of Sandwich Beams, Plates and Shells*, John Wiley and Sons, New York.

15.9. Allen, H.G. (1969) *Analysis and Design of Structural Sandwich Panels*, Pergamon Press, Oxford.

15.10. Zenkert, D. (1995) *An Introduction to Sandwich Construction*, Chameleon Press Ltd., London.
15.11. Zenkert, D. (1997) *The Handbook of Sandwich Constructions*, EMAS Publishing.

15.6 Problems

15.1. Consider a foam core sandwich panel composed of two identical faces and a foam core. Each face is a cross-ply laminate of $[0°, 90°, 90°, 0°]$ construction, with each ply being 0.25 mm (0.01″) thick. The core is made of Klegecell foam of density 1.0 lb/ft^3 and shear modulus G_c of 10,000 psi, $\nu_c = 0.4$, and is 25.4 mm (1″) thick. The stiffness matrix quantities for the boron/epoxy faces are:

$$Q_{11} = 2.43 \times 10^5 \text{ MPa } (35.32 \times 10^6 \text{ psi})$$
$$Q_{22} = 2.43 \times 10^4 \text{ MPa } (3.532 \times 10^6 \text{ psi})$$
$$Q_{12} = 7.30 \times 10^3 \text{ MPa } (1.06 \times 10^6 \text{ psi})$$
$$Q_{66} \quad 1.034 \times 10^4 \text{ MPa } (1.5 \times 10^6 \text{ psi})$$

and the boron/epoxy density is ρ_f □0.0721 lb/in^3.

(a) Using Equations (15.1) through (15.4), what are the $[A]$ and $[D]$ stiffness matrices for this sandwich panel?
(b) Using the simpler equations analogous to (15.6) and (15.7), what are the $[A]$ and $[D]$ stiffness matrices for this sandwich panel?
(c) What is the largest percentage difference between (a) and (b) above of any component in either matrix?
(d) If the panel is subjected to in-plane tension in the x-direction (i.e. the 0° direction) such that the faces and core are equally strained, what percentage of the load is carried by the faces, and what percentage is carried by the core, using the results of (a) above.

15.2. Consider a mid-plane symmetric sandwich panel, wherein the faces are made of a unidirectional Kevlar 49/epoxy composite with the 1-axis coincident with the x-axis. The properties are given in Table 15.2, and ν_{12} 0.34. This sandwich panel has faces 1 mm (0.04 inches) thick, and the core thickness is 25.4 mm (1 inch).
(a) Provide the values of the $[A]$ matrix and the $[D]$ matrix.
(b) How does this panel compare with that of Problem 15.1 which geometrically is the same, i.e., compare the properties of a sandwich with isotropic faces with one of faces of a unidirectional composite?

15.3. In selecting face materials, quite often one selects a material which has the highest specific strength, defined here as σ_{11} / ρ, or the highest specific stiffness defined here as E_{11} / ρ. In Table 15.1 and 15.2
(a) Which material has the highest specific strength? Which has the lowest?
(b) Which material has the highest specific stiffness? Which has the lowest?

15.4. Consider a rectangular sandwich hull plate on the flat bottom of a ship as a rectangular plate under a uniform lateral distributed loading, p_0, from the water pressure, and clamped along all edges. The steel sandwich faces are 0.125″ thick, and the foam core depth $\boldsymbol{h_c = 2.5''}$. The panel is 4 feet wide and 6 feet long, and the ship draws 14 feet of water, where sea water weighs 64 lb./ft.^{3}. The steel face properties are given in Table 15.2. What is the maximum stress in the sandwich panel? What is the maximum deflection? Use Table 3.4.

15.5. A rectangular wing panel component, $16'' \times 12''$, is made of a foam core sandwich with identical aluminum faces $(E = 10 \times 10^6 \text{ psi}, \nu = 0.3, \sigma_{\text{all.}} = \pm 20{,}000 \text{ psi})$. The panel is considered to be simply supported on all four edges. With a face thickness of 0.064″ and a core depth of 1.25″, if a maximum design pressure of 20 psi is reached,

(a) What and where is the maximum stress?
(b) What and where is the maximum deflection?
Use Table 3.1.

15.6. A flat portion of a wind tunnel measuring $30'' \times 54''$ is subjected to a uniform pressure of 20 psi. If the sandwich faces are steel (see Table 15.2) and the foam core depth is 1.0″, what face thicknesses are needed if all four sides of the panel are:

(a) Simply supported? (Use Table 3.1)
(b) Clamped? (Use Table 3.4)
The panel cannot be overstressed and the maximum deflection cannot exceed 0.125″.

15.7. A portion of the cover of a hovercraft is approximated by a rectangular sandwich plate measuring 8 feet by 4 feet in planform, and is simply supported on all four edges. It is subjected to a uniform lateral pressure of 20 psi. Assume that in the design $h_c = 20t_f$

(a) How thick must the faces be if made of 6061-T6 aluminum (see Table 15.2) to not be overstressed?
(b) What is the maximum deflection in (a) above?
(c) How much will the panel weigh if $\rho_{\text{Al}} = 0.116$ lb./in.^{3}, $\rho_c = 15$ lb./ft.^{3}?
(d) How thick must the faces be if made of the steel in Table 15.2 to not be overstressed?
(e) What is the maximum deflection in that case?
(f) How much will the panel weight if $\rho_{\text{st}} = 0.283$ lb./in.^{3}, $\rho_c = 15$ lb./ft.^{3}?
(g) Which design will weigh less?

15.8. A designer must design a rectangular plate cover over an opening measuring 9 feet by 3 feet. The maximum design load uniform pressure is 10 psi. If the steel faced sandwich of Table 15.2 is use and assume $h_c = 20t_f$, what will the panel weight if all four edges are:

(a) Simply supported?
(b) Clamped?
You may use Tables 3.1 and 3.4.

15.9. Consider a mid-plane asymmetric sandwich panel, wherein the two faces differ in thickness and stiffness quantities as shown below

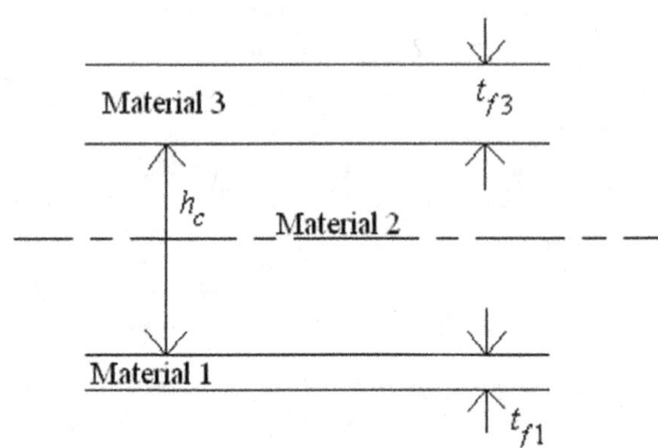

Assume $t_f << h_c$, that the core material contributes nothing to the stiffnesses discussed here, and that both faces are specially orthotropic.
Determine the A_{ij}, B_{ij} and D_{ij} stiffness analogous to (15.6) and (15.7) in terms of the material properties and sandwich geometries shown above.

Table 15.1. Quasi-Isotropic Composite Properties.

Material	Configuration	V_f (%)	E (10^6 psi)	ρ (lb/in^3)	σ_{max} (10^3 psi)
C/epoxy	Cross Ply	58	12.04	0.0555	55.1
B/epoxy	Cross Ply	60	15.37	0.0721	55.1
SiC/T_i 6Al-4V	Woven	35	27.0	0.1492	245
2024-T3 Al			10.88	0.0973	50.0
Ti			16.5	0.160	128.0
E glass/epoxy	Cross Ply	57	3.12	0.0710	82.0
Welton Steel			30.45	0.283	126.9

Table 15.2. Unidirectional Composite Properties.

Material	Elastic Moduli			Ultimate Strength			Density
	Axial E_{11}	Transverse E_{22}	Shear G_{12}	Axial tens. σ_{11}	Trans. Tens. σ_{22}	Shear τ_{12}	
High strength GR/epoxy	20 (138)	1.0 (6.9)	0.65 (4.5)	220 (1517)	6 (41)	14 (97)	0.057 (1.57)
High modulus GR/epoxy	32 (221)	1.0 (6.9)	0.7 (4.8)	175 (1206)	5 (34)	10 (69)	0.058 (1.60)
Ultra high modulus GR/epoxy	44 (303)	1.0 (6.9)	0.95 (6.6)	110 (758)	4 (28)	7 (48)	0.061 (1.68)
Kevlar 49/epoxy	12.5 (86)	0.8 (5.5)	0.3 (2.1)	220 (1517)	4 (28)	6 (41)	0.050 (1.38)
S glass/ epoxy	8 (55)	1.0 (14)?	0.5 (3.4)	260 (1793)	6 (41)	10 (69)	0.073 (2.00)
Steel	30 (207)	30 (207)	11.5 (79)	60 (414)	60 (414)	35 (241)	0.284 (7.83)
Aluminum 6061-T6	10.5 (72)	10.5 (72)	3.8 (26)	42 (290)	42 (290)	28 (193)	0.098 (2.70)

*Moduli in Msi (GPa); Stress in Ksi (MPa); Density in lb/in^3 (g/cm^3)

CHAPTER 16

ELASTIC INSTABILITY (BUCKLING) OF SANDWICH PLATES

16.1 General Considerations

As stated previously, structures usually fail in one of four ways:

- overstressing (strength critical structure)
- over deflection (stiffness critical structure)
- resonant vibration
- buckling.

For many cases, because sandwich structures (compared to monocoque structures) minimize stresses, are extremely stiff, and have high fundamental natural vibration frequencies, care must be taken to insure that unanticipated buckling does not undermine a structural design.

In monocoque structures for given plate dimensions, material, boundary conditions, and a given load type (in-plane compression, in-plane shear), only one buckling load will result in actual buckling. This is the lowest eigenvalue of a countable infinity of such eigenvalues. All other eigenvalues exist mathematically, but only the lowest value has physical significance. This differs from natural frequencies in which several eigenvalues can be very important.

For the simplest cases, for columns and isotropic plates, an introduction was given in Chapter 6. While philosophically the simple examples cover the topic of buckling; more complex structures can have several types of buckling instabilities, any one of which can destroy the structure.

Historically, there have been four major textbooks dealing primarily with elastic stability or buckling. These are authored by Timoshenko and Gere [6.1], Bleich [6.2], Brush and Almroth [12.1] and Simitses [12.2]. A new text by Jones [6.4] will supplement these four. Although these texts deal primarily with structures other than sandwich, the solutions can be applied by using the appropriate flexural stiffnesses.

16.2 The Overall Buckling of an Orthotropic Sandwich Plate Subjected to In-Plane Loads - Classical Theory

From previous developments, it was seen that for a plate there are five equations associated with the in-plane stress resultants N_x, N_y and N_{xy} and the in-plane displacements they cause, namely u_0 and v_0. For the isotropic rectangular plate, see (2.50)-(2.54), and the isotropic circular plate, see (5.15), (5.16) and (5.25) through (5.27). For a composite material plate, the in-plane equilibrium equations are given by Equations

(11.6) and (11.7). From Equation (10.66), for the case of mid-plane symmetry ($B_{ij} = 0$) and no thermal or moisture considerations it is seen that the in-plane constitutive equations are:

$$N_x = A_{11}\varepsilon_x^0 + A_{12}\varepsilon_y^0 + 2A_{16}\varepsilon_{xy}^0 \tag{16.1}$$

$$N_y = A_{12}\varepsilon_x^0 + A_{22}\varepsilon_y^0 + 2A_{26}\varepsilon_{xy}^0 \tag{16.2}$$

$$N_{xy} = A_{16}\varepsilon_x^0 + A_{26}\varepsilon_y^0 + 2A_{66}\varepsilon_{xy}^0. \tag{16.3}$$

For this case, all of the equations of Section 12.2 apply, simply by using the sandwich stiffness properties for the D_{ij} that are found in Section 15.1.

Likewise, for the mid-plane symmetric panel, the six governing equations involving $M_x, M_y, M_{xy}, Q_x, Q_y$ and w, are given by Equations (11.9), (11.11), (11.12), (11.17), (11.18), and (11.19), the latter three neglecting the D_{16} and D_{26} terms. One can see there is no coupling between in-plane and lateral action for the plate with mid-plane symmetry. Yet it is well knows and often observed that in-plane loads do cause lateral deflections through buckling, which is usually disastrous.

The answer to the paradox is that in the above discussion only <u>linear</u> elasticity theory is considered, while the physical event of buckling is a non-linear problem. For brevity, the development of the non-linear theory will not be included herein because it is included in so many other texts, such as those cited in Section 16.1.

The results of including the terms to predict the advent or inception of buckling for the beam and plate are, modifying Equation (11.26), shown previously as (12.4),

$$\begin{aligned} D_1\frac{\partial^4 w}{\partial x^4} + 2D_3\frac{\partial^4 w}{\partial x^2\partial y^2} + D_2\frac{\partial^4 w}{\partial y^4} = p(x,y) + N_x\frac{\partial^2 w}{\partial x^2} \\ + 2N_{xy}\frac{\partial^2 w}{\partial x\partial y} + N_y\frac{\partial^2 w}{\partial y^2} \end{aligned} \tag{16.4}$$

where clearly there is a coupling between the in-plane loads and the lateral deflection. For overall buckling of a sandwich panel the D_1, D_2 and D_3 flexural stiffnesses are given by (15.1) through (15.8)

It should be noted that the buckling loads, like the natural frequencies, are independent of the lateral loads, which will be disregarded in what follows. However, in actual structural analysis, the effect of lateral loads, in combination with the in-plane loads could cause overstressing and failure before the in-plane buckling load is reached. However, the buckling load is still independent of the type or magnitude of the lateral load, as are the natural frequencies. Incidentally, common sense dictates that if one is designing a structure to withstand compressive loads, with the possibility of buckling being the failure mode, one had better design the structure to be mid-plane symmetric, so that $B_{ij} = 0$. Otherwise the bending-stretching coupling would likely cause overstressing before the buckling load is reached.

Looking now at (16.4) for the buckling of the composite plate under an in-plane compressive load N_x only, and ignoring $p(x,y)$ Equation (16.4) becomes:

$$D_1 \frac{\partial^4 w}{\partial x^4} + 2D_3 \frac{\partial^4 w}{\partial x^2 \partial y^2} + D_2 \frac{\partial^4 w}{\partial y^4} - N_x \frac{\partial^2 w}{\partial x^2} = 0. \tag{16.5}$$

Again, one may assume the buckling mode for a sandwich plate to be that of the Navier solution for the case of the plate simply supported on all four edges:

$$w(x,y) = \sum_{m=1}^{\infty} \sum_{n=1}^{\infty} A_{mn} \sin\frac{m\pi x}{a} \sin\frac{n\pi y}{b}. \tag{16.6}$$

Substituting (16.6) into (16.5) it is seen that the equation is satisfied only when N_x has certain values, namely the critical values, $N_{x\,\mathrm{cr}}$,

$$N_{x\,\mathrm{cr}} = -\frac{\pi^2 a^2}{m^2}\left[D_1\left(\frac{m}{a}\right)^4 + 2D_3\left(\frac{m}{a}\right)^2\left(\frac{n}{b}\right)^2 + D_2\left(\frac{n}{b}\right)^4\right]. \tag{16.7}$$

Again several things are clear: Equation (16.5) is a homogeneous equation, so this is an eigenvalue problem and therefore one cannot determine the value of A_{mn}; and again only the lowest value of $N_{x\,\mathrm{cr}}$ is of any importance. However, it is not clear which value of m and n results in the lowest critical buckling load. All values of n appear in the numerator for this case of all edges being simply supported, so $n = 1$ is the necessary value. But m appears several places, and depending upon the value of the flexural stiffnesses D_1, D_2 and D_3, and the length to width ratio, i.e., the aspect ratio, of the plate, a/b, it is not clear which value of m will provide the lowest value of $N_{x\,\mathrm{cr}}$. However, for a given plate this is easily determined computationally.

What about the buckling loads of composite material sandwich plates with boundary conditions other than simply supported? It is seen that all combinations of beam vibrational mode shapes are applicable for plates with various boundary conditions. These have been developed by Warburton [8.3] and all derivatives and integrals of those functions catalogued conveniently by Young and Felgar [3.1, 3.2] for easy use. Likewise they can be used instead of Equation (16.6) to obtain solutions of Equation (16.5) to determine the critical buckling load per unit width, $N_{x\,\mathrm{cr}}$.

Using classical plate theory the treatment of overall buckling for sandwich plates with other boundary conditions is given in Section 16.5.2 below.

The buckling loads calculated in this Section do not include transverse shear deformation effects, and are therefore only approximate – but they are useful for preliminary design, because of their relative simplicity. If transverse shear deformation were included, the buckling loads are lower than those calculated with classical theory.

Therefore the buckling loads calculated, neglecting transverse shear deformation, are not conservative. Transverse shear deformation effects will now be investigated.

16.3 The Buckling of Honeycomb Core Sandwich Panels Subjected to In-Plane Compressive Loads

In sandwich plates and panels the core is usually either a honeycomb, a foam or solid core, a truss core or web core. The honeycomb core is treated in this Section.

Consider a rectangular sandwich panel of length a (the load direction), width b, face thickness t_f, core depth h_c, core cell wall thickness t_c, and diameter d of a circle inscribed in the cell as shown in Figures 16.1 and 16.2 for the hexagonal-cell honeycomb core. A sketch analogous to Figure 16.2 could also be drawn for the square-cell honeycomb. It is assumed for this study that the core is composed of an isotropic material of shear modulus G_c and modulus of elasticity E_c. If the core is orthotropic, the properties normal to the plane of the panel for E_c and G_c are used. Consider the faces to be composed of identical composite materials that are balanced about their own planes of symmetry with no unwanted couplings, i.e., B_{ij} 0 and all $(\ )_{16}$ and $(\ )_{26}$ terms equal to zero. It is assume that the in-plane load in the x-direction is uniform and has the value of N_x (load/unit width) in compression, so $\overline{N}_x$ □ $-N_x$ for convenience.

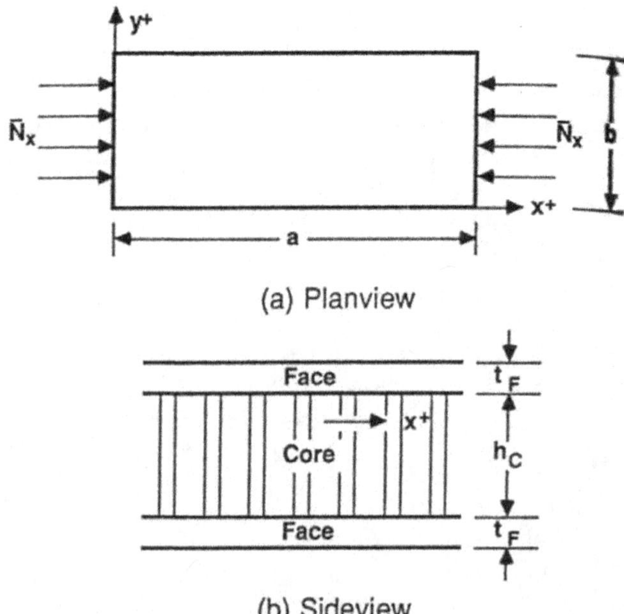

Figure 16.1. Sandwich panel with in-plane compressive load in the x-direction. (Reprinted from References [1.7] and [10.1].)

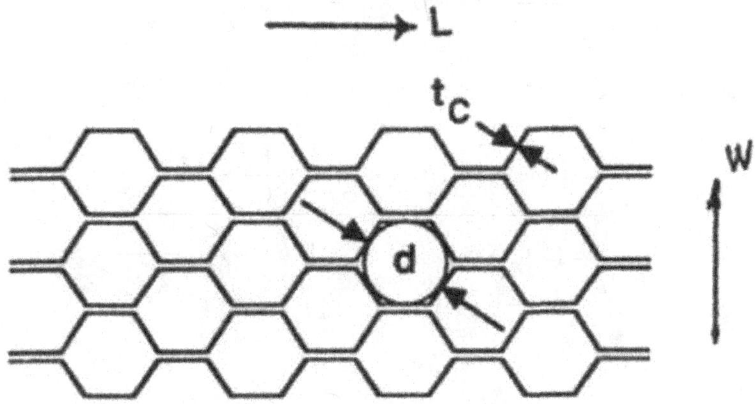

Honeycomb Core in Planview

Figure 16.2. Honeycomb core in plan view. (Reprinted from References [1.7] and [10.1].)

16.3.1 Face Stresses

If it is assumed that the in-plane loads are resisted only by the faces, not the honeycomb core. Therefore, the applied compressive face stress in the load direction is written as

$$\sigma_x = N_x / 2t_f \tag{16.8}$$

For an applied load per unit width, N_x, the face stress σ_x is, of course, restricted to some prescribed maximum value to prevent overstressing.

The honeycomb sandwich panel shown in Figure 16.1 can be overstressed according to Equation (16.8), but can also buckle in one of several modes, any one of which will render the panel to be useless. These modes of buckling are overall instability, core shear instability, face wrinkling, and face dimpling (monocell buckling). These are discussed in turn.

16.3.2 Overall Instability

The equation to use for the compressive face stress for overall buckling of the subject panel that includes the effects of transverse shear deformation is given by [16.1] as follows, where the bar over the modulus indicates that for stresses above the proportional limit, a plasticity reduction factor should be used.

$$\sigma_{cr} = \frac{\pi^2}{4(1-\nu_{xy}\nu_{yx})}\sqrt{E_{fx}E_{fy}}\frac{h_c^2}{b^2}K, \text{ for } V_x \le k_1 B_1 r \tag{16.9}$$

In this equation, the coefficient K is given by

$$K = \frac{B_1C_1 + 2B_2C_2 + \dfrac{C_3}{B_1} + A\left(\dfrac{V_y}{C_4} + V_x\right)}{1 + (B_1C_1 + B_3C_2)\dfrac{V_y}{C_4} + \left(\dfrac{C_3}{B_1} + B_3C_2\right)V_x + \dfrac{V_yV_xA}{C_4}} \tag{16.10}$$

where

$$A = C_1C_3 - B_2C_2^2 + B_3C_2\left(B_1C_1 + 2B_2C_2 + \frac{C_3}{B_1}\right) \tag{16.11}$$

$$B_1 = \sqrt{\frac{D_y}{D_x}}, \quad B_3 = \frac{D_y\nu_{xy} + 2D_{xy}}{\sqrt{D_xD_y}} = 2B_3 + B_1\nu_{xy}, \quad B_3 = \frac{D_{xy}}{\sqrt{D_xD_y}} \tag{16.12}$$

$$V_y = \frac{\pi^2}{b^2}\frac{\sqrt{D_xD_y}}{U_{yz}}, \quad V_x = \frac{\pi^2}{b^2}\frac{\sqrt{D_xD_y}}{U_{xz}} \tag{16.13}$$

$$U_{xz} = G'_{cx}h_c, \quad U_{yz} = G'_{cy}h_c \tag{16.14}$$

In Equation (16.13) the V_i quantities are seen to be particular ratios of the sandwich-panel flexural stiffness D, to the transverse shear stiffness U, hence the V quantities are called the transverse shear flexibility parameters. The D quantities relate to face properties while the U quantities are core properties. The effective core moduli, G'_{cx} and G'_{cy} are defined in Equation (16.27) and (16.28).

The constants C_1 through C_4 in Equations (16.10) and (16.11) are associated with the boundary conditions and are listed as follows where n is the number of buckling waves in the direction of the compressive loading.

(1) All edges simply supported (Note that this gives results identical to Equation (16.7))

$$C_1 = C_4 = \frac{a^2}{n^2b^2}, \quad C_2 = 1, \quad C_3 = \frac{n^2b^2}{a^2} \tag{16.15}$$

(2) Loaded edges simply supported, other edges clamped

$$C_1 = \frac{16}{3}\frac{a^2}{n^2 b^2},\quad C_2 = \frac{4}{3},\quad C_3 = \frac{n^2 b^2}{a^2},\quad C_4 = \frac{4}{3}\frac{a^2}{n^2 b^2} \tag{16.16}$$

(3) Loaded edges clamped, other edges simply supported

$$\begin{aligned} C_1 &= C_4 = \frac{3}{4}\frac{a^2}{b^2} \quad \text{for } n = 1 \\ C_1 &= C_4 = \frac{1}{(n^2+1)}\frac{a^2}{b^2} \quad \text{for } n \geq 2 \\ C_2 &= 1,\quad C_3 = \frac{(n^2 + 6n^2 + 1)}{(n^2+1)}\frac{b^2}{a^2} \end{aligned} \tag{16.17}$$

(4) All edges clamped

$$\begin{aligned} C_1 &= 4C_4 = \frac{4a^2}{b^2} \quad \text{for } n = 1 \\ C_1 &= 4C_4 = \frac{16}{3(n^2+1)}\frac{a^2}{b^2} \quad \text{for } n \geq 2 \\ C_2 &= \frac{4}{3},\quad C_3 = \frac{(n^4 + 6n^2 + 1)}{(n^2+1)}\frac{b^2}{a^2} \end{aligned} \tag{16.18}$$

Obviously, if the critical face stress given by Equation (16.9) is higher than the allowable compressive stress of the face material, the panel will be overstressed before there is an overall buckling problem.

16.3.3 Core Shear Instability

Referring to the expressions for overall stability, if the value of V_x is increased through increasing the panel bending stiffnesses (D_x and D_y) or decreasing the core transverse shear stiffness (U_{xz}), the value of K in Equation (16.10) is decreased. There exists a value of V_x that causes K to equal $1/V_x$. This value depends both on the boundary conditions and the effective shear moduli of the core (G'_{cx}, G'_{cy}), that are defined in Equations (16.27) and (16.28). At this particular value of $K = 1/V_x$, K is independent of the length-to-width ratio a/b and n is infinite. For values of V_x greater than this value, $K = 1/V_x$, which is true for a great number of practical sandwich panels because they have high values of V_x. Under these conditions, the critical stress can be written as follows:

$$\sigma_{cr} = \frac{G'_{cx} h_c}{2t_f}, \quad \text{for } V_x \geq k_1 B_1 r \tag{16.19}$$

This value of critical stress is called the core-shear instability stress and cannot be exceeded for any given sandwich construction. It is seen that Equation (16.19) is independent of the panel length, width, and boundary conditions. Core shear instability is illustrated in Figure 16.3.

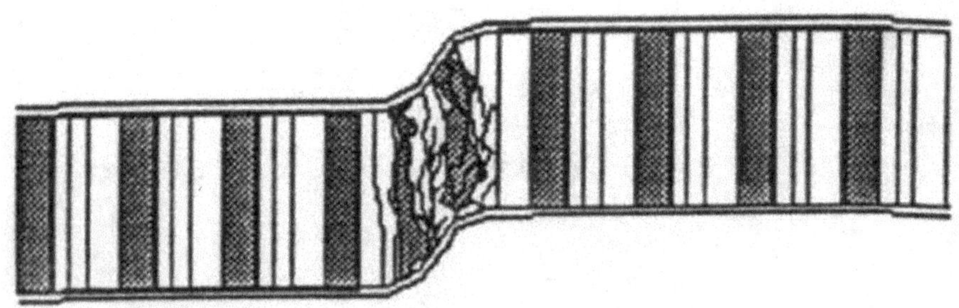

Figure 16.3. Core shear instability. (Reprinted from References [1.7] and [10.1].)

The particular value of V_x at which $K = 1/V_x$, is the value for which the critical stresses for overall buckling and core shear instability are equal. This value is given in Equation (16.20) and is dependent on the boundary conditions and the effective shear moduli of the core, G'_{cx} and G'_{cy}, defined in Equations (16.27) and (16.28). The values for the boundary condition factors k_1 are listed in Table 16.1.

$$V_x = k_1 B_1 r \quad \text{where here } r = G'_{cx} / G'_{cy} \tag{16.20}$$

More specifically this can be written as follows, where $\overline{E}$ refers to a modulus multiplied by a suitable plasticity reduction factor, when stresses exceed the proportional limit. Otherwise, the modulus of elasticity E is used.

$$\frac{\pi^2}{2(1-\nu_{xy}\nu_{yx})k_3} \frac{t_f h_c d}{b^2 t_c} \frac{\overline{E}_{fx}}{G_c} = k_1 r \tag{16.21}$$

Table 16.1. Boundary Condition Factors k_i for Various Edge Conditions.

Boundary Condition	k_1
All edges simply supported	1
Loaded edges simply supported, other edges clamped	¾
Loaded edges clamped, other edges simply supported	1
All edges clamped	¾

16.3.4 Face Wrinkling Instability

Wrinkling occurs across many cells of the honeycomb core and, under the loading conditions described here, extends across the width of the plate, but is localized in the direction of the applied load; that is, the wrinkle is essentially a short wavelength buckle, as shown in Figure 16.4. Heath [16.2] derived an expression for this mode of instability for the case of isotropic materials as

$$\sigma_{cr} = \left[\frac{2}{3}\frac{t_f}{h_c}\frac{E_c' E_f}{(1-\nu^2)}\right]^{\frac{1}{2}} \tag{16.22}$$

Figure 16.4. Face wrinkling instability. (Reprinted from References [1.7] and [10.1].)

Heath defined E_c' incorrectly in his paper, but Hemp [16.3] clarified the point in an earlier paper. The face-wrinkling stability equation for isotropic faces given by Equation (16.22) can be modified for anisotropic materials to be:

$$\sigma_{cr} = \left[\frac{2}{3}\frac{t_f}{h_c}\frac{E_c'\sqrt{E_{fx}E_{fy}}}{(1-\nu_{xy}\nu_{yx})}\right]^{\frac{1}{2}} \tag{16.23}$$

An earlier equation for face wrinkling was developed by Hoff and Mautner [16.4], and given by

$$\sigma_{cr} = c(E_{fx}E_{cx}'E_{cxz}')^{1/3} \tag{16.24}$$

for a honeycomb sandwich panel where E_{cx}' and G_{cxz}' are given by Equations (16.27) and (16.28), and c is a constant usually 0.5, 0.6, or 0.65. Note that here the critical stress depends on material properties only. It is not clear whether one should use Equation (16.23) or (16.24). The analyst/designer can be conservative by using the equation which gives the lowest critical stress for the particular case being studied.

16.3.5 Monocell Buckling or Face Dimpling

In honeycomb core sandwiches, a fourth type of instability occurs because the faces over one cell can buckle as a small plate supported by the cell walls. Methods of analysis developed at the Forest Products Laboratory used an empirical equation having the form of the plate-buckling equation with the numerical coefficient determined by empirical means. The result for an isotropic face material is as follows, where the subscript T denotes the tangent modulus.

$$\sigma_{cr} = \frac{2E_{f_T}}{(1-\nu^2)}\left(\frac{t_f}{d}\right)^2$$

For anisotropic faces, the expression can be written as

$$\sigma_{cr} = \frac{2(E_{fx_T}E_{fy})^{1/2}}{(1-\nu_{xy}\nu_{yx})}\left(\frac{t_f}{d}\right)^2 \tag{16.25}$$

16.3.6 Core Properties

Mechanical properties of honeycomb core used in the previous equations are called "effective" and are designated with a prime, because they are properties associated with the core acting as a homogeneous material having these "effective" properties. They are functions of the core material's properties, ρ_c, E_c, G_{cx}, and G_{cy}, the core wall thickness t_c, and the cell size *d*. To truly optimize the structure for minimum weight, it is advantageous to relate the effective properties back to fundamental geometry and material properties. For foam or solid core sandwich panels there is no need for effective properties and the actual mechanical properties are used for ρ_c, E_c, G_{cx}, and G_{cy}.

For hexagonal-cell honeycomb core having some double walls as shown in Figure 16.2, the properties were developed by Kaechele [16.5], and shown below. These effective core properties can be related to the geometry and actual material properties as follows, where ρ_c, E_c, and G_c are the weight density, the shear modulus and the modulus of elasticity of the core material itself:

$$\rho_c' = k_2(t_c/d)\rho_c \tag{16.26}$$

$$G_{cx}' = k_3(t_c/d)G_c \tag{16.27}$$

$$G_{cy}' = k_4(t_c/d)G_c \tag{16.28}$$

$$E_c' = k_2(t_c/d)E \tag{16.29}$$

It is seen from Equations (16.27) and (16.28) that here

$$r = G_{cx}'/G_{cy}' = k_3/k_4 \tag{16.30}$$

For the hexagonal cell construction of Figure 16.2 as well as other types of honeycomb core, the values of $k_2, k_3,$ and k_4 are given in Table 16.2 according to Kaechele [16.5] and MIL HDK-23 [16.1]. For other honeycomb configurations, the values of these constants can be easily derived using the methods of Kaechele [16.5]. A fairly inclusive listing of commercially available honeycomb core properties are given in Appendix 1 of [10.1].

Table 16.2. Values of k_2, k_3 and k_4 for Various Honeycomb Constructions.

Type of Construction	k_2	k_3	k_4
Hexagonal (Kaechele)	8/3	5/3	1
Hexagonal (MIL HDBK-23)	8/3	4/3	8/15
Square cell (Kaechele)	2	1	1
Square cell (MIL HDBK-23)	2	1	1

16.3.7 Plasticity Effects

The extension of elastic buckling theory to account for the buckling of structures at stresses above the proportional limit of the material has been widely studied mostly for ductile metallic materials. Many investigations have used the elastic equations, wherein Young's modulus E has been multiplied by a plasticity reduction factor η. However, there is considerable difference of opinion about a correct form for η. These expressions range in complexity and it is not at all clear which expression has more merit. For structural optimization, if the compressive stress-strain curve of the face material has a proportional limit, then for stresses above the proportional limit, all values of E_{fx} can be replaced by $\overline{E}_{fx}$ where

$$\overline{E}_{fx} = \eta E_{fx} \tag{16.31}$$

One often uses a plasticity reduction factor in which $\overline{E}_{fx} = (E_{fx_T} E_{fx})^{1/2}$ where the subscript T denotes the tangent modulus at that face stress. This will require an iteration to match face stress with tangent modulus. It is unlikely that for the in-plane compression in the x-direction, the stresses in the y-direction will cause deviation from the elastic value E_y, but this could also be modified. For many composite materials usually the moduli are linear to failure, so this subsection can be ignored.

16.3.8 Weight Relationship

To obtain a weight per unit area of the honeycomb sandwich structure, the following may be used:

$$W = 2t_f \rho_f + \rho_c' h_c + W_{ad} \tag{16.32}$$

The weight of the adhesive or other joining material, W_{ad}, cannot be easily related to the variables discussed earlier and is dependent upon the material, method of joining, fabrication techniques, and skill and temperament of the personnel. Since in many cases this is a small fraction of the weight and, because of the factors involved, it will not be specified further and need not be accounted for in the comparisons to select the optimum geometry and materials. However, care should be used to include it when comparing structures employing no adhesive or with other types of construction.

16.3.9 Analysis and Design Methods

For a honeycomb sandwich panel (such as that shown in Figure 16.1) to withstand an in-plane compressive load N_x, in force per unit width, the materials and component sizes must be sufficient to insure that overstressing [Equation (16.8)], overall buckling [Equation (16.9)], core shear instability [Equation (16.19)], face wrinkling [Equation (16.23) or (16.24)], and face dimpling [Equation (16.25)] will not occur. If any of the first four occur the panel is useless. If face dimpling occurs, it may not cause structural failure, but peeling of any coating such as paint could occur, the surfaces may be "unsightly," a boundary layer could be tripped from laminar to turbulent flow, or permanent core crushing in that vicinity may occur. These should be avoided.

16.4 The Buckling of Solid-Core or Foam-Core Sandwich Panels Subjected to In-Plane Compressive Loads

16.4.1 Face Stresses

For foam core sandwich panels and many other solid core panels it can be assumed that all in-plane loads, and bending loads as well, are resisted by the faces only. Therefore, in this case Equation (16.8) is used to determine the face stresses in terms of the applied load per unit width, N_x.

16.4.2 Overall Buckling

The equation for the overall buckling of an anisotropic composite sandwich panel with a solid core subjected to an in-plane compressive load is given by Equations (16.9) through (16.14). However, in this case the actual mechanical properties of the core, G_{cx} and G_{cy} are used in Equation (16.14).

16.4.3 Core Shear Instability

Core shear crimping or core shear instability will occur at a face stress lower than that of overall panel buckling when

$$V_x \geq k_1 B_1 r \tag{16.33}$$

The critical stress value on core shear instability is given by

$$\sigma_{cr} = \frac{G_{cx} h_c}{2t_f} \tag{16.34}$$

It is seen that overall panel buckling and core shear instability will occur at the same face stress value when $V_x = k_1 B_1 r$, or more specifically when

$$\frac{\pi^2}{2(1-\nu_{xy}\nu_{yx})} \frac{\overline{E}_{fx}}{G_{cx}} \frac{h_c t_f}{b^2} = k_1 r \tag{16.35}$$

Thus Equations (16.9) and (16.34) completely describe the conditions of the simultaneous overall buckling of the panel and core shear instability without the complexities of using Equations (16.10) through (16.18) with the lengthy determination of K.

16.4.4 *Face Wrinkling*

In addition to overall panel buckling and core shear instability, a short wavelength buckling can occur if the faces are thin and can be described by

$$\sigma_{cr} = \left[\frac{2}{3} \frac{t_f}{h_c} \frac{E_c (E_{fx} E_{fy})^{1/2}}{(1-\nu_{xy}\nu_{yx})} \right]^{\frac{1}{2}} \tag{16.36}$$

Of course the Hoff-Mautner equation is still used also

$$\sigma_{cr} = C[E_{fx} E_{cx} G_{cxz}]^{1/3} \tag{16.37}$$

where the constant C is 0.5, 0.6 or 0.65 by various users. Again, it is seen that with this equation the critical strain is dependent on material properties only. Plantema [15.8] uses 0.82 for the constant C in Equation (16.35), and Dreher [16.6] says that this corresponds well with the experimental data. He states emphatically that $C = 0.5$ does not correspond with his test results.

16.4.5 *Weight Relationship*

The weight of the solid-core sandwich panel per unit planform area, w is given by

$$W = 2t_f \rho_f + \rho_c h_c + W_{ad} \tag{16.38}$$

where ρ_i are the weight density of the materials involved and W_{ad} is the non-analytic weight per unit planform area of the adhesive bonding the face to the core.

16.4.6 Analysis and Design Methods

For a foam- or solid-core sandwich panel to withstand an in-plane compressive load N_x, in force per unit width, the materials and components thicknesses must be sufficient to insure that overstressing [Equation (16.8)], overall buckling [Equation (16.9)], core shear instability [Equation (16.34)], and face wrinkling [Equations (16.36) or {16.37)] will not occur, because any one of the above will cause panel failure. With the satisfactory design, the panel weight is determined from Equation (16.38). As discussed previously, foam cores available today comprise an almost continuous value of shear modulus G and density ρ, usually a linear relationship between them.

16.5 Buckling of a Truss-Core Sandwich Panel Subjected to Uniaxial Compression

Another type of sandwich construction for panels subjected to uniaxial compression involves a corrugated core, sometimes referred to as single-truss core. Several promising and clever manufacturing methods have been devised to make unique use of fiber-reinforced polymer matrix composites for this type of construction. At least one of these involves weaving the filaments of the face material and core material together at the junctions, thus increasing the structural integrity of the joint and avoiding the joining problems associated with conventional construction.

The analysis methods developed herein are applicable to both metallic and composite material construction and account for material orthotropy in both the face and core. They are also applicable to panels at elevated or lowered temperature, under steady state and nearly uniform temperatures. Only the stress-strain curve is necessary for each temperature under consideration.

Consider the flat corrugated-core (truss-core) sandwich panel cross-section idealized in Figure 16.5 for the panel shown in Figure 16.1(a). For a given material system there are four geometric parameters to consider; namely the core depth, h_c, the web thickness, t_c, the face thickness, t_f, and the angle the web makes with a line normal to the faces, θ.

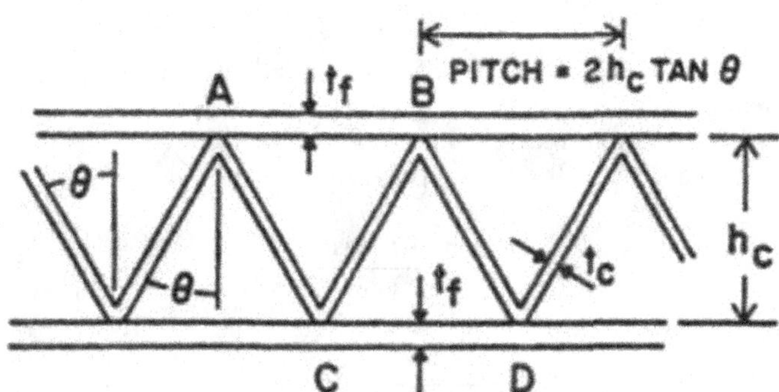

Figure 16.5. Cross-section of corrugated-core sandwich panel. (Reprinted from References [10.1], [16.12] and [16.13].)

This panel will fail if any of the following five events occur: overstressing of the face, overstressing of the core, overall panel instability, local face buckling, or web buckling. It can be shown that face wrinkling, which can occur in honeycomb sandwich construction will not occur in this type of construction, because it can be shown that local face buckling and web buckling will invariably occur at lower values of the applied load. Likewise, core shear instability, in the sense of shear crimping in honeycomb sandwich construction cannot occur for the same reason. Hence, there are three modes of instability and four geometric parameters.

Unlike honeycomb and solid- or foam-core sandwich construction, with a truss-core sandwich, it is intended that the core will carry or resist a portion of the in-plane compressive load. Thus, it is necessary to define some elastic and geometric quantities before proceeding.

16.5.1 Elastic and Geometric Constants

These are determined from those given in more general form by Libove and Hubka [16.7]. The core area per unit width and the area moment of inertia of the core per unit width are given by

$$\overline{A}_c = \frac{t_c}{\sin\theta} \tag{16.39}$$

$$\overline{I}_c = \frac{t_c h_c}{12\sin\theta} = \frac{\overline{A}_c h_c^2}{12} \tag{16.40}$$

The extensional stiffness of the plate per unit width in the x-direction, A_{11}, is given by the following for isotropic materials,

$$A_{11} = E_c \overline{A}_c + 2E_f t_f \tag{16.41}$$

where E_c and E_f are the compressive moduli of elasticity of the core and face material, respectively. Therefore, Equation (16.41) is A_{11} for the truss-core panel in the language of composite theory.

The transverse shear stiffness per unit width, in the x-direction and the y-direction, respectively, are found to be, in the notation of [16.7]

$$D_{q_x} = \frac{G_c t_c \cos\theta}{\tan\theta} \tag{16.42}$$

$$D_{q_y} = \frac{E_c t_c}{(1-\nu_c^2)} \cos^2\theta \sin\theta \tag{16.43}$$

The latter expression agrees with that derived by Anderson [16.8]. Thus, Equations (16.42) and (16.43) are A_{55} and A_{44} for the truss-core panel [see Equation (10.71)].

Lastly, the moment of inertia per unit width of the faces, considered as membranes, with respect to the sandwich middle surface, is seen to be, as before,

$$\overline{I}_f = \frac{t_f h_c^2}{2} \tag{16.44}$$

16.5.2 Overall Instability

The best expression describing the overall instability of a corrugated-core sandwich panel composed of isotropic materials under uniaxial compressive loads is derived by Seide [16.9] NACA TN2679 (which is used in ANC-23 in slightly modified form [16.1])

$$N_x = \frac{\pi^2 E_f \overline{I}_f K}{b^2} \tag{16.45}$$

where K is the buckling coefficient derived and plotted in Figures 2 and 4 of Reference [16.9]. It is given as a function of length to width ratio (a/b), for the cases of the unloaded edges simply-supported and clamped, for various values of the transverse shear flexibility parameter, V, defined as

$$V = \frac{\pi^2 E_f \overline{I}_f}{b^2 D_{q_y}} + \frac{\pi^2(1-\nu_c^2)}{2\cos^2\theta \sin\theta}\left(\frac{t_f}{t_c}\right)\left(\frac{h_c}{b}\right)^2\left(\frac{E_f}{E_c}\right) \tag{16.46}$$

The buckling coefficient K for this type of construction has the same general characteristics as that of a flat homogeneous plate; namely, that for $a/b > 1$, successive minima occur for increasing numbers of half sine waves in the loading direction, each minimum having the same value of K. Hence, this minimum value can be taken as the lower bound for all panels where $a > b$. Therefore, Figure 3 and 5 of Reference [16.9] can be used. These figures make use of the ratio $E_c\bar{I}_c / E_f\bar{I}_f$, which for this construction is given by

$$\frac{E_c\bar{I}_c}{E_f\bar{I}_f} = \frac{E_c}{E_f}\frac{t_c}{t_f}\frac{1}{6\sin\theta} \tag{16.47}$$

For example, for panels with the unloaded edges simply supported the buckling coefficient K is given as follows:

$$K = \frac{\left[\dfrac{1}{\beta}+\beta^2\right]}{(1-\nu_f^2)+2(1+\nu_f)\beta^2}+\frac{E_c\bar{I}_c}{E_f\bar{I}_f\beta^2}$$

where here $\beta = a/b$.

There is no published analytical expression describing the overall instability of corrugated-core sandwich panels utilizing orthotropic materials and for in-plane compressive loads. However, it is not difficult to deduce the form by observing the difference in the analogous expression for honeycomb-core sandwich panels for isotropic and orthotropic materials.

From the isotropic expressions on pages 53, 82, an 96 of ANC-23 [16.1], it is seen that when the overall instability expressions are written for orthotropic materials E_f is replaced by $\sqrt{E_{fx}E_{fy}}$ when flexural properties are involved, while E_f is replaced by E_{fx} in extensional property expressions. It is therefore deduced that for corrugated-core panels utilizing orthotropic materials using Equations (16.44), (16.45) and the above, the critical load per unit width is

$$N_x = \frac{\pi^2\sqrt{E_{fx}E_{fy}}\,t_f h_c^2 K}{2b^2} \tag{16.48}$$

It is also hypothesized that Figures 3 and 5 of Reference [16.10] may be used to determine K if the following expressions are used instead of Equations (16.46) and (16.47):

$$V = \frac{\pi^2(1-\nu_{xy}\nu_{yxc})}{2\cos^2\theta\sin^2\theta}\left(\frac{t_f}{t_c}\right)\left(\frac{h_c}{b}\right)^2\frac{\sqrt{E_{fx}E_{fy}}}{\sqrt{E_{cx}E_{cy}}} \tag{16.49}$$

$$\frac{E_c \bar{I}_c}{E_f \bar{I}_f} = \frac{\sqrt{E_{cx}E_{cy}}}{\sqrt{E_{fx}E_{fy}}}\frac{t_c}{t_f}\frac{1}{6\sin\theta} \tag{16.50}$$

16.5.3 Face Plate Instability

From Figure 16.5 it is seen that each plate element of the face from A to B can buckle due to the axial loading N_x. Since the support conditions at A and B, the unloaded edges, are not known precisely, it is conservative to assume a simply support. Since for almost all constructions the panel length a is greater than the distance AB, the buckling coefficient K is taken as four. Anderson [16.11] discusses the effect of more complex buckling modes due to the interaction between face and core elements. However, he shows that, at most, the buckling coefficient would be 4.21 for simultaneous buckling of face and core elements. Hence the value of four appears very realistic as a conservative value. The face plate instability equation can therefore be written in terms of the quantities given in Figure 16.5 as

$$\sigma_{cr} = \frac{\pi^2 E_f}{12(1-\nu_f^2)}\frac{t_f^2}{h_c^2}\frac{1}{\tan^2\theta} \tag{16.51}$$

where σ_f is the critical stress in the face.

Utilizing the critical stress expression for an orthotropic plate given by Timoshenko and Gere [6.1], and using the terminology of Figure 16.5, the expression can be written as

$$\sigma_{f\,cr} = \frac{\pi^2 E_{0f} t_f^2}{12h_c^2(1-\nu_{xyf}\nu_{yxf})\tan^2\theta} \tag{16.52}$$

where

$$2E_{0i} = \sqrt{E_{ix}E_{iy}} + \nu_{yxi}E_{ix} + 2G_{xyi}(1-\nu_{xyi}\nu_{yxi}) \quad (i = c, f) \tag{16.53}$$

16.5.4 Web-Plate Instability

Similar to the above, the plate instability equation for a web element can be written as

$$\sigma_{wcr} = \frac{\pi^2 E_c}{3(1-\nu_c^2)}\frac{t_c^2}{h_c^2}\cos^2\theta \tag{16.54}$$

where σ_{wcr} is the critical stress in the web element. Likewise, for orthotropic composite materials

$$\sigma_{wcr} = \frac{\pi^2 E_{0c} t_c^2 \cos^2\theta}{3h_c^2(1-\nu_{xyc}\nu_{yxz})} \tag{16.55}$$

where E_{oc} is given by Equation (16.53).

16.5.5 Applied Load-Face Stress Relationship

By enforcing equal axial strains in the core and face to insure compatibility in the overall construction, the following important relationship is easily derived for the stress in the core material as a function of the face stresses,

$$\sigma_c = \sigma_f \frac{E_{cx}}{E_{fx}} \tag{16.56}$$

Then since $N_{xf} = 2\tau_f \sigma_f$ and $N_{xc} = \sigma A_c$, the face stress can be written in terms of the applied load per unit width $N_x = N_{xf} + N_{xc}$:

$$\sigma_f = \frac{N_x}{\left(\frac{E_c}{E_f}\frac{t_c}{\sin\theta} + 2t_f\right)} \tag{16.57}$$

For orthotropic composite materials the modulus values to use in Equations (16.56) and (16.57) are those in the load direction.

Both Equations (16.51) and (16.52) hold only when stresses in the face and core are both below the proportional limit of each material. Above the proportional limit an iterative procedure would be needed to insure compatibility in determining an analogous relation to Equation (16.56). If both core and face materials are the same $\sigma_c = \sigma_f$ and the procedures which follow hold above the proportional limit if a suitable plasticity reduction factor is used with the modulus of elasticity.

16.5.6 Weight Relationship

From Figure 16.5 it is seen that

$$W = 2\rho_f t_f + \rho_c \overline{A}_c + W_{ad} \tag{16.58}$$

where ρ_c and ρ_f are the weight density of the core and face material, respectively, and W and W_{ad} are the weight per unit planform area of the panel and the weight of the adhesive or other material used to join face and core, respectively.

16.5.7 Analysis and Design Methods

For the truss core panel subjected to a uniform in-plane compressive load per unit width, N_x, the face stress is given by Equation (16.57) and the core stress by Equation (16.56). These stresses must be maintained at or below the allowable compressive stress for the face material and core material, respectively. For orthotropic materials the properties in the load direction should be used (the x-direction). It is also necessary that, with a specific applied load N_x, neither overall buckling [Equations (16.45) or (16.48)] nor face plate buckling [Equations (16.51) or (16.52)] nor web plate buckling [Equations (16.54) or (16.55)] occur.

Once satisfied of the structural integrity, Equation (16.58) is used to determine the panel weight.

Under in-plane compressive loads, because the core in truss-core sandwich panels does resist a portion of the load, and because the construction has such a high flexural stiffness, truss-core sandwich panels can efficiently resist higher loads than honeycomb solid- or foam-core sandwich panels. As a secondary advantage, this type of construction can be used as a heat exchanger or a liquid storage container.

16.6 Elastic Stability of a Web-Core Sandwich Panel Subjected to a Uniaxial Compressive In-Plane Load

16.6.1 Introduction

Consider a flat web-core sandwich panel, generalized to include some arbitrary angle θ, as shown in Figure 16.6. The overall geometry and loading is given in Figure 16.1(a). There are five geometric variables; namely, the core depth h_c, the web thickness t_c, the face thickness t_f, the angle the web makes with a line normal to the faces θ, and the distance between web elements d_f.

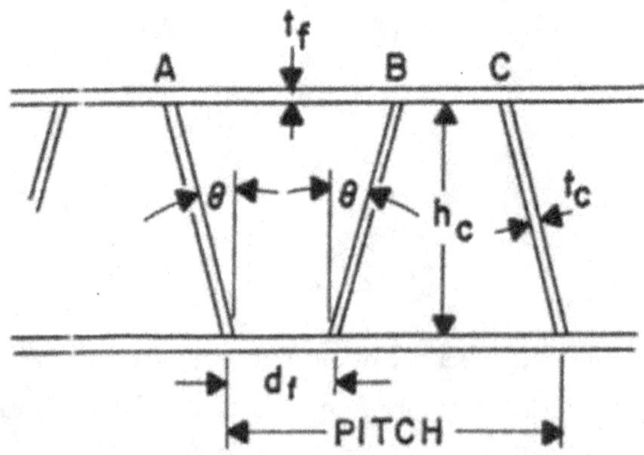

Figure 16.6. Generalized web-core sandwich panel. (Reprinted from References [1.7] and [10.1].)

The panel is considered to fail if any of the following instabilities occur: overall panel instability, local face buckling in the region from *A* to *B*, local face buckling in the region *B* to *C*, and web element buckling (see Figure 16.6). In each case, the expression used to mathematically describe the instabilities is the best available in the literature. Overstressing can occur if the stresses in either the face or core exceed established allowable stresses for the materials used. Hence, there are five geometric variables and four modes of failure

16.6.2 Elastic and Geometric Constants

The elastic and geometric constants for the web-core construction can be determined from those given in more general form by Libove and Hubka [16.7]. The core area per unit width and the moment of inertia of the core about the centroidal axis per unit width are given by

$$\overline{A}_c = t_c h_c/(d_f + h_c \tan\theta)\cos\theta \tag{16.59}$$

$$\overline{I} = t_c h_c^3/12\cos\theta(d_f + h_c \tan\theta) = \overline{A}_c h_c^2/12 \tag{16.60}$$

The transverse shear stiffness of the core, per unit width in the *x*-direction (D_{qY}) of an element of the sandwich panel cut by two *y-z* planes is seen to be negligible, due to the lack of structural continuity of the web core. Hence, following Libove and Hubka [16.7], as well as Seide [16.9].

$$\boldsymbol{D}_{qY} \to 0 \tag{16.61}$$

Hence, the transverse shear flexibility parameter in the *y*-direction is given by

$$V_y = \pi^2 E_f \overline{I}_f / b^2 D_{qY} \to \infty \tag{16.62}$$

This is not to say that the construction as a whole has no transverse-shear stiffness, but rather that the stiffness that does exist depends upon the faces to provide the continuity. In fact Seide discusses this case and states that for the case of $D_{qX} = \infty$ and $\boldsymbol{D}_{qY} = 0$, the compressive buckling load is finite, rather than being equal to zero, and varies with plate aspect ratio.

It should be noted that when $\theta \neq 0$ and $d_f = 0$, the construction has a continuous core and is called corrugated core or truss core, with the result that $\boldsymbol{D}_{qY} \neq 0$, and there is a specific value of θ for which the weight is a minimum. This construction is treated in Section 16.5.

The area moment of inertia per unit width of the faces considered as membranes, with respect to the sandwich middle surface, per unit width, is seen to be

$$\overline{I}_f = t_f h_c^2 / 2 \tag{16.63}$$

16.6.3 Applied Load-Face Stress Relationship

For a given N_x, the face stress and the core stress are given by Equations (16.56) and (16.57) where $\overline{A}_c$ is given by Equation (16.59).

16.6.4 Overall Panel Instability

Just as in Section 16.5.2 earlier, the best expression applicable to the overall instability of a web-core sandwich panel composed of isotropic materials under in-plane compressive loads is derived by Seide [16.9] and given as follows:

$$N_x = \pi^2 E_f \overline{I}_f K / 2b^2 \tag{16.64}$$

where K is the buckling coefficient derived and plotted in Figure 2 and 4 of Reference [16.9] for various boundary conditions. For this construction, with simply supported unloaded edges, K can be given explicitly as

$$K = \frac{[(1/\beta) + \beta^2]}{1 - \nu_f^2 + 2(1 + \nu_f)\beta^2} + \frac{E_c \overline{I}_c}{E_f \overline{I}_f \beta^2} \tag{16.65}$$

where here $\beta = a/b$.

There is no published analytical expression describing the overall panel instability of web-core sandwich panels utilizing orthotropic materials subjected to uniaxial compressive loads. However, as in previous sections, it is not difficult to deduce the form of the equation by observing the differences in the analogous expressions for honeycomb-core sandwich panels for isotropic and orthotropic materials.

The remainder of this subsection is identical to Section 16.5.2, but is repeated here for completeness. From the expressions in Reference [16.1], it is seen that when the overall instability expressions are written for orthotropic materials E_f is replaced by $(E_{fx} E_{fy})^{1/2}$ when flexural properties are considered, while E_f is replaced by E_{fx} when extensional properties are involved. It is therefore deduced that for web-core panels utilizing orthotropic materials, Equations (16.64) becomes, utilizing Equation (16.63),

$$N_x = \pi^2 (E_{fx} E_{fy})^{1/2} t_f h_c^2 K / 2b^2 \tag{16.66}$$

It is also hypothesized that Figures 2 and 4 of Reference [16.10] may be extended to find the buckling coefficient K for the orthotropic construction if $E_c \overline{I}_c / E_f \overline{I}_f$ is determined by

$$\frac{E_c\bar{I}_c}{E_f\bar{I}_f} = \frac{(E_{cx}E_{cy})^{1/2}t_c h_c}{6(E_{fx}E_{fy})^{1/2}t_f(d_f + h_c\tan\theta)\cos\theta} \qquad (16.67)$$

Likewise Equation (16.62) may be extended to orthotropic construction by replacing

$$(1-\nu_f^2) \text{ by } (1-\nu_{xyf}\nu_{yxf}), \text{ and } (1+\nu_f) \text{ by } (1+\nu_{xyf})$$

16.6.5 Face Plate Instability

From Figure 16.6 it is seen that the face elements *A-B* and *C-D* may each undergo an elastic instability under in-plane compressive loads. Since the unloaded edge supports are not known precisely, it is conservative to assume a simple support. Since for almost all panels, the panel length a is much greater than the width, which is the distance A to B or C to D, the buckling coefficient is taken as 4.

For the faces made of an orthotropic material the expression for the critical face stress given by Timoshenko and Gere [6.1] is used. In terms of the quantities defined in Figure 16.6 the critical stress can be written as follows for the region A to B:

$$\sigma_{f1} = \frac{\pi^2 E_{0f}t_f^2}{3(1-\nu_{xyf}\nu_{yxf})(d_f + 2h_c\tan\theta)^2} \qquad (16.68)$$

and in the region B to C:

$$\sigma_{f2} = \frac{\pi^2 E_{0f}t_f^2}{3(1-\nu_{xyf}\nu_{yxf})d_f^2} \qquad (16.69)$$

where $\boldsymbol{E_{0i}}$ $(\boldsymbol{i = c, f})$ is given by Equation (16.53).

16.6.6 Web-Plate Instability

Similar to the above, the plate instability equation for a web element composed of an orthotropic material can be written as

$$\sigma_c = \frac{\pi^2 E_{0c}t_c^2\cos^2\theta}{3(1-\nu_{xyc}\nu_{yxc})h_c^2} \qquad (16.70)$$

16.6.7 Applied Load-Face Stress Relationship

For the construction of Figure 16.6, it is seen that the load per unit width, N_x, is related to the face and web stresses, σ_f and σ_c, by the following relationship:

$$N_x = \sigma_c \overline{A}_c + 2\sigma_f t_f \tag{16.71}$$

By equating the axial strains in the core and face to insure compatibility in the overall construction, the following relationship is easily derived:

$$\sigma_c = \sigma_f E_{cx} / E_{fx} \tag{16.72}$$

Thus from Equations (16.71), (16.72), and (16.59), the load per unit width N_x is related to the face stress as follows:

$$N_x = \sigma_f \{[E_{cx} t_c h_c / E_{fx}(d_f + h_c \tan\theta] + 2t_f\} \tag{16.73}$$

Both Equations (16.72) and (16.73) hold only when stresses in the face and core are below the proportional limit of each material, where Hooke's Law applies. Above the proportional limit of either material an iterative procedure would be needed to insure compatibility in determining an analogous relationship to Equation (16.62), employing some reduced moduli $\overline{E}_{ci}$ and $\overline{E}_{fi}$ involving a plasticity reduction factor.

If both core and face materials are the same, $\sigma_f = \sigma_c$, then the same procedures apply for stresses above the proportional limit if a suitable plasticity reduction factor is used with the modulus of elasticity.

16.6.8 Weight Relation

From Figure 16.6, it is seen that

$$W = 2\rho_f t_f + \rho_c \overline{A}_c + W_{ad} \tag{16.74}$$

where ρ_c and ρ_f are the weight densities of the core and face material, respectively, and W and W_{ad} are the panel weight per unit of planform area and the weight of the adhesive or other material used to join face to core, respectively.

At the outset, independent of the material system, it is clear that from Equations (16.68) and (16.69) that for minimum weight construction of the web-core sandwich

$$\theta = 0^\circ \tag{16.75}$$

This is intuitively obvious. The result is that all other expressions used for optimization are simplified, and the construction shown in Figure 16.6, with $\theta = 0^\circ$ now assumes the familiar web-core configuration.

16.6.9 Design and Analysis

For the web-core panels subjected to an in-plane compressive load per unit width, N_x, it is clear that components must be sized such that neither the faces nor the web elements are overstressed [Equations (16.71) and (16.72)]. Also, overall buckling [Equations (16.64) through (16.67)], face-element buckling [Equations (16.68) through (16.69)], and web-plate buckling [Equation (16.70)] cannot be allowed. Finally the panel weight can be determined from Equation (16.74).

16.7 Buckling of Honeycomb Core Sandwich Panels Subjected to In-Plane Shear Loads

If the honeycomb core sandwich panel discussed in Section 16.3 and shown in Figure 16.1 is subjected to in-plane shear loads, the following equations in this section can be used for their design and analysis [10.1]. Again, the honeycomb sandwich panel faces can be overstressed, or any of four buckling modes can occur.

16.7.1 Applied Load-Face Stress Relationship

Since the honeycomb core does not take any of the applied in-plane shear load, the applied load face stress relationship is

$$\frac{N_{xy}}{2t_f} = \frac{N_{yx}}{2t_f} = \sigma_{xy} = \sigma_{yx} \tag{16.76}$$

16.7.2 Overall Panel Buckling

The overall panel buckling equation is

$$N_{xy} = \frac{\pi^2\sqrt{\overline{D}_x\overline{D}_y}}{b^2}L_{cr} \tag{16.77}$$

where the following the terminology of the original source document the buckling coefficient is L_{cr}; where, as seen earlier in the text, i.e., Equation (15.7),

$$\overline{D}_x = \frac{\overline{E}_{fx}h_c^2 t_f}{2(1-\nu_{xy}\nu_{yx})}, \quad \overline{D}_y = \frac{\overline{E}_{fy}h_c^2 t_f}{2(1-\nu_{xy}\nu_{yx})}$$

and where, if the stresses are above the proportional limit, the following plasticity-reduction factor may be used, where the subscript T refers to the tangent modulus.

$$\overline{E}_{fx} = \sqrt{E_{fx_T}E_{fx}}, \quad \overline{E}_{fy} = \sqrt{E_{fy_T}E_{fy}}$$

L_{cr} is a buckling coefficient defined by curves in Reference [16.10]. Overall buckling will occur when

$$V_x \leq k_1 S_{xi} \tag{16.78}$$

where V_x and k_1 have been defined previously in Section 16.3 and S_{xi} is defined as

$$S_{xi} = \frac{\left[\frac{1-\bar{\nu}}{2}\right]4r + (1-r)^2}{\left[\frac{1-\bar{\nu}}{2}\right](1+r^2)\left[r + \frac{b^2}{a^2}\right]} \tag{16.79}$$

where $\bar{\nu}$ in this expression is defined as

$$\bar{\nu} = \sqrt{\nu_{xy}\nu_{yx}}$$

and in this equation,

$$r = V_y / V_x = G'_{cx} / G'_{cy} = k_3 / k_4 \tag{16.80}$$

16.7.3 Core Shear Instability

Core shear instability occurs when $V_x \geq k_1 S_{xi}$, and is described by

$$N_{xy} = h_c \sqrt{G'_{cx} G'_{cy}} \tag{16.81}$$

16.7.4 Face Wrinkling

The appropriate equation for the critical face stress for face wrinkling is:

$$\sigma_{xy} = \left[\frac{2t_f}{3h_c}\frac{E'_c\sqrt{\bar{E}_{fx}\bar{E}_{fy}}}{(1-\nu_{xy}\nu_{yx})}\right]^{\frac{1}{2}} \tag{16.82}$$

16.7.5 Monocell Buckling

The equation to use to determine the critical face stress for monocell buckling is:

$$\sigma_{xy} = \frac{2\sqrt{E_{fxT}E_{fyT}}}{(1-\nu_{xy}\nu_{yx})}\left(\frac{t_f}{d}\right)^2 \tag{16.83}$$

where here, the authors of Equation (16.83) have employed the tangent modulus for the use of this equation with face stresses that are above the material's proportional limit.

16.7.6 Analysis and Design

Again, the weight equation is given by Equation (16.32). At this point any honeycomb-core panel can be designed and analyzed for specified in-plane shear loads. The faces must be sized that the allowable shear stress does not exceed Equation (16.76). Also, the panel sizes must preclude overall buckling [Equations (16.77) through (16.80)], core shear instability [Equation (16.81)], face wrinkling [Equation (16.82)], and face dimpling [Equation (16.83)]. Obviously, if the face material is isotropic then $E_{fx} = E_{fy}$.

16.8 Buckling of a Solid-Core or Foam-Sandwich Panel Subjected to In-Plane Shear Loads

Again the expression for overall buckling is given by Equation (16.77). This can occur when Equation (16.78) holds where for this case

$$V_x = \frac{\pi^2}{b^2}\frac{\sqrt{\bar{E}_{fx}\bar{E}_{fy}}}{G_{cx}}\frac{h_c t_f}{2(1-\nu_{xy}\nu_{yx})} \le k_1 S_{x1} \tag{16.84}$$

The core shear instability equation and the face-wrinkling equations are given by Equations (16.81) and (16.82), but in this case effective core properties $(\ )'$ must be replaced by the actual core material properties for the solid-core or foam-core materials.

Thus, the applied load-face stress relations and the weight equation are

$$N_{xy} = 2t_f\sigma_{xy} \tag{16.85}$$

$$W = 2t_f\rho_f + \rho_c h_c + W_{ad} \tag{16.86}$$

any solid-core sandwich panel can now be analyzed and designed for in-plane shear loads.

16.9 Buckling of a Truss-Core Sandwich Panel Subjected to In-Plane Shear Loads

16.9.1 Introduction

Consider the flat corrugated-core sandwich panel cross section of Figure 16.5. For a given material system, there are four geometric parameters with which to optimize; namely the core depth h_c, the web thickness t_c, the face thickness t_f, and the angle the web makes with a line normal to the faces θ. The overall panel to be considered is shown in planform in Figure 16.7. This panel of width b and length a is subjected to in-plane shear loads per unit edge distance N_{xy} and N_{yx} (lb/in).

In addition to overstressing, the panel is considered to fail if any of the following instabilities occur: overall instability, shear instability of the faces, and shear instability in the web. Thus, there are three modes of instability, and four geometric variables. Since panels in which the faces and cores utilize different orthotropic materials are the most general materials system, it is convenient to derive all expressions for that situation.

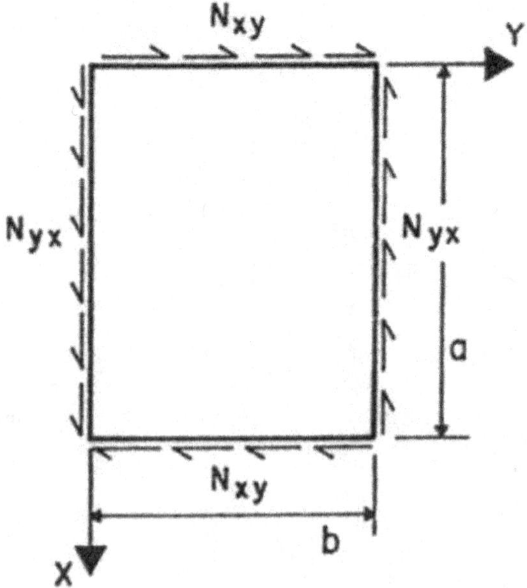

Figure 16.7. Planform view of panel. (Reprinted from References [10.1] and [16.13].)

16.9.2 Overall Stability

The overall stability of a truss-core sandwich panel under in-plane shear loading is give in ANC-23 [16.1] as well as other places as

$$\tau_f^{cr} = \frac{(E_{fy}^3 E_{fx})^{1/4}}{(1-\nu_{xyf}\nu_{yxf})}\left(\frac{h_c}{b}\right)^2 \bar{j} \tag{16.87}$$

where τ_f^{cr} is the critical shear stress in the face of the sandwich and where $\bar{j}$ is a buckling coefficient given by Vinson and Shore [16.12], related to the buckling coefficient j.

The coefficient j, found in [16.1], for orthotropic panels with simply-supported edges whose axes of elastic symmetry are parallel to the edges is given in Figure 16.8. In this figure, j is plotted as a function of B_2 and $1/r$ where

$$B_2 = \frac{2G_{xyf}(1-\nu_{xyf}\nu_{yxf}) + E_{fy}\nu_{xyf}}{\sqrt{E_{fx}E_{fy}}}, \quad \frac{1}{r} = \frac{b}{a}\left(\frac{E_{fx}}{E_{fy}}\right)^{\frac{1}{4}} \tag{16.88}$$

16.9.3 Face-Plate Instability

Looking at Figure 16.5, it is seen that each plate element of the faces from A to B can buckle due to the applied shear loads N_{xy} and N_{yx}. Since the support condition of the plate element along the edges depicted by A and B are not known precisely, it is conservative to assume that they are simply-supported edges. For such a case, the governing equation is given by Timoshenko and Gere [6.1] and others, for an orthotropic plate whose axes of elastic symmetry are parallel to the edges, of thickness h and width b,

$$\tau_{cr} = \frac{k}{3}\frac{(E_y^3 E_x)^{1/4}}{(1-\nu_{xy}\nu_{yx})}\left(\frac{h}{b}\right)^2 \tag{16.89}$$

In this expression k is a coefficient plotted in Reference [6.1] (Figure 9.42) as a function of two parameters: β and $1/\theta$. If one looks at the original figures, it is obvious that is $\beta = 1/r$ and $1/\theta = B_2$, then Figure 9.42 of Reference [6.1] and Figure 16.8 herein are identical. Hence, the k of Equation (16.89) is identical to j of Figure 16.8 under these conditions.

From Figure 16.5 it is seen that for the face-plate instability Equation (16.89) can be written as

$$\tau_{f\,cr} = \frac{k}{12}\frac{(E_{fy}^3 E_{fx})^{1/4}}{(1-\nu_{xyf}\nu_{yxf})}\frac{t_f^2}{h_c^2\tan^2\theta} \tag{16.90}$$

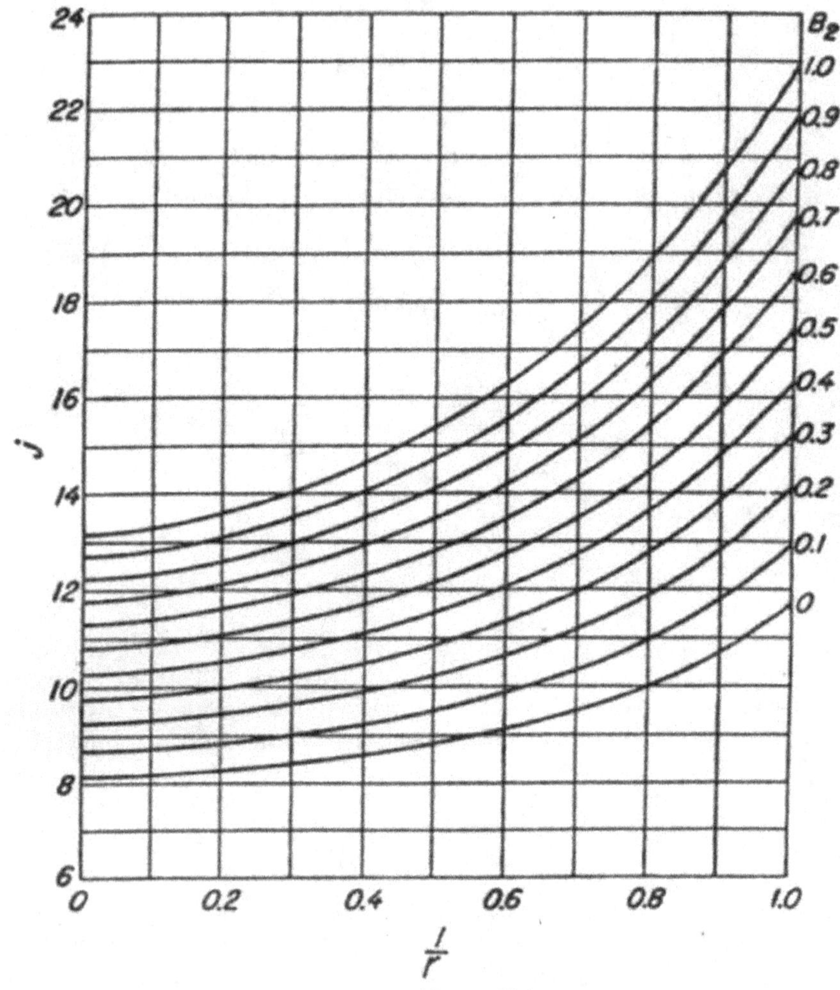

Figure 16.8. Buckling coefficient.

where $k_f = j$ is given by Figure 16.8, in which for this plate element B_2 is given by Equation (16.88) and

$$\frac{1}{r} = \frac{2h_c \tan\theta}{a}\left(\frac{E_{fx}}{E_{fy}}\right)^{\frac{1}{4}} \tag{16.91}$$

16.9.4 Web-Plate Instability

Likewise, the local plate elements of the triangulated core can become unstable due to shear stresses induced into the core by the shearing of the faces. Again, the conservative assumption is made here that the web elements are simply supported along the edges A and C depicted in Figure 16.5.

Referring to Equation (16.90) and the geometry of Figure 16.5, it is seen that the expression describing the web-plate instability can be written as

$$\tau_{f\,cr} = \frac{k_c \; [E_{yc}^3 E_{xc}]^{1/4}}{3\;(1-\nu_{xyc}\nu_{yxc})}\frac{t_c^2}{h_c^2}\cos^2\theta \tag{16.92}$$

where $k_c = j$ is found from Figure 16.8, where here

$$B_2 = \frac{2G_{xyc}(1-\nu_{xyc}\nu_{yxc})+E_{cy}\nu_{xyc}}{\sqrt{E_{cx}E_{cy}}} \quad \text{and} \quad \frac{1}{r} = \frac{h_c}{a\cos\theta}\left(\frac{E_{cx}}{E_{cy}}\right)^{\frac{1}{4}} \tag{16.93}$$

16.9.5 Applied Load-Face Stress Relationship

Looking at the construction shown in Figure 16.7 along the edges at $x = 0$ or $x = a$, the shear resultant N_{yx} is primarily resisted by the two faces. Even if the core elements are bonded or otherwise fastened to some edge fixture through which the shear N_{yx} is transmitted, little load will be introduced into the core web plates directly. Hence, the applied load-face stress relationship is taken to be

$$\frac{N_{yx}}{2t_f} = \frac{N_{xy}}{2t_f} = \tau_f \tag{16.94}$$

This is not to imply that loads are not introduced into the core elements by the faces, as shown below.

16.9.6 Core Stress-Face Stress Relationship

Consider the repeated unit of the triangulated-core construction shown in Figure 16.9. Due to the shearing deformations of the faces, shearing deformations occur in the core since the web element and the face are bonded or otherwise connected to their junction along db, gh, etc.

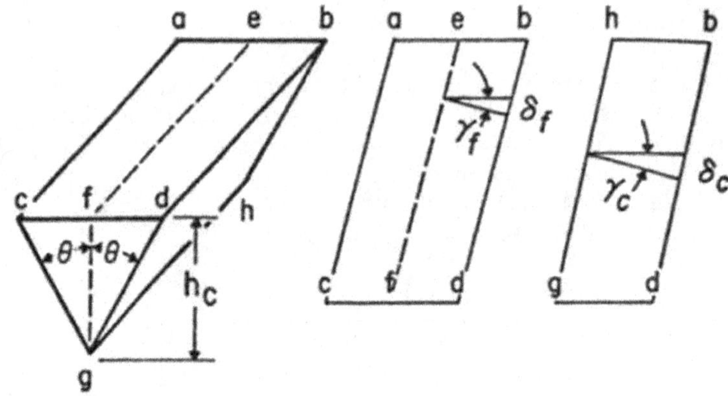

Figure 16.9. Triangulated-core construction. (Reprinted from Reference [16.13].)

The following relationships are valid in the elastic range, where ε_{xyi} $(i = c, f)$ are the shearing strains and the other symbols are given in Figure 16.5, 16.6 and 16.9.

$$\gamma_1 = 2\varepsilon_{xyi} = \frac{\tau_i}{G_{xyi}}, \quad \gamma_f = \frac{\delta_f}{h_c \tan\theta}, \quad \gamma_c = \frac{\delta_c}{h_c}\cos\theta$$

For compatibility of deformations, $\delta_c = \delta_f$ hence

$$\tau_c = \frac{G_{xyc}}{G_{xyf}}\tau_f \sin\theta \tag{16.95}$$

Note that $\tau_c < \tau_f$ when the same materials are used for faces and core.

16.9.7 Weight Relationship

The weight relationship from Figure 16.5 is repeated here for completeness,

$$W = 2\rho_f t_f + \rho_c \overline{A}_c + W_{ad} = 2\rho_f t_f + \frac{\rho_c t_c}{\sin\theta} + W_{ad} \tag{16.96}$$

where ρ_f and ρ_c are the weight density of the face and core materials, respectively; W_{ad} is the weight in $\mathrm{lb/in^2}$ of planform area of the adhesive or any other material used to join face and core; W is the total weight in $\mathrm{lb/in^2}$ of planform area of the entire panel; and $\overline{A}_c = t_c / \sin\theta$.

16.10 Buckling of a Web-Core Sandwich Panel Subjected to In-Plane Shear Loads

16.10.1 Overall Buckling

For the web-core sandwich subjected to in-plane shear loads, it is seen that all the applied load is taken by the face components, and none of the applied shear can be induced into the web elements because they are perpendicular to the faces when $\theta = 0°$ (see Figure 16.6).

The overall panel in-plane shear instability is given as follows, following Libove and Hubka [16.7], and Vinson and Shore [16.12]

$$\tau_{f\,\mathrm{cr}} = \frac{(E_{fy}^3 E_{fx})^{1/4}}{1-\nu_{xyf}\nu_{yxf}}\left(\frac{h_c}{b}\right)^2 \frac{1}{1+4\left[\dfrac{K_m(V=0)}{K_m}-1\right]} j \tag{16.97}$$

where $\tau_{f\,\mathrm{cr}}$ is the critical shear stress in the face, and where for this construction

$$K_m = \frac{A}{B_1 C_1 + B_3 C_2 + V_x A} \tag{16.98}$$

$$K_m(V=0) = B_1 C_1 + 2B_2 C_2 + \frac{C_3}{B_1} \tag{16.99}$$

$$A = C_1 C_3 - B_2^2 C_2^2 + B_3 C_2\left(B_1 C_1 + 2B_2 C_2 + \frac{C_3}{B_1}\right) \tag{16.100}$$

$$B_1 = (E_{fy} E_{fx})^{1/2} \tag{16.101}$$

$$B_2 = \frac{2G_{xyf}(1-\nu_{xyf}\nu_{yxf}) + E_{fy}\nu_{xyf}}{(E_{fx}E_{fy})^{1/2}} \tag{16.102}$$

$$B_3 = \frac{G_{xyf}(1-\nu_{xyf}\nu_{yxf})}{(E_{fx}E_{fy})^{1/2}} \tag{16.103}$$

C_1 through C_4 are given in Reference [16.1] for various boundary conditions and constructions. For the web-core construction with in-plane shear loading ($n = 1$) the constants are

All edges simply supported:

$$C_1 = C_4 = a^2/b^2, \; C_2 = 1, \; C_3 = b^2/a^2 \tag{16.104}$$

$x = 0$, a simply supported; $y = 0$, b clamped:

$$C_1 = \frac{16}{3}\frac{a^2}{b^2}, \; C_2 = \frac{4}{3}, \; C_3 = b^2/a^2, \; C_4 = \frac{4}{3}\frac{a^2}{b^2} \tag{16.105}$$

$x = 0$, a clamped; $y = 0$, b simply supported:

$$C_1 = C_4 = \frac{3}{4}\frac{a^2}{b^2}, \; C_2 = 1, \; C_3 = 4\frac{b^2}{a^2} \tag{16.106}$$

All edges clamped:

$$C_1 = 4C_4 = 4\frac{a^2}{b^2}, \; C_2 = \frac{4}{3}, \; C_3 = 4\frac{b^2}{a^2} \tag{16.107}$$

The coefficient j is obtained from Reference [6.1] for panels with simply supported edges, given herein as Figures 16.8, where for this construction, B_2 is given by Equation (16.102) above and

$$1/r = (b/a)(E_{fx}/E_{fy})^{1/4} \tag{16.108}$$

16.10.2 Face-Plate Buckling

From Figure 16.5, it is seen that each face plate can buckle between each web element due to the applied shear loads N_{xy} and N_{yx}. Since the support conditions along the face element at the web element are not known precisely, it is conservative to assume simply-supported edges. For that case Timoshenko and Gere [6.1] provide a stability equation for an orthotropic plate, which can be written as

$$\tau_{f\,cr} = \frac{k_f}{3}\frac{(E_{fy}^3 E_{fx})^{1/4}}{(1-\nu_{xyf}\nu_{yxf})}\frac{t_f^2}{d_f^2} \tag{16.109}$$

where it can be shown that $k_f = j$ of Figure 16.8, B_2 is given by Equation (16.102) and $1/r$ is

$$\frac{1}{r} = \frac{d_f}{a}\left(\frac{E_{fx}}{E_{fy}}\right)^{\frac{1}{4}} \tag{16.110}$$

Note that for $\theta = 0^\circ$, when the panel is subjected to in-plane shear loads, the web plates are unloaded. If $\theta \neq 0^\circ$, then some load is introduced into the web plates analogous to the truss-core construction discussed earlier.

16.10.3 Applied Load-Face Stress Relationship

Looking at the construction of Figure 16.5, it is seen that in-plane shear resultants N_{xy} and N_{yx} are resisted by the two faces. Hence, the applied load-face stress relationship is seen to be, again,

$$\frac{N_{xy}}{2t_f} = \frac{N_{yx}}{2t_f} = \tau_f \tag{16.111}$$

16.10.4 Weight Relationship

The weight of the web-core sandwich panel per unit planform area is given by

$$W - W_{ad} = 2\rho_f t_f + \rho_c \overline{A}_c = 2\rho_f t_f + \frac{\rho_c t_c h_c}{d_f} \tag{16.112}$$

where W_{ad} is the weight of the adhesive or other bonding material.

16.11 Other Considerations

Of course, the theory and analysis becomes far more complex if the sandwich panel is curved, has dissimilar faces, initial imperfections, and dynamic loading. Librescu, Hause, and Camarda [16.14] developed methods of analysis for giving accurate predictions for static and dynamic behavior subjected to complex mechanical and thermal loads in the pre-buckled and post-buckled ranges, and made comparisons with available experimental data.

Smidt [16.15] also investigated curved sandwich panels because of their applications to high-speed boats, containers, tanks and aircraft. His experimental research was compared to finite element solutions he obtained. He comments that in Equation (16.54) the coefficient can vary from 0.5 to 0.8.

Also, Laine and Rio [16.16] found that for foam-core sandwich panels typical of those used in ship construction, loaded up to 60% of their elastic-buckling load, creep (that they define as more than 10% influence on deformation) is significant. They therefore advocate either the use of a factor of safety of 2 on the critical buckling load, or introduce a creep law into the structural analysis.

The vibration of sandwich plates and panels can be studied by the use of vibration equations and solutions derived for isotropic and composite plates if the proper stiffness quantities are used. Recommended reading includes papers by Meunier and Shenoi [16.17-16.19].

16.12 References

16.1. Anon (1955) Materials, Properties and Design Criteria Part II, Sandwich Construction for Aircraft, *Military Handbook 23* (ANC-23), Department of the Air Force Research and Development Command; Department of the Navy, Bureau of Aeronautics; and Department of Commerce, Civil Aeronautics Administration, 2nd Edition.

16.2. Heath, W.G. (1960) Sandwich Construction, Part 2: The Optimum Design of Flat Sandwich Panels, *Aircraft Engineering*, Vol. 32, August, pp. 230-235.

16.3. Hemp, (1948) On a Theory of Sandwich Construction, *College of Aeronautics Report*, No. 15.

16.4. Hoff, N.J. and Mautner, S.E. (1945) Buckling of Sandwich Type Panels, *Journal of the Aeronautical Sciences*, Vol. 12, No. 3, pp. 285-297.

16.5. Kaechele, L.E. (1957) Minimum Weight Design of Sandwich Panels, *USAF Project Rand Research Memorandum*, RM 1895, AD-133011, March.

16.6. Dreher, G. (1992) Stability Failure of Sandwich Structures, in *Sandwich Constructions 2 – Proceedings of the Second International Conference on Sandwich Construction*, Gainesville, Florida, Editors, D. Weissman-Berman and K-A. Olsson, EMAS Publications, United Kingdom.

16.7. Libove, C. and Hubka, R.E. (1951) Elastic Constants for Corrugated-Core Sandwich Plates, *NACA TN 2289*.

16.8. Anderson, M.S. (1959) Optimum Proportions of Truss Core and Web-Core Sandwich Plates Loaded in Compression, *NASA TN D-98*, September.

16.9. Seide, P. (1952) The Stability Under Longitudinal Compression of Flat Symmetric Corrugated-Core Sandwich Plates with Simply Supported Loaded Edges and Simply Supported or Clamped Unloaded Edges, *NACA TN 2679*.

16.10. Kuenzi, E.W., Ericksen, W.S. and Zahan, J., Shear Stability of Flat Panels of Sandwich Construction, *Forest Products Laboratory Report 1560*, Rev. N62-13403.

16.11. Anderson, M.S. (1959) Local Instability of the Elements of a Truss-Core Sandwich Plate, *NASA TR R-30*.

16.12. Vinson, J.R. and Shore, S. (1967) Structural Optimization of Flat Corrugated Core Sandwich Panels Under In-Plane Shear Loads and Combined Uniaxial Compression and In-Plane Shear Loads, *Naval Air Engineering Center Report NAEC-ASC-1110*, July.

16.13. Vinson, J.R. (1986) Minimum Weight Triangulated (Truss) Core Composite Sandwich Panels Subjected to In-Plane Shear Loads, in *Composites 86 – Recent Advances in Japan and the United States*, Eds. K. Kawata, S. Umekawa and A. Kobayashi, *Proceedings of the Third Japan-U.S.*

Conference on Composite Materials, Japan Society of Composite Materials, Publishers, Tokyo, pp. 663-671

16.14. Librescu, L., Hause, T. and Camarda, C.J. (1997) Geometrically Nonlinear Theory of Initially Imperfect Sandwich Curved Panels Incorporating Nonclassical Effects, *AIAA Journal*, Vol. 35, No. 8, August, pp. 1393-1403.

16.15. Smidt, S. (1996) Testing of Curved Sandwich Panels and Comparison with Calculations Based on the Finite Element Method, in *Sandwich Construction 3 – Proceedings of the Third International Conference on Sandwich Construction*, Southampton, Great Britain, Editor, H.G. Allen, EMAS Publications, United Kingdom, pp. 665-680.

16.16. Laine, C. and Rio, G. (1992) Buckling of Sandwich Panels Used in Shipbuilding: Experimental and Theoretical Approach, in *Sandwich Construction 2 – Proceedings of the Second International Conference on Sandwich Construction*, Gainesville, Florida, Editors, D. Weissman-Berman and K-A. Olsson, EMAS Publications, United Kingdom, pp. 243-252.

16.17. Meunier, M. and Shenoi, R.A. (1999) Free Vibration Analysis of Composite Sandwich Plates, *Journal of Mechanical Engineering Science, Part C.*

16.18. Meunier, M. and Shenoi, R.A. (1999) Dynamic Behavior of Composite Sandwich Structures Modified Using HSOT, *Proceedings of the European Conference on Computational Mechanics*, Munich, September.

16.19. Meunier, M. and Shenoi, R.A. (2001) Dynamic Analysis of Composite Sandwich Plates With Damping Modeled Using High-Order Shear Deformation Theory, *Composite Structures*, pp. 243-254.

16.13 Problems

16.1. For the sandwich panel of Problem 15.1,

(a) What is the overall buckling load, $N_{x\,cr}$ (lb/in) the sandwich plate can withstand, using classical plate theory?

(b) What is the overall buckling load $N_{x\,cr}$ (lb/in) the plate can withstand including transverse shear deformation effects?

(c) What is the critical load $N_{x\,cr}$ for core-shear instability?

(d) What is the critical load $N_{x\,cr}$ for face wrinkling?

16.2. If the sandwich panel of Problem 16.1 were clamped on all four edges, solve 16.1(b), (c) and (d) again for this case.

16.3. Consider a foam-core sandwich panel that is $16'' \times 16''$ in planform dimensions, simply supported on all four edges. Each face is made of cross-ply Kevlar 49/epoxy composite whose properties are given in Table 17.1. The face stacking sequence is $[\mathbf{0^\circ/90^\circ/90^\circ/0^\circ}]$, wherein each ply is 0.0055 inches thick. The foam core has $\boldsymbol{G_c} = \mathbf{15{,}000}$ psi, and ρ_c 2 lb/ft^3, and is 3/4 inches thick.

(a) Determine each element of the $[A]$, $[B]$, and $[D]$ stiffness matrices for the sandwich.

(b) Could any of the perturbation techniques of Section 11.8 be used to solve problems for this plate?

(c) If the sandwich plate is subjected to an in-plane compressive load in the x-direction, what is the critical buckling load per unit edge distance, $N_{x\,cr}$ using classical-plate theory?
(d) What is the mode of buckling?
(e) If transverse shear deformation effects are included, what is the overall critical buckling load, $N_{x\,cr}$?
(f) What is the fundamental natural frequency in Hz, using classical plate theory? Equation (13.4) can be used.
(g) What is the answer to (f) if transverse shear deformation effects are included? Equation (13.15) can be used.

16.4. Consider a hexagonal-cell honeycomb core sandwich panel, simply supported on all four edges, subjected to an in-plane compressive load, N_x (lb/in). The face material is made of unidirectional T300/934 graphite epoxy, whose properties are given in Table 17.1. The core is aluminum whose properties are $\boldsymbol{E} = \mathbf{10 \times 10^6}$ psi, and $\nu = 0.3$. The dimensions of the panel are:

$a = 48$ in.	$t_f = 0.066$ in.	$\boldsymbol{h_c} = \mathbf{0.50}$ in.
$b = 30$ in.	$\boldsymbol{t_c} = \mathbf{0.001}$ in.	$d = 0.50$ in.

(a) At what face stress will face dimpling occur?
(b) Would face dimpling ever occur in this panel under this load situation? Why?
(c) At what face stress will face wrinkling occur?
(d) For this panel, which will have the lower buckling stress, overall panel buckling (i.e. $V_x \le k_1 B_1 r$) or core-shear instability (i.e. $V_x \ge k_1 B_1 r$)?
(e) At what face stress will core-shear instability occur?
(f) Of all modes of failure, how will this sandwich panel fail, when subjected to an increasing in-plane compressive loading?
(g) Therefore what is the greatest in-plane compressive load N_x (lb/in of width) that this sandwich panel can withstand?
(h) What is the greatest in-plane compressive load, (pounds) that this sandwich panel can withstand without failing?
(i) What is the total weight of the panel if the density of the faces is 0.055 lb/in^3 and the density of the aluminum is 0.100 lb/in^3?

16.5. Regarding Problem 16.4, consider a monocoque laminated panel of the same size made of the unidirectional T300/934 graphite/epoxy, that is $2t_f = 2 \times 0.066'' = 0.132''$ thick, simply supported on all four edges. Thus this panel has the same face material, but no core.
(a) What in-plane buckling load per unit width, $N_{x\,cr}$ can this monocoque panel withstand before buckling?
(b) What is the face stress at that load?
(c) What is the panel weight per unit planform area?
(d) What thickness would be necessary to have the same buckling load $N_{x\,cr}$ as the sandwich panel considered in Problem 16.4 above?

(e) What would the monocoque panel weigh in *d* above? What is the comparison of the weight of this panel to the sandwich panel of 16.4 above?

16.6. A designer is considering various alternatives in the design of a rectangular sandwich panel measuring 60 inches by 30 inches in planform. The panel must withstand an in-plane compressive load in the longer direction of $N_x = -600$ lb/in. Consider the foam core to weigh 15 lb/in^3, and $h_c = 25t_f$.

(a) For a sandwich with aluminum faces ($E = 10\times10^6$ psi, $\nu = 0.3$, $\rho = 0.1$ lb/in^3, $\sigma_{all} = \pm 30{,}000$ psi). To simply support this panel or all four edges will require 20 lb of support structure; to clamp it on all four edges will require 40 lb of support structure. For the plate to not be overstressed or buckle, which design will result in less system weight?

(b) One alternative is to use magnesium faces ($E = 6.5\times10^6$ psi, $\nu = 0.3$, $\rho = 0.065$ lb/in^3, $\sigma_{all} = \pm 30{,}000$ psi). Would a simply supported or clamped magnesium-faced system weigh less than the better aluminum-faced sandwich system?

(c) What about steel faces ($E = 30\times10^6$ psi, $\nu = 0.3$, $\rho = 0.283$ lb/in^3, $\sigma_{all} = \pm 60{,}000$ psi); could a steel-faced sandwich result in weight savings?

CHAPTER 17

STRUCTURAL OPTIMIZATION TO OBTAIN MINIMUM WEIGHT SANDWICH PANELS

17.1 Introduction

It is sometimes sufficient to be able to design and analyze sandwich panels or any other structure to insure its structural integrity. However, it is very desirable to be able to design a sandwich panel that not only successfully resists the applied load but is also of minimum weight. A design can proceed the search for the minimum weight solution by trial and error through examining all of the possible combinations of materials and thicknesses for each sandwich element. This can produce a minimum weight design for (only) those options examined. However, even with excellent insight and intuition, this random walk approach is time consuming, and is not guaranteed to provide the absolute minimum weight panel.

Fortunately, more rational methods have been developed so that the absolute minimum weight can be found analytically for honeycomb, solid, foam, truss-, and web-core rectangular sandwich panels. The methods apply to panels which are subjected to either an in-plane compressive load, N_x, or to in-plane shear loads, N_{xy}. Each optimization uses the best available equations (established through long use and given in Chapter 16 earlier) to depict the various failure modes. When, and if, new and better equations are formulated, they can be used to develop the same optimization procedures as those presented here.

In each case, the age-old principle of the "weakest link in the chain" is applied. The philosophy is that with any sandwich panel there are several failure modes, each independent of the others, and any one of which will result in the panel's failure. Each failure mode is thus a "link" in the sandwich "chain" that resists the applied load, and failure of any one "link" means failure of the whole "chain". Logically, the most efficient chain is one in which each link fails simultaneously. In the sandwich, each "link" has an associated weight varying directly with its load carrying ability. Therefore, the minimum weight panel ("chain") is found by ensuring that the failure modes ("links") occur simultaneously. Conversely, one could say that if a particular mode of failure is significantly greater or higher than the others, then the material associated with this excess strength or buckling resistance could either be removed or could be reallocated to bring all modes closer to equality. This extra material is truly "dead weight" (and cost) as the overall panel will fail via some other mode before ever reaching the load associated with the stronger mode of failure.

In performing the optimization, unique expressions are obtained for each geometric variable of the minimum weight panel. Additionally, a "universal relationship" is found which relates the applied load index to unique values of face and core element stresses for given materials and boundary conditions. The methods developed are also applicable to panels at any steady state temperature if the material properties are known at that temperature. In order to select the best material systems to achieve minimum weight for a given load, material property figures of merit are also found; these figures of merit are tabulated for many of the current material systems. The optimization methods have been found to be extremely useful for the following reasons:

1. Unique values of each geometric variable are found in order to obtain the minimum weight panel for a specified load index, material system, panel dimensions, and boundary conditions.
2. Various material systems can be compared for best materials selection.
3. The best stacking sequence can be determined for laminated composite face materials.
4. The optimum sandwich panel weight can be compared with other types of sandwich, reinforced panel, or monocoque architectures.
5. Weight penalties for non-optimal construction can be determined rationally.

Much of the text of this chapter parallels that of Vinson [10.1, Chapter 10].

17.2 Minimum Weight Optimization of Honeycomb Core Sandwich Panels Subjected to a Unidirectional Compressive Load [17.2]

For this type of panel and load, there can be four buckling failure modes: overall buckling, Equations (16.9) through (16.18); core shear instability, Equation (16.19); face wrinkling instability, Equations (16.22) and (16.23); and monocell buckling, Equation (16.24). For the same construction there are four dependent variables for which the analyst/designer can specify, namely, the face thickness, t_f, the core depth, h_c, the core cell size, d, and the core wall thickness, t_c. In addition the applied load is related to the face stress by (16.8). Therefore for any panel size, boundary conditions, and given face and core material, unique values of each dependent variable are determined by the philosophy expressed earlier through solving four equations with four unknowns. In addition it is found that there exists one and only one face stress, σ_0, for a given load index (N_x/b), that will result in minimum weight. If the face is stressed below or above this optimum value the sandwich panel will weigh more.

It was shown that overall buckling and core shear instability occur at the same stress under the conditions given by (16.20) and (16.21), that is,

$$V_x \quad k_1 B_1 r .$$

Equations (16.21), (16.19), (16.23), (16.25) and (16.9) yield the universal relationship for optimum honeycomb sandwich panels with orthotropic facings and core under uniaxial compression,

$$\left(\frac{k_2}{k_3 k_1 r}\right)^{\frac{1}{2}}\left(\frac{N_x}{b}\right) \quad \frac{2(3)^{1/2}(1-\nu_{xy}\nu_{yx})}{\pi}\left(\frac{G_c}{E_c}\right)^{\frac{1}{2}} \frac{\sigma_0^2}{E_{fx}^{3/4}E_{fy}^{1/4}} \tag{17.1}$$

It must be noted that, in (17.1), σ_0, the optimum face stress, must be limited to some predetermined allowable stress (usually defined by the yield strength or ultimate strength dividing by appropriate factors of safety for each such that the lower value of the two is the allowable stress); hence, there is an upper bound on the load that can be carried by the panel for a given material system. Therefore, when given a set of boundary conditions (denoted by k_1) and any type of honeycomb core construction (denoted by k_2, k_3, and r), it is seen that the universal relationship relates the load index (N_x/b) to the optimum face stress where other items in the equations are material properties only. It is independent of geometric variables and it establishes a unique optimum face buckling stress, σ_0, for each value of the load index.

Explicit values of the unique optimum geometric variables as functions of the load index (N_x/b) are determined to be, from [17.1] and [17.2],

$$h_c = b\left(\frac{2}{\pi}\right)^{\frac{3}{4}}\left(\frac{k_2}{3k_3}\right)^{\frac{1}{8}}(1-\nu_{xy}\nu_{yx})^{1/4}(k_1 r)^{3/8}\left(\frac{E_c}{G_c}\right)^{\frac{1}{8}}\frac{E_{fy}^{1/16}}{E_{fx}^{5/16}}\left(\frac{N_x}{b}\right)^{\frac{1}{4}} \tag{17.2}$$

$$d = \frac{b2^{1/4}}{\pi^{3/4}}\left(\frac{3k_3k_1 r}{k_2}\right)^{\frac{3}{8}}(1-\nu_{xy}\nu_{yx})^{1/4}\left(\frac{G_c}{E_c}\right)^{\frac{3}{8}}\frac{E_{fxT}^{1/4}E_{fy}^{1/16}}{E_{fx}^{9/16}}\left(\frac{N_x}{b}\right)^{\frac{1}{4}} \tag{17.3}$$

$$t_c = b\left[\frac{3}{2k_2k_3}\frac{1}{G_cE_c}\sqrt{\frac{E_{fxT}}{E_{fx}}}\right]^{\frac{1}{2}}\left(\frac{N_x}{b}\right) \tag{17.4}$$

$$t_f = b\left(\frac{3k_3k_1 r}{k_2}\right)^{\frac{1}{4}}\left[\frac{1-\nu_{xy}\nu_{yx}}{2\pi}\right]^{\frac{1}{2}}\left(\frac{G_c}{E_c}\right)^{\frac{1}{4}}\frac{(N_x/b)^{1/2}}{E_{fx}^{3/8}E_{fy}^{1/8}} \tag{17.5}$$

Through the use of the universal relationship given in (17.1), the explicit values of the optimum geometric variables in terms of optimum critical stresses, σ_0, are written as

$$h_c = b\left(\frac{2}{\pi}\right)(k_1 r)^{1/2}(1-\nu_{xy}\nu_{yx})^{1/2}\frac{\sigma_0^{1/2}}{E_{fx}^{1/2}} \tag{17.6}$$

$$d = \frac{b}{\pi}\left[\frac{6(1-\nu_{xy}\nu_{yx})k_3k_1 r}{k_2}\right]^{\frac{1}{2}}\left(\frac{G_c}{E_c}\right)^{\frac{1}{2}}\frac{E_{fxT}^{1/4}\sigma_0^{1/2}}{E_{fx}^{3/4}} \tag{17.7}$$

$$t_c = \frac{b3\sqrt{2}}{\pi k_2}(k_1 r)^{1/2}(1-\nu_{xy}\nu_{yx})\frac{E_{fxT}^{1/4}\sigma_0^2}{E_c E_{fx} E_{fy}^{1/4}} \tag{17.8}$$

$$t_f = b\left(\frac{3k_3k_1r}{k_2}\right)^{\frac{1}{2}}\frac{(1-\nu_{xy}\nu_{yx})}{\pi}\left(\frac{G_c}{E_c}\right)^{\frac{1}{2}}\frac{\sigma_0}{E_{fx}^{3/4}E_{fy}^{1/4}}. \tag{17.9}$$

It should be noted that should if a designer/analyst decides that face dimpling can be ignored because the monocell buckling does not necessarily mean panel failure, then equating the three remaining buckling stresses provides the same values for the core depth, Equation (17.6), for the face thickness, Equation (17.9), and a ratio for $(\boldsymbol{t_c / d})$ equal to Equation (17.8) divided by (17.7). So either way the optimum minimum weight panel remains the same.

The expressions for the weight per unit planform area for the optimum panel as a function of optimum face stress, σ_0, and the load index $(\boldsymbol{N_x / b})$ can be found by substituting the above expressions into (16.32). The results are expressed both in terms of the optimum face stress and the applied load index for ease of use.

$$\begin{aligned} W = {} & \left(\frac{3k_3k_1r}{k_2}\right)^{\frac{1}{2}}\frac{2(1-\nu_{xy}\nu_{yx})b}{\pi} \\ & \times\left(\frac{G_c}{E_c}\right)^{\frac{1}{2}}\frac{\sigma_0}{E_{fx}^{3/4}E_{fy}^{1/4}}\left[\rho_f+\rho_c\frac{k_2}{k_3}\frac{\sigma_0}{G_C}\right]+W_{ad} \end{aligned} \tag{17.10}$$

$$\begin{aligned} W = {} & \left(\frac{3k_3k_1r}{k_2}\right)^{\frac{1}{4}}\left[\frac{2(1-\nu_{xy}\nu_{yx})}{\pi}\right]^{\frac{1}{2}} \\ & \times\left(\frac{G_c}{E_c}\right)^{\frac{1}{4}}\frac{b(N_x/b)^{1/2}}{E_{fx}^{3/8}E_{fy}^{1/8}}\left[\rho_f+\rho_c\frac{k_2}{k_3}\frac{\sigma_0}{G_C}\right]+W_{ad}. \end{aligned} \tag{17.11}$$

Several very important and interesting conclusions can be drawn from (17.11) about the weight of optimized honeycomb sandwich panels under uniaxial compressive loads. It is seen that in the selection of the materials to use in these optimum panels, the core material with the highest ratio of G_c / ρ_c will result in minimum weight. For the selection of a facing material, the material that has the highest ratio of $E_{fx}^{3/8}E_{fy}^{1/8} / \rho_f(1-\nu_{xy}\nu_{yx})$ for the particular applied load index $(\boldsymbol{N_x / b})$ will result in the panel of lowest weight. For optimized honeycomb sandwich panel construction, the ratio of core weight to face weight is

$$\frac{W_c}{W_f} = \frac{\rho_c}{\rho_f} \frac{k_2}{k_3} \frac{\sigma_0}{G_c}. \tag{17.12}$$

Therefore, it is obvious that, for a construction employing the same material in both faces and honeycomb core, the great majority of the weight is in the faces. Also note that the ratio of the core weight to face weight for an optimum panel is independent of boundary conditions and the orthotropy factor r. In (17.11), it is seen that the panel weight varies as the one-fourth power of the boundary condition factor k_1. For a panel that is simply supported on the unloaded edges, $k_1 = 1$. In a panel that is clamped on the unloaded edges, $k_1 = \frac{3}{4}$. Therefore, the ratio of the weight of an optimum panel simply supported on the unloaded edges to the weight of an optimum panel clamped along the unloaded edges is 1.0745. This is a slight over-simplification, but it can be concluded that for a given load index an optimum honeycomb sandwich panel simply supported on the unloaded edges weighs no more than 1.0745 times the weight of the optimum panel with unloaded edges clamped.

This result has major implications. Foremost, it implies that, in almost all cases of sandwich panels subjected to in-plane compression, the optimizations should be conducted for simply supported boundary conditions on all edges. (Note that the boundary conditions on the loaded edges have no effect for this loading). Also:

1. Such an optimization will result in the panel weighing at most 7.45% over an optimum panel whose unloaded edges are clamped. Choosing a simply supported panel is thus conservative as far as all dimensions selected, enabling the panel to have additional structural integrity even when the edges are clamped or partially clamped.
2. In actual construction, the auxiliary structural elements required to make the unloaded edges clamped would possibly offset the potential saving of 7.45% in weight resulting from the clamped boundary condition. In the final analysis, the totally clamped panel assembly would weigh more.
3. It is virtually impossible to insure a truly clamped edge; hence, most panel edge conditions are somewhere between the conditions of fully simply supported and totally clamped. The choice of simple support conditions for the optimization is therefore conservative as well as rational.

For a honeycomb core sandwich panel subjected to an in-plane compressive load, independent of the boundary condition, the "best" face material is determined from (17.11) to be

$$FM = \frac{E_x^{3/8} E_y^{1/8}}{\rho_f (1-\nu_{xy}\nu_{yx})^{1/2}}. \tag{17.13}$$

It is interesting to use this Factor of Merit (FM) to some of the material systems currently available. The values in Table 17.1 are for 70 °F, and because the Poisson's ratios are rarely given tables of mechanical properties for the new materials, it is assumed below that $(1-\nu_{xy}\nu_{yx}) \approx 1$. It is seen clearly that in this application composite materials are better than any all-metallic construction. It is also clear that graphite/epoxy and boron/aluminum composites clearly result in the minimum weight construction. Even boron/epoxy is quite competitive.

Table 17.1. Face Material Comparison, Based on Factor of Merit for an Optimized Honeycomb Core Sandwich Panel Subjected to an In-Plane Uniaxial Compressive Load.

Material	Configuration	V_f (%)	E_x (msi)	E_y (msi)	ρ (lb/in^3)	σ_{ult} ksi[a]	FM[b]	Ref.
T300/934	Unidir.	60	23.69	1.7	0.0555	105	63.09	1.7
C/epoxy	Cross-ply	58	12.04	12.04	0.0555	55.1(T)	62.50	17.2
T300/5208	Unidir.	72	22.2	1.58	0.0555	110	61.01	1.7
T300/5208	Unidir.	60	21.9	1.53	0.0555	164	60.46	1.7
B/2024Al	--	49	33.93	22.04	0.0916	220(T)	60.24	17.2
B/6061Al	--	49	33.79	22.04	0.0916	513.3	60.14	17.2
$B(B_4C)/Al$	--	49	34.95	18.85	0.0918	210.3(T)	59.62	17.2
B/5052Al	--	49	33.35	21.03	0.0916	168.2(T)	59.50	17.2
B/110Al	--	49	33.35	19.72	0.0916	323.3	59.03	17.2
B/3002Al	--	49	32.48	20.74	0.0916	365.4	58.81	17.2
B/6061	Unidir.	50	32.0	20.00	0.0915	250.0	58.33	17.2
AS/3501	--	67	20.2	1.30	0.0555	209.9	57.47	1.7
AS1/3501-6	--	--	18.85	1.52	0.0555	246.0	57.10	17.2
B/Al	--	48	30.02	21.03	0.0918	221.9(T)	57.09	17.2
Borsic/Al	--	45	31.03	20.16	0.0926	190.0(T)	57.03	17.2
B/epoxy	--	67	31.18	3.05	0.0740	468.4	56.43	17.2
B/epoxy	--	67	30.3	2.80	0.0740	362.6	55.24	1.7
B/epoxy	Cross-ply	60	15.37	15.37	0.0721	55.1(T)	54.37	17.2
T300/2500	--	--	17.55	1.17	0.0555	--	53.80	17.2
$\alpha Al_2O_3/M_8$	Cont. fiber	50	30.02	15.08	0.1009	275.5	49.89	17.2
Kevlar49/epoxy	Cross-ply	60	5.80	5.80	0.0505	--	47.72	17.2
SiC/6061Al	--	48	31.63	18.06	0.112	--	46.74	17.2
S glass/xp-251	--	67	8.29	2.92	0.0555	170.0	45.53	1.7
$\gamma Al_2O_3/Al$	Cont. fiber	50	21.75	15.95	0.1045	203.0	42.91	17.2
$\gamma Al_2O_3/Al-5Cu$	Cont. fiber	50	21.75	14.50	0.1045	319.0	42.41	17.2
SiC/Ti	Woven, ISO	39.5	28.70	28.70	0.1432	228(T)	37.41	17.2
$\gamma Al_2O_2/Al$	Cont. fiber	50	31.9	20.16	0.1172	406	36.14	17.2
SiC/Ti-6Al-4V	Woven, ISO	35	27.0	27.0	0.1492	245(T)	34.83	17.2
2024 Al	ISO	--	10.88	10.88	0.0973	--	33.88	17.2
Titanium	ISO	--	17.4	17.4	0.1600	145	26.07	17.2
E glass/epoxy	Cross-ply	57	3.12	3.12	0.0710	82.0	24.87	17.2
Welton 80 steel	ISO	--	30.45	30.45	0.283	126.9	19.50	17.2

[a] Compressive Strength: (T) denotes tensile properties available only. [b] Units of FM are 10^2 in^2 / lb$^{1/2}$.

The above comparison does not indicate maximum loads that the panel may carry. To investigate that, knowledge of maximum allowable stresses must be available and

(16.8) must be used to determine the maximum load index, N_x/b, to which the optimum panel may be subjected. Figure 17.1 illustrates that point, where the maximum allowable stresses are taken to be the ultimate compressive strength and a 2024 aluminum core is arbitrarily chosen for the honeycomb core. It is seen that T300/934 is the lowest weight material, but limited to a load index of $\boldsymbol{N_x/b = 719}$ psi. For higher load levels, boron/aluminum is the most efficient material to a load index of 5840 psi and boron/epoxy the best to a load index of 8740 psi. Beyond that load index, other panel architectures are perhaps required.

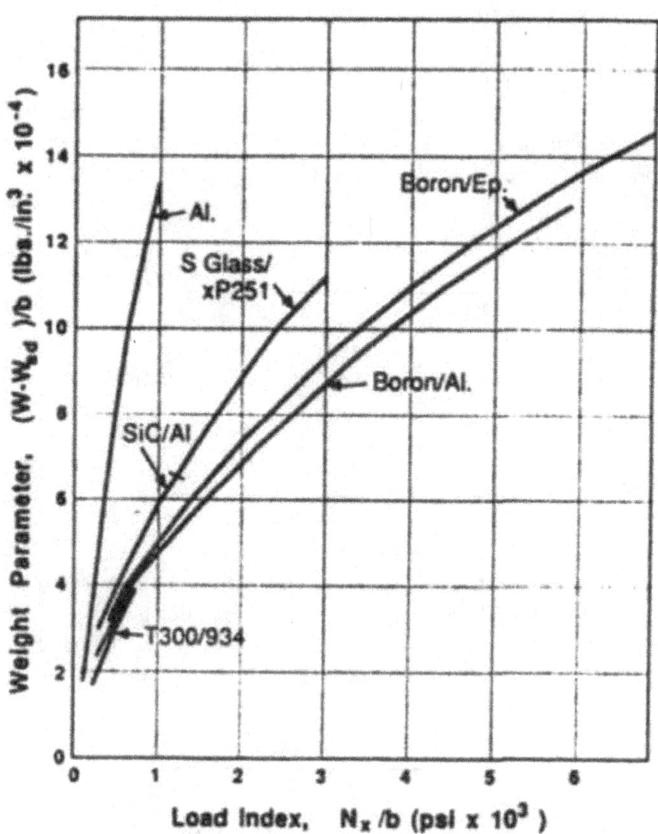

Figure 17.1. Weight parameter as a function of load index for a honeycomb core sandwich panel under an in-plane compressive load [17.1].

In all of the above, it is assumed the materials are elastic to the ultimate stress. If ductility permits and the stress-strain curve are known for stresses higher than the yield point, a plasticity reduction factor can be employed. Finally, it should be remembered that in any laminated face construction, in which B_{ij} $(\;)_{16}$ $(\;)_{26}$ 0, E_x and E_y can always be obtained by utilizing the appropriate stiffness matrix values, discussed in Section 15.1.

17.3 Minimum Weight Optimization of Foam Core Sandwich Panels Subjected to a Unidirectional Compressive Load

For this type of sandwich panel and loading, there are three buckling failure modes: overall buckling, Equations (16.9) through (16.18); core shear instability, Equation (16.34); and face wrinkling, Equation (16.36). In this case of foam or solid core one uses the actual core properties in (16.14), namely

$$U_{xz} = G_{cx} h_c, \quad U_{yz} = G_{cy} h_c \tag{17.14}$$

For the optimum minimum weight construction, overall panel buckling and core shear instability occur simultaneously, i.e., when $V_x = k_1 B_1 r$, or more specifically, as in (16.35),

$$\frac{\pi^2}{2(1-\nu_{xy}\nu_{yx})} \frac{\overline{E}_{fx}}{G_{cx}} \frac{h_c t_f}{b^2} = k_1 r. \tag{17.15}$$

For the optimized construction the applied load-optimum face stress relationship is found to be

$$\frac{1}{(k_1 r)^{1/2}} \left(\frac{N_x}{b} \right) = \frac{6(1-\nu_{xy}\nu_{yx})^{3/2} \sigma_0^{7/2}}{\pi E_c \overline{E}_{fx} E_{fy}^{1/2}} \tag{17.16}$$

The geometric variables and the core shear stiffness that results in minimum weight design are as follows because of the linear relationship between core shear modulus and core density, as shown in Figure 17.4 below, G_{cx} is treated as a variable in the optimization:

$$t_f = \frac{3(1-\nu_{xy}\nu_{yx})^{3/2} b (k_1 r)^{1/2} \sigma_0^{5/2}}{\pi E_c \overline{E}_{fx} E_{fy}^{1/2}} \tag{17.17}$$

$$h_c = \frac{2}{\pi}(1-\nu_{xy}\nu_{yx})^{1/2} b (k_1 r)^{1/2} \frac{\sigma_0^{1/2}}{\overline{E}_{fx}^{1/2}} \tag{17.18}$$

$$G_{cx} = \frac{3(1-\nu_{xy}\nu_{yx})\sigma_0^3}{E_c \sqrt{\overline{E}_{fx} E_{fy}}}. \tag{17.19}$$

The above expressions could be written in terms of the applied load index load (N_x/b) by using (17.16) but the expressions are lengthier.

The optimized panel weight in terms of applied load is

$$\frac{W-W_{ad}}{b} = \frac{6^{2/7}(k_1 r)^{1/7}(1-\nu_{xy}\nu_{yx})^{3/7}(N_x/b)^{5/7}}{\pi^{2/7}E_c^{2/7}\overline{E}_{fx}^{2/7}E_{fy}^{1/7}} \times\left[\rho_f+\rho_c\frac{E_c\overline{E}_{fx}^{1/2}E_{fy}^{1/2}}{3(1-\nu_{xy}\nu_{yx})\sigma_0^2}\right]. \tag{17.20}$$

From (17.20) it is seen that the best face material to use is the one with the largest value of

$$FM = \frac{\overline{E}_{fx}^{2/7}E_{fy}^{1/7}}{\rho_f}. \tag{17.21}$$

Note that the face modulus in the load direction is barred to indicate the use of a plasticity reduction factor if the stresses used for the design are above the proportional limit. For this loading the face stress in the y-direction will never exceed the yield strength, hence E_{fy} is not barred.

17.4 Minimum Weight Optimization of Truss Core Sandwich Panels Subjected to a Unidirectional Compressive Load

For the optimization in this case there are three modes of buckling and four geometric variables discussed in Section 16.5. As a result, the approach is to solve for three of the variables in terms of the fourth variable, utilizing the three buckling equations, and the load-stress equation. Then substitute these expressions into the weight equation such that the panel weight is a function of the core and face material properties, the load index and the chosen variable θ. With five equations and six unknowns, a sixth equation is obtained by setting the derivative of the weight equation with respect to θ equal to zero to obtain the value of that variable which will provide minimum weight. In the case of the faces and core made of the same orthotropic material, the weight equation in terms of the load index is written as

$$\frac{W-W_{ad}}{b} = \left[\frac{3(1-\nu^2)}{K}\right]^{\frac{1}{4}}\left[\frac{N_x/b}{\overline{E}}\right]^{\frac{1}{2}}\frac{\rho\,[1+4\sin^2\theta]^{3/4}}{\pi\,\sin\theta\,(\cos\theta)^{1/2}}. \tag{17.22}$$

Setting the derivative of $(W-W_{ad})/b$ with respect to θ equal to zero it is found that for the optimum construction, where the core and faces are of the same material, isotropic or anisotropic,

$$\sin^2\theta = 2/7 \quad \text{or} \quad \theta = 32.4°. \tag{17.23}$$

For core and faces of different materials the lengthy results are given in [17.3]. The load index–optimum face stress relationship is

$$\frac{N_x}{b} = \frac{45}{2}\frac{(1-\nu_{xy}\nu_{yx})^{1/2}\sigma_0^2}{\pi^2 K^{1/2} E_0^{1/2}(E_x E_y)^{1/4}}. \tag{17.24}$$

where E_0 is defined by (16.53). The unique values for the geometric variables are:

$$\frac{t_f}{b} = \frac{6(1-\nu_{xy}\nu_{yx})^{1/2}\sigma_0}{\pi^2 K^{1/2} E_0^{1/2}(E_x E_y)^{1/4}}, \tag{17.25}$$

$$\frac{t_c}{b} = 3\left(\frac{7}{2}\right)^{\frac{1}{2}}\frac{(1-\nu_{xy}\nu_{yx})^{1/2}\sigma_0}{E_0^{1/2}(E_x E_y)^{1/4} K^{1/2}} \tag{17.26}$$

$$\frac{h_c}{b} = \frac{1}{(E_x E_y)^{1/4}}\left(\frac{15}{2}\frac{\sigma_0}{\pi^2 K}\right)^{\frac{1}{2}}. \tag{17.27}$$

The optimized panel weight is given by

$$\frac{W - W_{ad}}{b} = 3\left(\frac{5}{2}\right)^{\frac{1}{2}}\frac{(1-\nu_{xy}\nu_{yx})^{1/4}\rho(N_x/b)^{1/2}}{K^{1/4}\pi E_0^{1/4}(E_x E_y)^{1/8}} \tag{17.28}$$

from which it is seen that the factor of merit to use in materials selection is

$$FM = \frac{E_0^{1/4}(E_x E_y)^{1/8}}{\rho}. \tag{17.29}$$

For the optimized case, using Figures 3 and 5 of [16.9] to obtain the precise value of the coefficient K, use

$$\frac{E_c \bar{I}_c}{E_f \bar{I}_f} \square \frac{7}{24} \tag{17.30}$$

and

$$V = \frac{21(1-\nu_{xy}\nu_{yx})\sigma_0}{(E_x E_y)^{1/2} K}. \tag{17.31}$$

and perform a simple iteration. Note that in the weight equation (10.37), K appears only as the 1/4 power.

If the truss core sandwich panel has the face and core of the same isotropic materials, then as noted before, the optimization relations hold also above the proportional limit, if a suitable plasticity reduction factor, η, is used, such that

$$\overline{E} \equiv \eta E. \tag{17.32}$$

All expressions will be written in the general form utilizing $\overline{E}$. One then simplifies (17.22) through (17.29) by letting $E_c = E_f = \overline{E}$, $\nu_c = \nu_f = \nu$, and $\sigma_c = \sigma_f \equiv \sigma$. The buckling coefficient K is a slowly varying function such that it can be considered constant, and can be determined later by a trial iteration. The unknown variables are $t_f, h_c, t_c, \theta, \sigma$ and $(W - W_{ad})$. For the case of the faces and core being the same isotropic materials, the "universal relation" relating load index to a unique stress value for any set of material properties, which results in minimum weight is:

$$\frac{N_x}{b} = \frac{45}{2}\frac{(1-\nu^2)^{1/2}\sigma_0^2}{\pi^2 \overline{E} K^{1/2}}. \tag{17.33}$$

Unique values of all other variables are now found in terms of the applied load index, N_x/b, and the material properties for the optimum, minimum weight isotropic construction.

$$\frac{t_f}{b} = 2\left(\frac{2}{5}\right)^{\frac{1}{2}} \frac{(1-\nu^2)^{1/4}(N_x/b)^{1/2}}{\pi \overline{E}^{1/2} K^{1/4}}, \tag{17.34}$$

$$\frac{t_c}{b} = \sqrt{\frac{7}{5}} \frac{(1-\nu^2)^{1/4}(N_x/b)^{1/2}}{\pi \overline{E}^{1/2} K^{1/4}}, \tag{17.35}$$

$$\frac{h_c}{b} = \left[\frac{5(N_x/b)}{2\overline{E}\pi^2(1-\nu^2)^{1/2} K^{3/2}}\right]^{\frac{1}{4}}, \tag{17.36}$$

$$\frac{W - W_{ad}}{b} = 3\left(\frac{5}{2}\right)^{\frac{1}{2}} \frac{(1-\nu^2)^{1/4}\rho(N_x/b)^{1/2}}{\pi \overline{E}^{1/2} K^{1/4}}. \tag{17.37}$$

Using (17.33) analogous expressions can be written for each variable in terms of the optimum face stress, as was done earlier.

It is interesting to note that when the same core and face materials are used, optimum angle, θ, is independent of load and materials used. Note also that in the optimum construction described above, for the isotropic case:

$$\frac{t_c}{t_f} = \sqrt{\frac{7}{8}} \tag{17.38}$$

and

$$\frac{W_c}{W_f} = \frac{7}{8}, \tag{17.39}$$

where W_c and W_f are the weight of the core and face per unit planform area, respectively.

The core of the optimized truss-core sandwich weighs nearly as much as the faces, but it must be remembered that the core also carries a significant share of the in-plane compressive load unlike in honeycomb and foam core sandwich constructions.

17.5 Minimum Weight Optimization of Web Core Sandwich Panels Subjected to a Unidirectional Compressive Load

Again, the philosophy of structural optimization is as follows: within the class of structures being studied and for each material system, a truly optimum structure is one that has a unique value for each geometric variable, that results in the minimum possible weight for a specified loading condition, and yet maintain its structural integrity. In this case, the optimum structure will have the characteristic that the panel will become unstable in all four buckling modes simultaneously.

The equations with which to optimize are: the four buckling equations, Equations (16.66), (16.68), (16.69) and (16.70); the applied load-face stress relationship, Equation (16.73); and the weight relationship, Equation (16.74). The known quantities for any optimization are N_x, a, b, and the material properties of both core and face material. The buckling coefficient K is a constant depending on the value of (16.67), and is given by (16.65). The unknown variables to be determined are $t_f, h_c, t_c, d_f, \theta, \sigma_f$ and $(W - W_{ad})$. With six equations and seven unknown variables, a seventh equation is obtained by placing the weight equations in terms of one variable only, and setting the derivative of weight with respect to that variable to zero as in the previous section. Thus, one finds the value of that variable, and subsequently the value of every other variable, which result in minimum weight.

At the outset, independent on the material system, it is clear that in executing the above optimization philosophy, with Equations (16.68) and (16.69) that for optimum construction of the web-core sandwich,

$$\theta = 0^\circ. \tag{17.40}$$

This is intuitively obvious because with $\theta = 0^\circ$, upper and lower face plated components have the same width. The result is that all other expressions used for optimization are simplified, and the construction shown in Figure 16.6, with $\theta = 0^\circ$ now assumes the familiar web-core configuration.

Proceeding with the optimization procedure for panels with faces and core of different orthotropic materials the following expressions are obtained for the optimum minimum weight construction. First, one obtains the "universal relationship" relating applied load index, N_x / b, to a unique value of face stress σ_{f0}, as a function of given material properties for the optimum construction

$$\frac{N_x}{b} = \frac{(6)^{1/2}(1-\nu_{xyf}\nu_{yxf})^{1/4}(1-\nu_{xyc}\nu_{yxc})^{1/4}}{(E_{0f}E_{0c})^{1/4}(E_{fx}E_{fy})^{1/4}\pi^2 K^{1/2}} \times \left(\frac{E_{cx}}{E_{fx}}\right)^{\frac{7}{4}}\left(\frac{\rho_f}{\rho_c}\right) R^{3/2}\sigma_{f0}^2 \tag{17.41}$$

Here, $R = 1+2(\rho_c/\rho_f)(E_{fx}/E_{cx})$ and K is determined from Figures 2 or 4 of [16.9] in which $V \rightarrow \infty$ and for the optimum construction

$$\frac{E_c\bar{I}_c}{E_f\bar{I}_f} = \frac{1}{6}\left(\frac{\rho_f}{\rho_c}\right)\left(\frac{E_{cx}E_{cy}}{E_{fx}E_{fy}}\right)^{\frac{1}{2}} \tag{17.42}$$

In a panel with all edges simply supported K is given as follows, using the above expression for $E_c\bar{I}_c / E_f\bar{I}_f$

$$K = \frac{\left[(1/\beta)+\beta^2\right]}{1-\nu_{xyf}\nu_{yxf}+2(1+\nu_{xyf})\beta^2} + \frac{E_c\bar{I}_c}{E_f\bar{I}_f\beta^2} \tag{17.43}$$

where $\beta = a/b$.

The significance of the universal relationship given by Equation (17.41) is that, given panel length, width, load, and materials, a panel designed for a face stress either higher or lower than that given by (17.41), will be heavier than one designed for the optimum face stress σ_{f0} given by (17.41). Making use of (17.41) then, the geometry of the panel for minimum weight construction can be written as:

$$\frac{h_c}{b} = \left\{2\left(\frac{\rho_f}{\rho_c}\right)\left(\frac{E_{cx}}{E_{fx}}\right)\frac{R\sigma_{f0}}{(E_{fx}E_{fy})^{1/2}\pi^2 K}\right\}^{\frac{1}{2}} \tag{17.44}$$

$$\frac{h_c}{d_f} = \left\{\left(\frac{\rho_f}{\rho_c}\right)^2\left(\frac{1-\nu_{xyf}\nu_{yxf}}{1-\nu_{xyc}\nu_{yxc}}\right)\left(\frac{E_{0c}}{E_{0f}}\right)\left(\frac{E_{fx}}{E_{cx}}\right)\right\}^{\frac{1}{4}} \tag{17.45}$$

$$\frac{t_c}{b} = \frac{6^{1/2}}{\pi^2 K^{1/2}}\frac{(1-\nu_{xyc}\nu_{yxc})^{1/2}}{E_{0c}^{1/2}(E_{fx}E_{fy})^{1/4}}\left(\frac{\rho_f}{\rho_c}\right)^{\frac{1}{2}}\left(\frac{E_{cx}}{E_{fx}}\right)R^{1/2}\sigma_{f0} \tag{17.46}$$

$$\frac{h_c}{d_f} = (\rho_f/\rho_c)(t_f/t_c) \tag{17.47}$$

$$(W - W_{ad})/b = 3\rho_f(t_f/b). \tag{17.48}$$

Other alternative expressions for the optimum construction are given in [17.4].

It can also be shown easily that even in this general material system, the ratio of the face weight W_f to the core weight W_c for the optimum construction is:

$$W_f/W_c = 2. \tag{17.49}$$

Note that this is independent of applied load, material system and panel boundary conditions.

The optimum construction for panels with faces and core of different isotropic materials can be obtained from Equations (17.41) through (17.49) by allowing $E_{ix} = E_{iy} = E_{0i} = E_i$ and $\nu_{xyi} = \nu_{yxi} = \nu_i$ where $i = c, f$.

The resulting expressions for the optimum construction for panels with faces and core of the same orthotropic material can be obtained from (17.41) through (17.49) by letting each quantity $(\ )_f = (\ )_c = (\ )$. The results are

$$\frac{N_x}{b} = \frac{9(2)^{1/2}(1-\nu_{xy}\nu_{yx})^{1/2}\sigma^2}{\pi^2 K^{1/2}E_0^{1/2}(E_x E_y)^{1/4}} \tag{17.50}$$

where K is determined from Figures 2 or 4 of [16.9] or, for a simply supported panel, from (17.45), in which

$$E_c\bar{I}_c/E_f\bar{I}_f \,\square\, \frac{1}{6}$$

Making use of (17.50) to obtain σ, the other geometric variables are obtained easily from the following:

$$h_c/b = \left\{6\sigma_0/\left[\pi^2 K(E_x E_y)^{1/2}\right]\right\}^{1/2} \tag{17.51}$$

$$d_f \quad h_c \tag{17.52}$$

$$\frac{t_c}{b} \quad \frac{3(2)^{1/2}(1-\nu_{xy}\nu_{yx})^{1/2}\sigma_0}{\pi^2 K^{1/2}(E_x E_y)^{1/2} E_0^{1/2}} \tag{17.53}$$

$$t_f / t_c \quad 1 \tag{17.54}$$

$$\frac{W - W_{ad}}{b} \quad \frac{\rho}{\sigma_0}\left(\frac{N_x}{b}\right) \quad 3\rho\frac{t_f}{b} \quad 3\rho\frac{t_c}{b}. \tag{17.55}$$

It is see that for a panel in which the same orthotropic material is used in both face and core, the optimum geometry results in face and core of the same thickness ($t_c \quad t_f$) and square cells ($d_f \quad h_c$)

It is seen from (17.50) and (17.55) that the factor of merit for selecting the face material is:

$$FM \quad E_0^{1/4}(E_x E_y)^{1/8} / \rho$$

Lastly, the expressions for optimum construction of panels with faces and core of the same isotropic material can be easily determined from (17.50) through (17.55) by letting $(\ )_x \quad (\ )_y \quad (\ )$. With this material system, since the material is isotropic and the face and core are equally stressed, the expressions can be employed for loads resulting in stresses above the proportional limit by utilizing a suitably defined plasticity reduction factor η, such that $\overline{E} \quad \eta E$. Thus, in the following, $\overline{E}$ is used to denote that the expressions are valid in the range of inelastic deformations.

$$N_x / b \quad 9(2)^{1/2}(1-\nu^2)^{1/2}\sigma_0^2 / (\pi^2 \overline{E} K^{1/2}) \tag{17.56}$$

$$h_c / b \quad d_f / b \,\square\, (6\sigma_0 / \sigma^2 K \overline{E})^{1/2} \tag{17.57}$$

$$t_f / b \quad t_c / b \quad 3(2)^{1/2}(1-\nu^2)^{1/2}\sigma_0 / (\pi^2 K^{1/2} \overline{E}) \tag{17.58}$$

$$\frac{W - W_{ad}}{b} \quad \frac{3(2)^{1/4}\rho(1-\nu^2)^{1/4}}{\pi K^{1/4} E^{1/2}}\left(\frac{N_x}{b}\right)^{\frac{1}{2}} \quad \frac{\rho}{\sigma_0}\left(\frac{N_x}{b}\right) \quad 3\rho\frac{t_f}{b} \quad 3\rho\frac{t_c}{b} \tag{17.59}$$

Again it is seen that the web and face have the same thickness and that the cells are square for the optimum construction. Note also that the isotropic material which has the highest ratio of $E^{1/2}/\rho$ is the material which will result in the lowest weight panel in the elastic range.

As an example of the use of the optimization procedures derived above, curves of the weight function $K^{1/2}(W - W_{ad})$ as a function of load index $(K^{1/2}N_x/b)$ are plotted in Figure 17.2 for the following materials:

1. 7075-T6 Aluminum (clad)
2. S994-181 HTS glass fabric, ERSB-0111 Resin
3. 143 glass fabric laminate with polyester resin (MIL-R-7575)
4. 143 glass fabric laminate with epoxy resin (MIL-R-9300)
5. 181 glass fabric laminate with epoxy resin (MIL-R-9300)
6. cross rolled beryllium
7. unidirectional boron fibers with epoxy resin
8. unidirectional Thornel-40 graphite fibers with epoxy resin
9. AISI 4340 steel, 200,000 psi yield strength

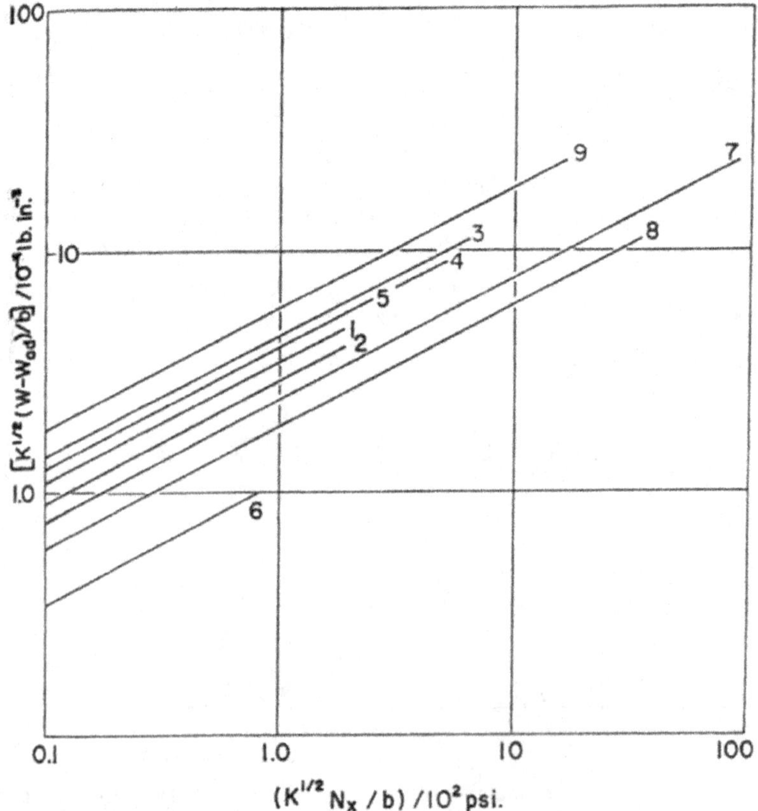

Figure 17.2. Weight parameter as a function of load index, web-core sandwich panel unidirectional compressive load. (Reprinted from Reference [10.1].)

In each case, the load index is taken only to an optimum stress equal to the yield strength of the material system. The factor $K^{1/2}$, appearing as it does in both the ordinate and abscissa, permits the weight comparison for various material systems regardless of boundary conditions and aspect ratio a/b.

It is clearly seen that the beryllium results in the lowest weight construction, but the yield strength is reached at a low value of load index. It is also clearly seen that not only do boron-epoxy and graphite-epoxy materials provide lower weight construction than glass-reinforced plastic systems, steel, and aluminum, but of perhaps equal significance, higher loads can be carried by graphite-epoxy and boron-epoxy panels of given size and edge support conditions than any other material system.

17.6 Minimum Weight Optimization of Honeycomb Core Sandwich Panels Subjected to In-Plane Shear Loads

In this optimization, the faces must be sized such that the allowable shear stress is not exceeded. Also, the panel geometry must preclude overall buckling, Equations (16.77) through (16.80); core shear instability, Equation (16.81); face wrinkling, Equation (16.82); and face dimpling, Equation (16.83). The optimization of the honeycomb core sandwich panel is then performed by equating the four buckling loads given by Equations (16.77) through (16.83). The results are as follows:

$$\left(\frac{k_2}{k_3 k_1 S_{xi}}\right)^{\frac{1}{2}}\left(\frac{N_{xy}}{b}\right) = \frac{2\sqrt{3}(1-\nu_{xy}\nu_{yx})}{\pi}\left(\frac{G_c}{E_c \overline{E}_{fx} \overline{E}_{fy}}\right)^{\frac{1}{2}}\sigma_0^2 \tag{17.60}$$

$$\frac{h_c}{b} = 2\left(\frac{k_3}{k_4}\right)^{\frac{1}{4}}(k_1 S_{xi})^{1/2}\left(\frac{1-\nu_{xy}\nu_{yx}}{\pi}\right)^{\frac{1}{2}}\frac{\sigma_0^{1/2}}{(\overline{E}_{fx}\overline{E}_{fy})^{1/4}} \tag{17.61}$$

$$\frac{d}{b} = \sqrt{6}\left(\frac{(1-\nu_{xy}\nu_{yx})}{\pi}\right)^{\frac{1}{2}}\left(\frac{G_c}{E_c}\right)^{\frac{1}{2}} \times \frac{(E_{fx_t}E_{fy_t})}{(\overline{E}_{fx}\overline{E}_{fy})^{1/2}}\sigma_0^{1/2}\left(\frac{k_3 k_1 S_{xi}}{k_2}\right)^{\frac{1}{2}} \tag{17.62}$$

$$\frac{t_c}{b} = 3\sqrt{2}\left[\left(\frac{k_3}{k_4}\right)^{\frac{1}{4}}\frac{(k_1 S_{xi})}{k_2}\right]^{\frac{1}{2}}\frac{(E_{fx_t}E_{fy_t})^{1/4}\sigma_0^2}{E_c(\overline{E}_{fx}\overline{E}_{fy})^{3/4}} \tag{17.63}$$

$$\frac{t_f}{b} = \left(\frac{3k_3 k_1 S_{xi}}{k_2}\right)^{\frac{1}{2}}\frac{(1-\nu_{xy}\nu_{yx})}{\pi}\left(\frac{G_c}{E_c}\right)^{\frac{1}{2}}\frac{\sigma_0}{(\overline{E}_{fx}\overline{E}_{fy})^{1/2}} \tag{17.64}$$

$$\left(\frac{W-W_{ad}}{b}\right) \left(\frac{3k_3k_1S_{xi}}{k_2}\right)^{\frac{1}{4}}\left[\frac{2(1-\nu_{xy}\nu_{yx})}{\pi}\right]^{\frac{1}{2}}\left(\frac{G_c}{E_c}\right)^{\frac{1}{4}}$$
$$\times\frac{(N_{xy}/b)^{1/2}}{(\overline{E}_{fx}\overline{E}_{fy})^{1/4}}\left[\rho_f+\frac{\rho_c k_2}{(k_3k_4)^{1/2}}\frac{\sigma_0}{G_c}\right] \tag{17.65}$$

Therefore for the face material selection the factor of merit is:

$$FM \square (\overline{E}_{fx}\overline{E}_{fy})^{1/4}/\rho \tag{17.66}$$

With this factor of merit a comparison of several materials is given in Table 17.2.

Table 17.2. Face Material Comparison, Based Upon the Factor of Merit, for an Optimized Honeycomb Core Sandwich Panel Subjected to In-Plane Shear Loads [10.1].

Material	Configuration	FM (10^3 in^2/lb$^{1/2}$)
C/epoxy	Cross-ply	62.53
B/2024Al		57.09
B/6061Al		57.03
B/5052Al		56.18
B/3002Al		55.61
B/1100Al		55.29
B (B_4C) /Al		55.18
B/Al		54.60
B/6061Al		54.57
B/epoxy	Cross-ply	54.37
Borsic/Al		54.03
Kevlar49/epoxy	Cross-ply	47.68
αAl_2O_3/Mg	Continuous fiber	45.72
T300/934	Unidirectional	45.39
αAl_2O_3/Al - SCu	Continuous fiber	44.04
T300/5208	Unidirectional	43.85
SiC/6061Al		43.57
T300/SP286		43.35
αAl_2O_3/Al	Continuous fiber	42.97
E glass/epoxy		42.75
B/epoxy		42.20
AS4/3501-6	Unidirectional	41.88
AS1/3501-6		41.68
Boron/epoxy		41.62
HS graphite/epoxy	Unidirectional	41.56
αAl_2O_3/Al	Continuous fiber	41.30
B/EP	Unidirectional	41.09
B/EP		41.02
Boron/EP		41.01
AS/3501		40.70
2024 Al		35.79

It can be shown [17.7] from Equation (17.66) that, for any laminated composite material, with laminae of continuous unidirectional fibers, used for the sandwich faces, the best stacking sequence to resist an in-plane shear load is a balanced cross-ply laminate.

17.7 Minimum Weight Optimization of Solid and Foam Core Sandwich Panels Subjected to In-Plane Shear Loads

To minimize the weight, the three buckling critical stresses are equated, and the results for the minimum weight panel are:

$$\left(\frac{1}{k_1 S_{xi}}\right)^{\frac{1}{2}}\left(\frac{N_{xy}}{b}\right) = \frac{6\sqrt{3}(1-\nu_{xy}\nu_{yx})^2\sigma_0^5}{\pi E_c^{3/2} G_{cy}^{1/2}\overline{E}_{fx}\overline{E}_{fy}} \tag{17.67}$$

Again, because of the continuous linear relationship of core shear modulus to core density for foam cores, G_{cx} is treated as a variable; therefore,

$$G_{cx} = \frac{9(1-\nu_{xy}\nu_{yx})^2\sigma_0^6}{G_{cy}E_c^2\overline{E}_{fx}\overline{E}_{fy}} \tag{17.68}$$

$$t_f = \frac{\sqrt{3}(1-\nu_{xy}\nu_{yx})^2 b(k_1 S_{xi})^{1/2}\sigma_0^4}{\pi E_c^2 G_{cy}^{1/2}\overline{E}_{fx}\overline{E}_{fy}} \tag{17.69}$$

$$h_c = \frac{2\sqrt{3}(1-\nu_{xy}\nu_{yx})b(k_1 S_{xi})^{1/2}\sigma_0^2}{\pi E_c^{1/2} G_{cy}^{1/2}\overline{E}_{fx}^{1/2}\overline{E}_{fy}^{1/2}} \tag{17.70}$$

The optimum panel weight equation is

$$\left(\frac{W-W_{ad}}{b}\right) = \frac{2^{1/5}3^{3/10}(1-\nu_{xy}\nu_{yx})^{2/5}(k_1 S_{xi})^{1/10}(N_{xy}/b)^{4/5}}{\pi^{1/5}E_c^{3/10}\overline{E}_{fx}^{1/5}G_{cy}^{1/10}} \times\left[\rho_f + \frac{\rho_c E_c(\overline{E}_{fx}\overline{E}_{fy})^{1/2}}{3(1-\nu_{xy}\nu_{yx})\sigma_0^2}\right], \tag{17.71}$$

from which it is seen that the best face material can be selected from

$$FM = (\overline{E}_{fx}\overline{E}_{fy})^{1/5}/\rho. \tag{17.72}$$

17.8 Minimum Weight Optimization of Truss-Core Sandwich Panels Subjected to In-Plane Shear Loads

Stated once again, the philosophy of the optimization for the truss core sandwich panel with in-plane shear loads is as follows: a truly optimum structure is one which has a unique value for each dependent variable within the class of structure being studied (triangulated core sandwich panel, for example), for each set of materials for each set of boundary conditions and is the minimum possible weight for a specified set of design loads and will maintain its structural integrity (no mode of failure will occur at a load less than the optimum design load). In this case the optimum (minimum weight) structure will have the characteristic that the panel will become unstable in all three buckling modes simultaneously.

The governing equations pertaining to this construction to be used in the optimization are given by (16.87), (16.89), (16.92), (16.94), (16.95) and (16.96). The known or specified quantities are the applied shear load per inch, N_{xy}, and the panel width b, which can be combined as the load index (N_{xy}/b); the material properties; and the panel boundary conditions. The buckling coefficient j, k_f and k_c are given in Figure 16.8 for any given set of variables, and hence are constants for the optimum construction being sought. The dependent variables with which to optimize the construction are the face thickness, t_f, the core depth, h_c, the web material thickness, t_c, the web angle, θ, the face stress, τ_f, the core stress, τ_c, and the weight, $W - W_{ad}$.

It is seen that there are six equations and seven unknowns. The seventh equation is obtained by placing the weight equation in terms of one convenient variable, taking the derivative of the weight equation with respect to this variable, and equating it to zero to obtain the unique value of that variable which results in a minimum weight structure. Subsequently, one can determine the value of all other variables for the optimum constructions. Manipulation of the equations listed above results in the weight equation involving only the dependent variable θ, as shown below

$$\frac{W - W_{ad}}{b} = \frac{3^{1/4}}{2}\frac{\rho_f}{(k_f \bar{j})^{1/4}}\frac{(N_{xy}/b)^{1/2}}{E_{sf}^{1/2}} \times \frac{4(\sin\theta)^{3/2} + \left(\frac{\rho_c}{\rho_f}\right)\left(\frac{G_{xyc}}{G_{xyf}}\right)^{\frac{1}{2}}\left(\frac{E_{sf}}{E_{sc}}\right)^{\frac{1}{2}}\left(\frac{k_f}{k_c}\right)^{\frac{1}{2}}}{(\sin\theta)(\cos\theta)^{1/2}} \tag{17.73}$$

where

$$E_{si} = [E_{iy}^{3} E_{ix}]^{1/4}/(1 - \nu_{xyi}\nu_{yxi}), \quad (i = c, f). \tag{17.74}$$

Taking the derivative of (17.73) with respect to θ, and equating it to zero results in a value of θ in terms of material properties and buckling coefficients which will result in minimum weight structure. This expression is:

$$2(\sin\theta)^{3/2}(\cos\theta)^2 + 2(\sin\theta)^{7/2} - \left(\frac{\rho_c}{\rho_f}\right)\left(\frac{G_{xyc}}{G_{xyf}}\right)\left(\frac{E_{sf}}{E_{sc}}\right)^{\frac{1}{2}}\left(\frac{k_f}{k_c}\right)^{\frac{1}{2}}\left[\cos^2\theta - \frac{1}{2}\sin^2\theta\right] = 0. \tag{17.75}$$

Note that the optimum web angle, defined by this equation is independent of the load and boundary conditions to which the panel is subjected.

A "universal relationship" may be obtained which relates the applied load index (N_{xy}/b) to a unique value of face shear stress, τ_{f0}, for any set of material properties, which will result in minimum weight panel. For a given load index (N_{xy}/b), a panel designed to have a face stress τ_f higher or lower than the value given by the following relationship will result in a panel which has a weight greater than can be achieved if this universal relationship is used.

$$\frac{N_{xy}}{b} = \frac{4\sqrt{3}\tan\theta}{k_f^{1/2}\bar{j}^{1/2}}\frac{\tau_{f0}}{E_{sf}} \tag{17.76}$$

where θ is obtained from (17.75) and E_{sf} from (17.74) and τ_{f0} is the optimum face stress.

The remaining geometric variable t_f, t_c, and h_c, as well as the weight equation can now be expressed in terms of the optimum face stress τ_{f0} obtained from (17.76), and the optimum angle θ determined by (17.75) above. For the case of faces and core of the same materials, these are:

$$\frac{t_f}{b} = \frac{2/3\tan\theta\tau_{f0}}{k_f^{1/2}\bar{j}^{1/2}E_s}, \quad \frac{h_c}{b} = \left(\frac{\tau_{f0}}{E_s\bar{j}}\right)^{\frac{1}{2}} \tag{17.77), (17.78}$$

$$\frac{t_c}{b} = \frac{\sqrt{3}}{k_c^{1/2}\bar{j}^{1/2}}\frac{(\sin\theta)^{1/2}}{\cos\theta}\frac{\tau_{f0}}{E_s} \tag{17.79}$$

$$\frac{W - W_{ad}}{b} = \frac{\sqrt{3}\rho\tau_{f0}}{k_f^{1/2}\bar{j}^{1/2}E_s}\,\frac{4(\sin\theta)^{3/2} + \left(\frac{k_f}{k_c}\right)^{\frac{1}{2}}}{\cos\theta\,(\sin\theta)^{1/2}} \tag{17.80}$$

Employing (17.76) the minimum weight panels for a given load index N_{xy}/b is given by

$$\frac{W - W_{ad}}{b} = \frac{3^{1/4}}{2}\frac{\rho}{k_f^{1/4}\bar{j}^{1/4}}\frac{(N_{xy}/b)^{1/2}}{E_s^{1/2}}\left(\frac{4(\sin\theta)^{3/2} + \left(\frac{k_f}{k_c}\right)^{\frac{1}{2}}}{\sin\theta\,(\cos\theta)^{1/2}}\right) \tag{17.81}$$

where E_x is given by (17.74).

Two conclusions are drawn. First, the best composite material to use in such construction is the one with the highest ratio of

$$FM = \frac{E_s^{1/2}}{\rho} = \frac{[E_y^3 E_x]^{1/8}}{\rho(1-\nu_{xy}\nu_{yx})^{1/2}}. \tag{17.82}$$

Secondly, the ratio of face weights to core weight for optimum construction is:

$$\frac{W_f}{W_c} = \frac{4(\sin\theta)^{3/2}}{(k_f/k_c)^{1/2}}. \tag{17.83}$$

In the case of a panel with the same face and core materials, then $k_f = k_c$, and from (17.75), $\theta = 28.4°$ and $W_f/W_c = 1.316$.

17.8.1 Certain Properties of the Buckling Coefficients $\bar{j}, k_f$, and k_c

It is advantageous to discuss certain characteristics of the coefficients $\bar{j}, k_f$, and k_c that result in significant simplifications to the design procedures. It is shown in the References of Chapter 16 and 17 that in all calculations of optimum constructions

$$\bar{j} = j \text{ and } k_f = k_c \tag{17.84}$$

17.8.2 Some Conclusions

For truss-core sandwich panels subjected to in-plane shear loads, in which both the faces and core are composed of the same isotropic or orthotropic materials, $\theta = 28.4°$ for minimum weight construction, whenever the edge restraint coefficients k_f and k_c are equal (the usual case). Under these conditions the weight ratio of face material to core material per unit planform area is 1.316.

For the case of core and face material being the same, the use of the orthotropic material which has the highest value of the following factor of merit will result in the least weight panel of specified geometry and loading. Since $(1-\nu_{xy}\nu_{yx}) \approx 1$ for most composites, it is seen that a factor of merit (FM) can be defined as:

$$FM = \frac{E_y^{3/8} E_x^{1/8}}{\rho} \tag{17.85}$$

For unidirectional composites, the fiber direction for this type construction should be in the y-direction. This is the same factor of merit as that found for the faces of an optimum honeycomb core subjected to in-plane compression [17.1, 17.2, 17.5]. In those references several dozen material systems were compared, and provide the wherewithal to select the best materials for the optimized structures of Figure 16.5.

However, to determine the limitations on the load index (N_{xy}/b) to prevent overstressing, Equations (17.76) and (17.80) must be plotted to make the final comparison, as in Figure 17.3. There, typical composite materials given in an Appendix of Reference [17.6] are plotted. There the maximum values of N_{xy}/b correspond to the maximum shear strengths given in Reference [17.6]; and some fraction of those values as a factor of safety would terminate each curve proportionately for safety. It is seen that among these materials, T300-934 graphite-epoxy provides not only the lowest weight construction but is also usable to the highest load index.

Weight as a Function of Load Index under In-plane Shear Loads

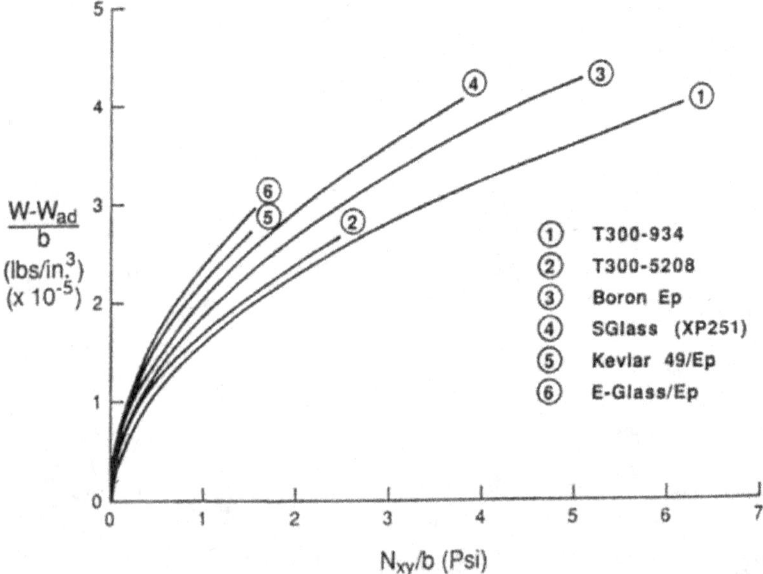

Figure 17.3. Weight as a function of load index under in-plane shear loads for a truss-core sandwich panel. (Reprinted from Reference [17.6].)

17.9 Minimum Weight Optimization of Web-Core Sandwich Panels Subjected to In-Plane Shear Loads

The equations to employ in the structural optimization of web-core sandwich panels subjected to in-plane shear loads are Equations (16.97), (16.109), (16.111) and (16.112). It should be noted that these equations can be used "as is" to design or analyze any non-optimum web-core panel under these loads.

It can be shown that for minimum weight, the critical faces stress, $\tau_{f\,\mathrm{cr}}$, for overall instability and face-element instability must be equal, and this optimized value can be denoted τ_{f0}.

It is also seen that due to the geometry of the web-core construction there is no buckling criterion for the web element because with $\theta = 0°$ no load can be introduced into the web plates where an in-plane shear load is applied to the panel. Therefore, the criteria for determining the web element thickness, t_c, is determined by strength alone.

Manipulating the equations enumerated above it is seen that

$$\frac{W-W_{ad}}{b} = \frac{\rho_f(N_{xy}/b)}{\tau_f} + \frac{\rho_c(1-\nu_{xyf}\nu_{yxf})(t_c/b)2\sqrt{3\tau_f^2}}{k_f^{1/2}\bar{j}^{1/2}[E_{fy}^3E_{fx}]^{1/4}(N_{xy}/b)} \tag{17.86}$$

where

$$\bar{j} = \frac{1}{1+4\left[\frac{K_m(V=0)}{K_m}-1\right]}j \tag{17.87}$$

This weight equation is expressed in terms of the face stress, τ_f, and the core wall thickness, t_c. As stated, t_c is not determined by buckling, and will temporarily be treated as a constant in (17.86). Now setting the derivative of (17.86) with respect to τ_f equal to zero, the following relationship provides the unique optimum value of τ_f for a minimum weight panel, denoted as τ_{f0}:

$$\frac{N_{xy}}{b} = \frac{2(3)^{1/4}}{k_f^{1/4}\bar{j}^{1/4}}\left(\frac{\rho_c}{\rho_f}\right)^{\frac{1}{2}}\frac{(1-\nu_{xyf}\nu_{yxf})^{1/2}}{[E_{fy}^3E_{fx}]^{1/8}}\left(\frac{t_c}{b}\right)^{\frac{1}{2}}\tau_{f0}^{3/2} \tag{17.88}$$

This is the "universal relationship" relating the applied load index to the optimum face stress, τ_{f0}, for a specific web-core element thickness, t_c, as yet undetermined. Substituting (17.88) into (17.86) results in

$$\frac{W - W_{ad}}{b} = \frac{(3)^{7/6}}{(2)^{1/3}} \frac{\rho_c^{1/3} \rho_f^{2/3}}{k_f^{1/6} \bar{j}^{1/6}} \frac{(1-\nu_{xyf}\nu_{yxf})^{1/3}}{[E_{fy}^3 E_{fx}]^{1/2}} \left(\frac{t_c}{b}\right)^{\frac{1}{3}} \left(\frac{N_{xy}}{b}\right)^{\frac{1}{3}}. \tag{17.89}$$

From (17.89) it is clear that the smaller the value of t_c, the lower the panel weight will be for given material systems, given panel geometry, and given load index (N_{xy}/b). Likewise from (17.88), for given materials systems, panel geometry, and load index, as t_c decreases the optimum face stress τ_{f0} increases. Therefore, it is concluded that minimum weight is obtained when the optimum face stress, τ_{f0}, equals the materials' allowable shear stress, τ_{all}, i.e.,

$$\tau_{f0} = \tau_{\text{all}}. \tag{17.90}$$

It should be noted that this result differs from all of the previously presented sandwich panel optimizations, where the optimum face stress for minimum weight construction was generally a unique stress below the maximum allowable value. That is, when the optimum face stress is the allowable stress, this is the upper bound for the applied load index, given by (17.88).

Now substituting (17.89) into (17.88) provides the optimum value of t_c for the minimum weight construction.

$$\frac{t_c}{b} = \left(\frac{\rho_f}{\rho_c}\right) \frac{k_f^{1/2} \bar{j}^{1/2} E_{sf} (N_{xy}/b)^2}{4\sqrt{3}\tau_{\text{all}}} \tag{17.91}$$

where $E_{sf} = \left[E_{fy}^3 E_{fx}\right]^{1/4} / (1-\nu_{xyf}\nu_{yxf})$.

The other variables for the minimum weight construction are:

$$(t_f/b) = (N_{xy}/b)/2\tau_{\text{all}} \tag{17.92}$$

$$(h_c/b) = (\tau_{\text{all}}/\bar{j}E_{sf})^{1/2} \tag{17.93}$$

$$(d_f/b) = \frac{k_f^{1/2}}{2\sqrt{3}} \frac{E_{sf}^{1/2}(N_{xy}/b)}{(\tau_{\text{all}})^{3/2}} \tag{17.94}$$

$$\frac{(W - W_{ad})}{b} = \frac{3\rho_f (N_{xy}/b)}{2\tau_{\text{all}}} \tag{17.95}$$

Other useful relations for the optimum construction are

$$(W_f / W_c) = 2 \tag{17.96}$$

where W_f and W_c are the weight of the face and core respectively, and

$$\frac{h_c}{d_f} = \left(\frac{\rho_f}{\rho_c}\right)\left(\frac{t_f}{t_c}\right) \tag{17.97}$$

Note that for the optimum construction, according to Equation (17.93) the core depth, h_c, is practically independent of the applied load. Note also that the weight of the optimum construction is independent of all material properties except the allowable shear stress and the density of the face material, and the weight varies linearly with the load index (N_{xy}/b). This differs markedly from the optimum constructions using other core geometries discussed earlier in Chapter 17.

Thus it is seen that to obtain a minimum weight web-core panel subjected to in-plane shear loads, the best face material to utilize is the one having the highest rate of τ_{all}/ρ_f. Thus the factor of merit to determine the best materials system to use is:

$$\text{Factor of Merit} = \frac{\tau_{\text{all}}}{\rho}. \tag{17.98}$$

Also, it can be shown that $\bar{j}$ and k_f are constants for the optimum construction and no iteration is needed.

The Factor of Merit, (17.98) is used to compare various materials systems. Such a comparison is found in Table 17.3 for several unidirectional composites for which shear strengths were available. It is seen that among these systems graphite-epoxy and boron aluminum look significantly better than the other materials.

Table 17.3. Face Material Comparison, Based on the Factor of Merit, for a Web-Core Sandwich Panel Subjected to In-Plane Shear Loads.

Rank	Material	V_f (%)	τ_u (ksi)	ρ (lb/in^3)	FM 10^5 in
1	T300/934	?	14.8	0.0555	2.667
2	E glass/Ep.	60	10.0	0.075	2.667
3	Boron 6061 Al	50	23.0	0.0915	2.515
4	AS/3501	67	13.5	0.0555	2.432
5	Boron/Ep.	?	18.0	0.0740	2.432
6	Kev. 49/3501	62	11.7	0.050	2.340
7	Hi Str. Gr./Ep.	60	12.0	0.057	2.105
8	Boron/Ep.	67	15.2	0.074	2.054
9	Braided FP/Al	17	19.9	0.1001	1.988
10	T300/SP-286	60	10.5	0.0555	1.842
11	Kev. 49/Ep.	60	9.0	0.050	1.800

17.10 Optimal Stacking Sequences for Composite Material Laminate Faces for Various Sandwich Panels Subjected to Various Loads

If one is given a particular composite material laminate, with its associated material properties, then the methods presented in Section 17.1 through 17.9 can be used to design a minimum weight panel. However, if the designer has the freedom to do so, this current section provides the means by which to select the particular stacking sequence to use with a given lamina of a filamentary composite in order to create a laminate for a minimum weight panel. The focus here is on the faces, as the majority of the weight of a sandwich panel is in the faces, because the faces take most of the in-plane loads.

It was seen earlier that the Factor of Merit for a honeycomb sandwich panel subjected to an in-plane compressive load is given by (17.13). Since the faces of the sandwich resist the applied compressive in-plane load prior to buckling, then for any face material, the best stacking sequence for this type of loading is one in which $A_{11}^3 A_{22}$ is a maximum, because the Factor of Merit involves $E_x^{3/8} E_y^{1/8}$. Because the term $(1-\nu_{xy}\nu_{yx})^{1/2}$ is approximately equal to one for most unidirectional composite laminae, it will not be included in this discussion. Therefore, to make $A_{11}^3 A_{22}$ a maximum for a laminate involving 0° and 90° plies, let N equal the total number of plies, S equal the number of 0° plies, and N-S the number of 90° plies. If h_k is the uniform ply thickness for the laminate then, from (10.58)

$$A_{11} = \frac{SE_{11}h_k}{(1-\nu_{12}\nu_{21})} + \frac{(N-S)E_{22}h_k}{(1-\nu_{12}\nu_{21})} \tag{17.99}$$

$$A_{22} = \frac{SE_{22}h_k}{(1-\nu_{12}\nu_{21})} + \frac{(N-S)E_{11}h_k}{(1-\nu_{12}\nu_{21})}. \tag{17.100}$$

So the Factor of Merit is proportional to

$$A_{11}^3 A_{22} = \left\{\left[SE_{11}+(N-S)E_{22}\right]^3 \times \left[SE_{22}+(N-S)E_{11}\right]\right\}\frac{h_k}{(1-\nu_{12}\nu_{21})}. \tag{17.101}$$

If one defines $\phi = E_{22}/E_{11}$, then the Factor of Merit (FM) is proportional to

$$\left[S+(N-S)\phi\right]^3\left[S\phi+(N-S)\right] \tag{17.102}$$

To maximize Equation (17.102), placing the derivative of FM with respect to S equal to zero results in

$$(4S-N)\phi+3N-4S=0 \tag{17.103}$$

or

$$S = \frac{1}{4}\frac{(3-\phi)N}{(1-\phi)} \qquad 0 \le S \le N \tag{17.104}$$

Equation (17.104) and Table 17.4 below shows that for $\phi = 0$, or E_{22} of essentially zero, the best stacking sequence is 75% 0° plies and 25% 90° plies. As ϕ increases, the percentage increases until when $\phi = 1/3$, $s = 1$. At $\phi \ge 1/3$, a unidirectional laminate is used to achieve minimum weight. Incidentally, that means for any metal matrix composite, a unidirectional composite is best when the loading is uniform in-plane compression, because in every practical case, $\boldsymbol{E_{22} > E_{11}}/3$.

Table 17.4. Optimum Stacking Sequence for the Laminated Face Material as a Function of ϕ for an Optimized Honeycomb Core Sandwich Panel Subjected to an Uniaxial Compressive Load.

$\phi = E_{22}/E_{11}$	S
0	0.75 N
0.1	0.806 N
0.2	0.875 N
0.3	0.96 N
0.333	N
>0.33	N

Correspondingly, for the case of a solid or foam-core sandwich panel subjected to in-plane compression, the Factor of Merit is given by (17.21).

As before, the Factor of Merit is proportional to

$$(E_x^2 E_y)^{1/7} \text{ or } (A_{11}^2 A_2)^{1/7} \tag{17.105}$$

Utilizing (17.99) and (17.100) the *FM* is proportional to

$$[\boldsymbol{S}+(\boldsymbol{N}-\boldsymbol{S})\boldsymbol{\phi}]^2[S\phi+(N-S)] \tag{17.106}$$

Setting the derivative of the above with respect to S equal to zero results in the optimum stacking sequence of

$$S = \frac{1}{3}\frac{(2-\phi)}{(1-\phi)}N \tag{17.107}$$

For the foam- or solid-core sandwich panel subjected to a uniform compressive in-plane load, it is seen from (17.106) and shown by Table 17.5 that for $\phi = 0$, two thirds of the plies should be 0°, and only after $\phi = 1/2$, i.e., $2E_2 \ge E_1$, should the laminate be unidirectional.

Table 17.5. Optimum Stacking Sequence for a Laminated Composite Face Material as a Function of ϕ for an Optimized Solid or Foam Core Sandwich Panel Subjected to an Uniaxially Compressive Load.

$\phi \equiv E_{22}/E_{11}$	S
0	0.667 N
0.1	0.704 N
0.2	0.75 N
0.3	0.810 N
0.4	0.889 N
0.5	N

Now examining (17.29) for the truss-core panel subjected to an in-plane compressive load, and the expression on page 359 for the web-core sandwich panel, one sees that they are identical. Remembering that $2E_0 = (E_xE_y)^{1/2} + \nu_{yx}E_x + 2G_{xy}(1-\nu_{xy}\nu_{yx})$, one sees that for the truss-core panel and for the web-core panel the FM is proportional to $(A_{11}A_{22})^{1/4}$. Therefore, for minimum weight

$$S = N/2 \tag{17.108}$$

Thus, the optimum stacking sequence in each case is cross-ply with the same number of 0° plies and 90° plies. Incidentally, using the same procedures as above, it can be shown that an angle ply laminate is never better than a cross-ply or unidirectional laminate to prevent buckling for any of the four sandwich architectures when the panel is subjected to an in-plane compressive load.

Turning now to the honeycomb core sandwich panel subjected to an in-plane shear load, it is seen that the factor of merit is given by (17.66) is proportional to $(E_xE_y)^{1/4}$. As a result the cross-ply laminate is best, as shown by (17.108). The same holds true for the solid- or foam-core panel subjected to in-plane shear loading.

For the truss-core sandwich panel subjected to in-plane shear loads, from (17.82), it is seen that the FM is proportional to

$$(E_xE_y^3)^{1/8} \tag{17.109}$$

By the same process used before it is seen that

$$S = \frac{1}{4}\frac{(1-3\phi)}{(1-\phi)}N \tag{17.110}$$

with the tabular values given in Table 17.6. It is seen that if $\phi = 0$, 75% of the fibers should be perpendicular to the flutes, and for $\phi \geq 1/3$, the best stacking sequence for the truss-core sandwich subjected to in-plane shear is to have all fibers (i.e. unidirectional) perpendicular to the flutes.

Table 17.6. Optimum Stacking Sequence for Laminated Composite Face Materials as a Function of ϕ for an Optimized Truss-Core Sandwich Panel Subjected to In-Plane Shear Loads.

$\phi = E_{22}/E_{11}$	S
0	0.25 N
0.1	0.194 N
0.2	0.125 N
0.3	0.036 N
0.333	0
>0.333	0

Finally the web-core sandwich subjected to in-plane shear loads differs from all the rest in that the factor of merit is strictly strength (not stiffness) dependent. Repeating (17.98),

$$FM = \tau_{all} / \rho. \tag{17.111}$$

Again for any of the four panel architectures subjected to in-plane shear loads, an angle ply laminate is never the best choice.

17.11 References

17.1. Vinson, J.R. (1985) Minimum Weight Composite Material Honeycomb Sandwich Panels Under Uniaxial Compression, *Transactions of the First European Conference on Composite Materials*, September.

17.2. Vinson, J.R. (1986) Optimum Design of Composite Honeycomb Sandwich Panels Subjected to Uniaxial Compression, *AIAA Journal*, Vol. 24, pp. 1690-1696.

17.3. Vinson, J.R. and Shore, S. (1967) Structural Optimization of Corrugated Core and Web Core Sandwich Panels Subjected to Uniaxial Compression, *Naval Air Engineering Center Report NAEC-ASC-1109*, May.

17.4. Vinson, J.R. (1986) Minimum Weight Web-Core Composite Sandwich Panels Subjected to In-Plane Compressive Loads, presented at the International Symposium on Composite Materials and Structures, Beijing, China, June 13-16.

17.5. Handel, P.I. and Vinson, J.R. (1988) Optimal Stacking Sequence of Composite Faces for Various Sandwich Panels and Loads to Attain Minimum Weight, *Proceedings of the 29th AIAA/ASME/ASCE/AHS/ASC Structures, Structural Dynamics and Materials Conference*, AIAA, pp. 999-1006.

17.12 Problems

17.1. Consider a honeycomb sandwich panel 20 inches long and 12 inches wide subjected to an in-plane compressive load in the longer direction of $N_x = -2000$ lbs./in. of width. The faces are made of boron/epoxy ($V_f = 67\%$) and the core is aluminum each with the following properties:

Boron/epoxy	Aluminum Honeycomb Core
$E_x = 30.3\times10^6$ psi	$E_c = 10.5\times10^6$ psi
$E_y = 2.8\times10^6$ psi	$\nu = 0.348$
$\gamma_{xy} = 0.21$	$G_c = 3.9\times10^6$ psi
$G_{xy} = 0.93\times10^6$ psi	$\rho_m = 0.101\ \text{lb/in}^3$
$\rho_f = 0.074\ \text{lb/in}^3$	
$\sigma_{\text{all}} = 362{,}600$ psi	

If the panel is simply supported on all four edges:

(a) What is the optimum face stress, σ_f?

(b) Is the panel overstressed?

(c) What are the optimum values of each dependent variable (in this case h_c, d, t_c, and t_f) for the honeycomb construction?

(d) If the boron/epoxy ply thickness is 0.0055 in., how many laminates are necessary to approximate the optimum construction?

(e) What is the weight of the panel per unit planform area?

17.2. Repeat Problem 17.1 for a foam core sandwich using the foam core of Problem 7.2.

17.3. Compare the results of Problems 17.1 and 17.2. Which core construction gives the lighter panel?

17.4. Consider a sandwich panel measuring $30'' \times 24''$ in planform are composed of 5052 aluminum faces whose properties are: $\boldsymbol{E = 10\times10^6}$ psi, $\nu = 0.3$, $\rho_w = 0.10\ \text{lb/in}^3$, $\boldsymbol{\sigma_{\text{all}} = 70{,}000}$ psi. The uniaxial compressive load per unit width, in the longer direction is $\boldsymbol{N_x = -4{,}000}$ lb/in. The sandwich plate is simply supported on all four edges. What are the optimum t_f, h_c, t_c, d, and $(W - W_{ad})$ for a honeycomb core also made of 5052 aluminum?

17.5. If the sandwich panel of Problem 17.4 has a rigid Klegecell foam core, what are the optimum values of t_f, h_i, G_c, and $(W - W_{ad})$ using Figure 17.4 to determine the weight of the rigid Klegecell foam?

17.6. If the sandwich panels of Problem 17.4 involved a truss-core also of 5052 aluminum, what are the optimum values of t_f, h_c, t_c, θ, and $(W - W_{ad})$ for this construction?

17.7. If the sandwich panel of Problem 17.4 involved a web-core was also made of 5052 aluminum, what are the optimum values of $t_f, h_c, t_c, d,$ and $(W - W_{ad})$ for this construction?

17.8. For Problems 17.4 through 17.7, compare the various constructions as to the weight, the various total panel thickness, and the various face thicknesses?

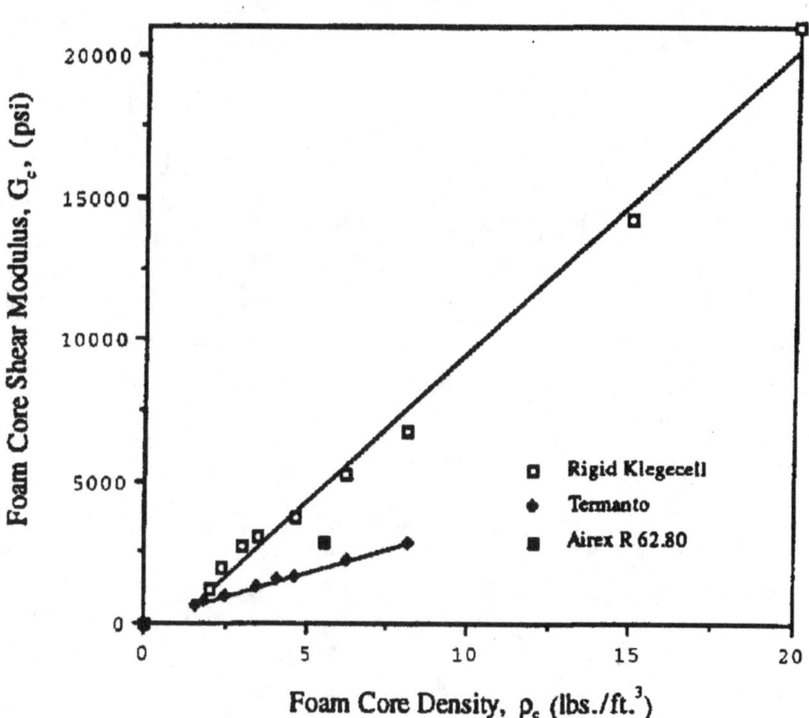

Figure 17.4. Typical foam core properties. (Reprinted from Reference [10.1].)

CHAPTER 18

PIEZOELECTRIC MATERIALS

18.1 Introduction

Recently, there has been increasing interest in using piezoelectric materials in advanced structures to transform them into "smart" or adaptive structures. To treat these developments adequately would require an entire text, well beyond the scope of this text; nevertheless, an introduction is included herein, hopefully, to provide the reader the capability of including piezoelectric effects into the structures comprising the first three parts of this book and enabling the reader to more easily follow the growing literature on this subject.

Smart, intelligent or adaptive materials are being used increasing due to their active interaction with their environment, and their increasing use in "smart" structures. Such materials include piezoelectric materials, electrostrictive materials, shape memory alloys, and electrorheological fluids. Only piezoelectric materials will be discussed briefly in Chapters 18 through 20 herein.

The use of piezoelectric materials in "smart" structures is increasing significantly. They can be used as sensors that recognize and measure the intensity of physical quantities such as strain in the structure and as such can be used as a structural health monitor to detect damage. They can also be used as actuators, where be responding to an applied voltage, they strain, and cause the "smart" structure in which they are imbedded to deform, or in the dynamic case cause it to excite or dampen vibration oscillations.

According to Larson [18.1], the ancient Greeks were the first to recognize the electrical features, particularly the static charges developed, in certain materials when rubbed. Another material electrical phenomenon, piezoelectricity, was named by Jacques and Pierre Curie more than one hundred years ago. In 1894 Voigt [18.2] rigorously stated the relationship between the material structure and the piezoelectric effects, namely that when a voltage is placed across a piezoelectric material, it generates a geometric change known as a converse piezoelectric effect. Depending on the material orientation and the poling direction, the material may elongate or shrink in different directions, or an angular distortion. Also, with the same material if the material is stressed due to a tensile, compressive or shear load, an electrical voltage results, and this is called a direct piezoelectric effect. Therefore, it is seen that piezoelectric materials can be used as actuators or sensors for a structure and could serve as both at different times.

Many materials exhibit a piezoelectric effect including Rochelle salt, quartz, tourmaline and barium titanate. As early as 1918, Langevin proposed a piezoelectric transducer for sonar during World War II. Prior to World War II lead zirconate titanate (PZT) was found by researchers at MIT to have a much higher piezoelectric response. Later, in 1969, Kawai determined that the polymer polyvinylidene fluoride (PVDF) was highly piezoelectric. Also in the 1960's it was found that human muscle and bone are also piezoelectric.

An excellent overview of recent activity in adaptive or smart structures is given in the early 1990's by Wada [18.3], especially regarding NASA future missions, by Wada, Fansom and Crawley [18.4] and Miura [18.5]. Analytical models of piezoelectric actuation of simple beams were treated by Forward [18.6], Baily and Hubbard [18.7], Crawley and de Luis [18.8], Burke and Hubbard [18.9] and Im and Atluri [18.10]. Piezoelectric plates were treated by Lee and Moon [18.11-18.13]; Crawley, Lazarus and Anderson [18.14, 18.15]; Wang and Rogers [18.16] and Pai and Nayfeh [18.17].

Two reference books are available for serious study of piezoelectric materials and layers, one by Nye [18.18] in 1972 and the other by Tiersten [18.19] in 1969.

The manufacturing of piezoelectric ceramics involves the intimate mixing of precise quantities of pure raw materials which are then heated to 1200°C to produce titanates and zirconates. The calcined material is milled to produce a fine powder, which after adding organic binders, is formed into prescribed shapes by pressing, casting and extrusion processes. The parts are then fired at temperatures up to 1350°C to produce dense polycrystalline ceramic components. Close tolerances of the parts is required which is achieved by diamond machining.

For subsequent polarization and operational use, metal electrodes must be applied to the ceramic. Fired-on silver or electrode nickel are the most satisfactory with respect to electrical conductivity, ease of soldering and good adhesion.

After the electrodes are applied the material is pole in a strong d.c. field in order to align the randomly oriented dipoles to produced the piezoelectric properties. The poling effect is permanent provided that the material is not subjected to high a.c. fields, to temperatures in the region of greater than half the Curie point or to very high mechanical stresses.

18.2 Piezoelectric Effect

When a piezoceramic element is stressed electrically by a voltage, its dimensions change. When it is stressed mechanically, it generates an electric charge. If the electrodes are not short circuited, a voltage associated with the charge appears.

A piezoceramic is therefore capable of acting either as a sensing or transmitting element or both. Since piezoceramic elements are capable of generating very high voltages, they are compatible with today's generation of solid-state devices.

Consider the element shown below:

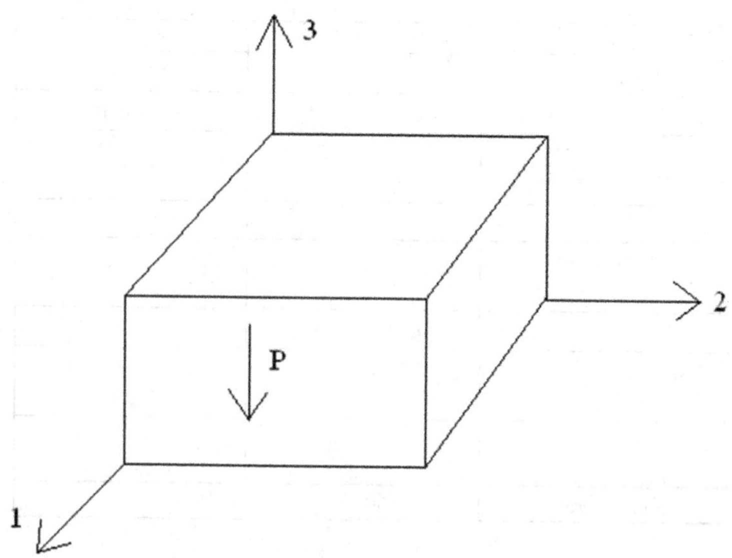

Figure 18.1. Piezoceramic element.

The polar or 3 axis is taken to be parallel to the direction of polarization within the ceramic. This direction is established during manufacturing by a high d.c. voltage that is applied between a pair of electroded faces to activate the material. The polarization vector P is represented by an arrow pointing from the positive to the negative poling electrode.

In shear operations, these poling electrodes are later removed and replaced by electrodes deposited on a second pair of faces. In this case, the 3 axis is not altered, but is them parallel to the electroded faces. Then the mechanical stress or strain is in shear and the subscript 5 is used as the second subscript.

Piezoelectric coefficients, d_{ij}, with double subscripts link electrical and mechanical quantities. The first subscript gives the direction of the electrical field associated with the voltage applied, or the charge or voltage produced. The second subscript gives the direction of the mechanical stress or strain

Since the 1980's the interest in the use of piezoelectric material for structural application mushroomed. The technical breakthrough has been the capability to manufacture thin layers of brittle piezoelectric materials and to successfully imbed them into a structural component. As a result intelligent structures have evolved including actively damped structures, smart fins, smart wings, active constrained layer damped structures, position controls for telescopes in complex structural platforms (NASA), and for structural health monitoring. The primary piezoelectric materials in use today are PZT and PVDF. Properties of PZT G1195 and PVDF are given below in Table 18.1 [18.2, 19.2].

Figure 18.1. Properties of PZT G1195 and PVDF.

Material	PZT G1195	PVDF
Density	7600 kg/m^3	1780 kg/m^3
Elastic Modulus ($Y^{E_{33}}$)	4.9×10^{10} N/m^2	0.2 N/m^2
Relative Dielectric Constant, K_3	1700	
K_1	1700	
Piezoelectric Strain Coefficient, d_{33}	360×10^{-12} m/Volt	285×10^{-12} m/Volt
d_{31}	-166×10^{-12} m/Volt	-122×10^{-12} m/Volt
Piezoelectric Voltage Coefficient, g_{33}	25×10^{-3} Volt m/N	24.9×10^{-3} Volt m/N
g_{31}	11×10^{-3} Volt m/N	-10.6×10^{-3} Volt m/N
Curie Temperature	360 °C	

18.3 References

18.1. Larson, P.H., Jr. (1994) The Use of Piezoelectric Materials in Creating Adaptive Shell Structures, Ph.D. Dissertation, Mechanical Engineering, University of Delaware, Spring.

18.2. Leibowitz, M. and Vinson, J.R. (1993) Intelligent Composites: Design and Analysis of Composite Material Structures Involving Piezoelectric Material Layers: Part B Experimental Demonstration of Damping Techniques in Simple Sandwich Structures, University of Delaware, Center for Composite Materials, CCM Report 93-01.

18.3. Wada, B.K. (1990) Adaptive Structures: An Overview, *Journal of Spacecraft and Rockets*, Vol. 3, No. 3, May-June, pp. 330-337.

18.4. Wada, B.K., Fansom, J.L. and Crawley, E.F. (1990) Adaptive Structures, *Journal of Intelligent Materials Systems and Structures*, Vol. 1, April, pp. 157-174.

18.5. Miura, K. (1992) Adaptive Structures Research at ISAS, *Journal of Intelligent Materials Systems and Structures*, Vol. 3, January, pp. 54-74.

18.6. Forward, R.L. (1981) Electronic Damping of Orthogonal Bending Modes in a Cylindrical Mast – Experimental, Vol. 18, No. 1, pp. 11-17.

18.7. Baily, T. and Hubbard, J.E. (1984) Distributed Piezoelectric – Polymer Active Vibration Control of a Cantilever Beam, *Journal of Guidance, Control and Dynamics*, Vol. 7, No. 4, pp. 437-442.

18.8. Crawley, E.F. and de Luis, J. (1987) Use of Piezoelectric Actuators as Elements of Intelligent Structures, *AIAA Journal*, Vol. 25, No. 10.

18.9. Burke, S. and Hubbard, J.E. (1987) Active Vibration Control of a Simply Supported Beam Using Spatially Distributed Actuators, *IEEE Control System Magazine*, Vol. 7, No. 6.

18.10. Im, S. and Atluri, S.N. (1989) Effects of Piezo-Actuator on a Finitely Deformed Beam Subjected to General Loading, *AIAA Journal*, Vol. 27, No. 12, December, pp. 1801-1807.

18.11. Lee, C.K. and Moon, F.C. (1989) Laminated Piezopolymer Plates for Torsion and Bending Sensors and Actuators, *Journal of the Acoustical Society of America*, Vol. 85, No. 6, June.

18.12. Lee, C.K. and Moon, F.C. (1990) Modal Sensor/Actuator, *ASME Journal of Applied Mechanics*, Vol. 57, June, pp. 434-441.

18.13. Lee, C.K. (1990) Theory of Laminated Piezoelectric Plates for Design of Distributed Sensor/Actuators, Part 1: Governing Equations and Reciprocal Relationships, *Journal of the Acoustical Society of America*, Vol. 87, No. 3, March, pp. 1144-1158.

18.14. Crawley, E.F. and Lazarus, K.B. (1989) Induced Strain Actuation of Isotropic and Anisotropic Plates, *AIAA 89-1326-CP*, pp. 1451-1461.

18.15. Crawley, E.F. and Anderson, E.H. (1990) Detailed Models of Piezoceramic Actuations of a Beam, *Journal of Intelligent Material Systems and Structures*, Vol. 1, January, pp. 4-25.

18.16. Wang and Rogers

18.17. Pai, P.F., Nayfeh, A.H. and Oh, K. (1992) A Nonlinear Theory of Laminated Piezoelectric Plates, *Proceedings of the 33rd AIAA/ASME/ASCE/AHS/ASC Structures, Structural Dynamics and Materials Conference, Part 3*, Dallas, April, pp. 577-585.

18.18. Nye, N.Y. (1972) *Physical Properties of Crystal: Their Representation by Tensors and Matrices*, Oxford University Press, Oxford, UK.

18.19. Tiersten, H.F. (1969) *Linear Piezoelectric Plate Vibration*, Plenum Press, NY.

CHAPTER 19

PIEZOELECTRIC EFFECTS

19.1 Laminate of a Piezoelectric Material

Consider a lamina of a laminate which is made of a piezoelectric material that is electrically anisotropic. For PZT the material is naturally piezoelectric, but for PVDF rolling and poling is necessary to make the material piezoelectric. The most general form of the piezoelectric strain constants for a plate can be written as follows [18.2, 19.1]:

$$d_{ij} = \begin{bmatrix} 0 & 0 & 0 & 0 & d_{15} & 0 \\ 0 & 0 & 0 & d_{24} & 0 & 0 \\ d_{13} & d_{23} & d_{33} & 0 & 0 & d_{36} \end{bmatrix} \tag{19.1}$$

where (i = 1,2,3) and (j = 1,2,3,4,5,6). For a thin piezoceramic, such as many laminae are, that are poled in the thickness direction, the only non-zero components are the $d_{13} = d_{31}$ and $d_{23} = d_{32}$. Thus, no shear strain components will be produced by a thin actuator.

Also piezoelectric stress resultants and piezoelectric stress couples can be defined that are completely analogous to thermal and hygrothermal resultants and couples, see (10.60), (10.61), (10.64) and (10.65). For a laminate these are defined as follows for i, j = 1,2,6:

$$N_j^{piezo} = \sum_{k=1}^{N} \left[\overline{EI}_j\right]_k (h_k - h_{k-1}) \tag{19.2}$$

$$M_j^{piezo} = \frac{1}{2}\sum_{k=1}^{N} \left[\overline{EI}_j\right]_k (h_k^2 - h_{k-1}^2) \tag{19.3}$$

In the above, in general

$$\left[\overline{EI}\right]_k = \begin{bmatrix} m^2 & n^2 & 2mn \\ n^2 & m^2 & -2mn \\ -mn & mn & m^2-n^2 \end{bmatrix} \begin{bmatrix} C_{11} & C_{12} & C_{16} \\ C_{21} & C_{22} & C_{26} \\ 0 & 0 & C_{66} \end{bmatrix} \begin{bmatrix} 0 & 0 & d_{13} \\ 0 & 0 & d_{23} \\ 0 & 0 & d_{63} \end{bmatrix} \begin{bmatrix} E_x \\ E_y \\ E_z \end{bmatrix} \tag{19.4}$$

where here again $m = \cos\theta$, $n = \sin\theta$ and θ is the angle between the x-y structural axes and the 1-2 material axes, defined as positive going for x^+ axis and the 1^+ axis. E_i (i = x, y, z) is the electrical field intensity vector.

For the laminate, (10.66) can be expanded to include the piezoelectric effects, so

$$\begin{bmatrix} N \\ M \end{bmatrix} = \begin{bmatrix} A & B \\ B & D \end{bmatrix} \begin{bmatrix} \varepsilon_0 \\ \kappa \end{bmatrix} - \begin{bmatrix} N^{\mathrm{T}} \\ M^{\mathrm{T}} \end{bmatrix} - \begin{bmatrix} N^{\mathrm{m}} \\ M^{\mathrm{m}} \end{bmatrix} - \begin{bmatrix} N^{\mathrm{piezo}} \\ M^{\mathrm{piezo}} \end{bmatrix} \tag{19.5}$$

Also, the components of the transverse shear resultants, $[Q]$, are defined as follows:

$$\begin{bmatrix} Q_x \\ Q_y \end{bmatrix} = \sum_{k=1}^{N} \left\{ \int_{h_{k-1}}^{h_k} [T]^{-1}[C_k][T] \begin{bmatrix} \bar{\alpha} + \dfrac{\partial w}{\partial x} \\ \bar{\beta} + \dfrac{\partial w}{\partial y} \end{bmatrix} \mathrm{d}z + \int_{h_{k-1}}^{h_k} [EI^*]_k \, \mathrm{d}z \right\} \tag{19.6}$$

where

$$[EI^*]_k = \begin{bmatrix} m & n \\ -n & m \end{bmatrix} \begin{bmatrix} C_{44} & C_{45} \\ C_{45} & C_{55} \end{bmatrix} \begin{bmatrix} 0 & d_{24} \\ d_{15} & 0 \end{bmatrix} \begin{bmatrix} E_x \\ E_y \end{bmatrix} \tag{19.7}$$

where E_i $(i = x, y)$ are components of the electrical field intensity vector, and $[C_k]$ is the C_{ij} matrix for the kth lamina. Thus with the integrated stress-strain relations (19.5) a piezoelectric plate, panel or beam structure can be analyzed by methods developed earlier in the text, because the piezoelectric effects are directly analogous to thermal and moisture effects.

When the piezoelectric material is thin in the 3 direction, if the surfaces of the cross sectional area are fully electroded, the electrical conditions are that $E_1 = E_2 = 0$ everywhere, and from the relation that the electrical field intensity vector E_i is derivable from a scalar potential ϕ, then $E_i = -\phi$. So for many beam and plate structures the piezoelectric material layers are so thin that the electrical field E_3 is obtained from

$$E_3 = \frac{\phi\left(\dfrac{h_k}{2}\right) - \phi\left(-\dfrac{h_k}{2}\right)}{h_k} = \frac{V}{h_k} \tag{19.8}$$

where V is the driving voltage, h_k is the thickness of the piezoelectric layer under consideration and ϕ is a scalar potential from which the electrical field vector E_i is derived.

To use energy methods, the expression for the kinetic energy and potential energy for a plate structure involving piezoelectric layers is given by Leibowitz and Vinson [19.1] as well as an example for the dynamic modeling of an elastic beam with piezoelectric actuator laminae on each side, using Hamilton's Principle. In this dynamic model, the piezoceramic and the beam are treated as an integral system and all of the natural and kinematic boundary conditions were included. Leibowitz has also developed methods for analyzing an active constrained layer beam involving the primary beam structure, soft core and piezoelectric actuator for various boundary conditions using Hamilton's Principle. This was a new (1993) concept in damping, one in which a piezoelectric material is embedded in the viscoelastic core such that it acts as an actuator to increase the shear strains in the viscoelastic material and thus increase the modal damping of the total laminate compared to the classical approach.

Because the governing equations for a plate including general anisotropy, mid-plane asymmetry and dynamic effects are so lengthy, suffice it to say that Equations (19.9) through (19.11) below clearly indicate how the right hand side of the governing plate equations given throughout this text are modified to include piezoelectric effects [see 19.2]. The resulting equations of motion are, for the classical plate theory, i.e., no transverse shear deformation [19.2].

$$\begin{aligned}
&A_{11}\frac{\partial^2 u_0}{\partial x^2}+2A_{16}\frac{\partial^2 u_0}{\partial x\partial y}+A_{66}\frac{\partial^2 u_0}{\partial y^2}+A_{16}\frac{\partial^2 v_0}{\partial x^2}+(\boldsymbol{A_{12}}+\boldsymbol{A_{66}})\frac{\partial^2 v_0}{\partial x\partial y}\\
&+A_{26}\frac{\partial^2 v_0}{\partial y^2}-B_{11}\frac{\partial^3 w}{\partial x^3}-3B_{16}\frac{\partial^3 w}{\partial x^2\partial y}-(\boldsymbol{B_{12}}+\boldsymbol{2B_{66}})\frac{\partial^3 w}{\partial x\partial y^2}-B_{26}\frac{\partial^3 w}{\partial y^3}\\
&\rho_m h\frac{\partial^2 u_0}{\partial t^2}+\frac{\partial N_1^P}{\partial x}+\frac{\partial N_6^P}{\partial y}
\end{aligned}\tag{19.9}$$

$$\begin{aligned}
&A_{16}\frac{\partial^2 u_0}{\partial x^2}+(\boldsymbol{A_{66}}+\boldsymbol{A_{12}})\frac{\partial^2 u_0}{\partial x\partial y}+A_{26}\frac{\partial^2 u_0}{\partial y^2}+A_{66}\frac{\partial^2 v_0}{\partial x^2}+2A_{26}\frac{\partial^2 v_0}{\partial x\partial y}\\
&+A_{22}\frac{\partial^2 v_0}{\partial y^2}-B_{16}\frac{\partial^3 w}{\partial x^3}-(\boldsymbol{2B_{66}}+\boldsymbol{B_{12}})\frac{\partial^3 w}{\partial x^2\partial y}-3B_{26}\frac{\partial^3 w}{\partial x\partial y^2}-B_{22}\frac{\partial^3 w}{\partial y^3}\\
&\rho_m h\frac{\partial^2 v_0}{\partial t^2}+\frac{\partial N_6^P}{\partial x}+\frac{\partial N_2^P}{\partial y}
\end{aligned}\tag{19.10}$$

$$\begin{aligned}
&B_{11}\frac{\partial^3 u_0}{\partial x^3}+3B_{16}\frac{\partial^3 u_0}{\partial x^2\partial y}+(\boldsymbol{B_{12}}+\boldsymbol{2B_{66}})\frac{\partial^3 u_0}{\partial x\partial y^2}+B_{26}\frac{\partial^3 u_0}{\partial y^3}+B_{16}\frac{\partial^3 v_0}{\partial x^3}\\
&+(\boldsymbol{B_{12}}+\boldsymbol{2B_{66}})\frac{\partial^3 v_0}{\partial x^2\partial y}+3B_{26}\frac{\partial^3 v_0}{\partial x\partial y^2}+B_{22}\frac{\partial^3 v_0}{\partial y^3}-D_{11}\frac{\partial^4 w}{\partial x^4}-4D_{16}\frac{\partial^4 w}{\partial x^3\partial y}\\
&-2(\boldsymbol{D_{12}}+\boldsymbol{D_{66}})\frac{\partial^4 w}{\partial x^2\partial y^2}-4D_{26}\frac{\partial^4 w}{\partial x\partial y^3}-D_{22}\frac{\partial^4 w}{\partial y^4}\\
&\rho_m h\frac{\partial^2 w}{\partial t^2}-f(x,y,t)+\frac{\partial^2 M_1^P}{\partial x^2}+\frac{\partial^2 M_2^P}{\partial y^2}+2\frac{\partial^2 M_6^P}{\partial x\partial y}
\end{aligned}\tag{19.11}$$

These are the equations of motion for an open circuit piezoelectric laminate under the influence of an externally applied electric field and lateral mechanical loading. If thermal and moisture effects are present, then simply add thermal and moisture stress resultants and couples to the above, analogous to the piezoelectric terms shown. In (19.11), if the plate is laminated, the $\rho_m h$ term is given by (13.8). Reference [19.2] also provides the changes in the governing equations if transverse shear deformation effects are included as well.

Piezoelectric ceramic materials (e.g. PZT) are available only in the form of small patches, although sheets of piezoelectric polymeric materials (PVDF) are available. The properties of each material differ significantly, as seen in Table 18.1.

Because the piezoceramics are only available in small patches, their use in plate or any other structures makes analytical solutions of governing differential equations with various boundary conditions impractical if not impossible. Therefore, energy principles are often used to obtain solutions.

Other useful references include the references of Chapter 18 and [19.3-19.6].

19.2 References

19.1. Leibowitz, M. and Vinson, J.R. (1993) The Use of Hamilton's Principle in Laminated Piezoelectric and Composite Structures, ASME-AD Vol. 35, Adaptive Structures and Material Systems, eds., G.P. Carman and E. Garcia, pp. 257-268.

19.2. Leibowitz, M. (1993) Studies on the Use of Piezoelectric Actuators in Composite Sandwich Constructions, Ph.D. Dissertation, University of Delaware.

19.3. Tzou, H.S. and Tseng, C.I. (1990) Distributed Piezoelectric Sensor/Actuator Design for Dynamic Measurement/Control of Distributed Parameter System: A Piezoelectric Finite Element Approach, *Journal of Sound and Vibration*, Vol. 138, No. 1, pp. 17-34.

19.4. Robbins, D.H. and Reddy, J.N. (1991) Analysis of Piezoelectrically Actuated Beam Using a Layer-Wise Displacement Theory, *Composites and Structures*, Vol. 41, No. 2, pp. 265-279.

19.5. Wang, B.T. and Rogers, C.A. (1991) Laminate Plate for Spatially Distributed Inducer Strain Actuators, *Journal of Composite Materials*, Vol. 25, April.

19.6. Fukada, E. and Sakurai, T. (1971) Piezoelectricity in Polarized Polyvinylidene Fluoride ( ) Fibers, *Polymer Journal*, Vol. 2, No. 5, pp. 656-662.

CHAPTER 20

USE OF MINIMUM POTENTIAL ENERGY TO ANALYZE A PIEZOELECTRIC BEAM

20.1 Introduction

The following example is given to analyze the simplest structure (a beam) in which a piezoelectric actuator is used, and because it is a "real world" problem in which two such beams are being used to actuate a "smart" fin [20.1].

Consider the beam shown in Figure 20.1 below. The host beam is of rectangular constant cross-section of length L. Piezoelectric actuators are adhesively bonded to the host beam from L_1 to L_2. Because of the geometry and material properties the adhesive must be included in the analysis.

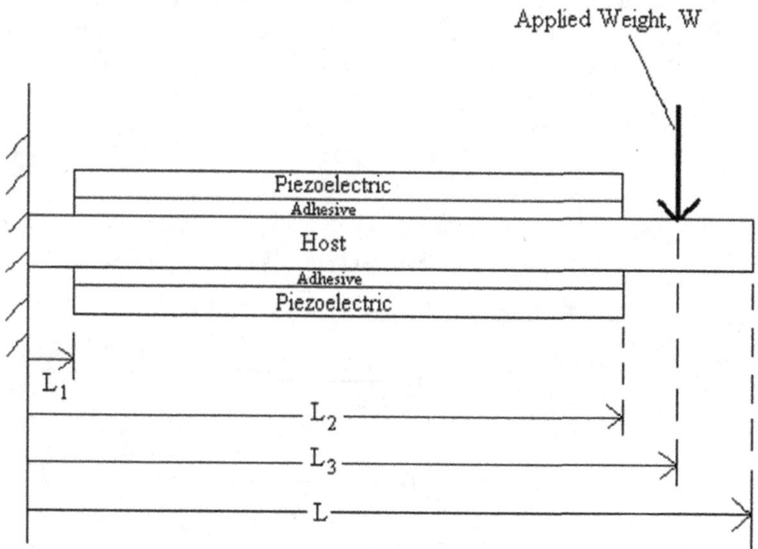

Figure 20.1. Piezoelectric beam.

Equation (14.2) is used to obtain the potential energy expression for the bending of the beam in Figure 20.1. Because it is a beam whose length is in the x-direction all $\partial(\ )/\partial y$ and v_0 terms are dropped. If mid-plane symmetric, all B_{ij} terms are dropped. Without extension in this example all u_0 terms are dropped. Using classical theory

$\bar{\alpha} = -\frac{dw}{dx}$. Because it is a beam, all remaining terms are multiplied by the width b, so that the $bD_{11} \square EI$, neglecting Poisson ratio effects, because it is a beam. Also the flexural stiffness for $0 \leq x \leq L_1$ is $(EI)_1$, $L_1 \leq x \leq L_2$ is $(EI)_2$ and $L_2 \leq x \leq L$ is $(EI)_3$. Beçause of the analogy between the thermal, moisture and piezoelectric stress couples let $M^T = M^P$. Lastly, the concentrated load applied at $x = L_3$ is accounted for. The result is Equation (20.1).

$$V = \frac{(EI)_1}{2}\int_0^{L_1}\left(\frac{d^2w}{dx^2}\right)^2 dx + \frac{(EI)_2}{2}\int_{L_1}^{L_2}\left(\frac{d^2w}{dx^2}\right)^2 dx + \frac{(EI)_3}{2}\int_{L_2}^{L}\left(\frac{d^2w}{dx^2}\right)^2 dx$$
$$+ Ww(L_3) + \int_{L_1}^{L_2} bM^P\left(\frac{d^2w}{dx^2}\right)dx \tag{20.1}$$

In this equation the last term represents the work done by the piezoelectric patches which have a constant moment effect M^P.

Since the beam is cantilevered at $x = 0$ and free at $x = L$, a deflection function (20.2) can be assumed which satisfies all of the boundary conditions at the ends of the beam structures.

$$w(x) = a_0x^2 + a_1x^3 \qquad 0 \leq x \leq L \tag{20.2}$$

If one now utilizes the fact that at $x = L_3$, a load W is applied in the negative z direction, then at $x = L_3$ the transverse shear resultant V and the beam bending moment M are:

$$V(L_3) = -(EI)_3\frac{d^3w(L_3)}{dx^3} = -W$$
$$M(L_3) = -(EI)_3\frac{d^2w(L_3)}{dx^2} = 0 \tag{20.3}$$

From (20.3) and (20.2), the constants a_1 and a_0 are found to be

$$a_1 = \frac{W}{6(EI)_3}$$
$$a_0 = \frac{-WL_3}{2(EI)_3} \tag{20.4}$$

Therefore the deflection $w(x)$ is seen to be

$$w(x) = \frac{W}{2(EI)_3}\left[-L_3x^2 + \frac{x^3}{3}\right] \tag{20.5}$$

For Minimum Potential Energy only the form is used hence one can assume the deflection to be

$$w(x) = A\left[-L_3x^2 + \frac{x^3}{3}\right] \tag{20.6}$$

where A is the unknown constant to be found. Substituting (20.6) into (20.1) the result is

$$\begin{aligned} V = {} & \frac{(EI)_1}{2}\int_0^{L_1} 4A^2\left[L_3^2 - 2L_3x + x^2\right]dx + \frac{(EI)_2}{2}\int_{L_1}^{L_2} 4A^2\left[L_3^2 - 2L_3x + x^2\right]dx \\ & \frac{(EI)_3}{2}\int_{L_2}^{L} 4A^2\left[L_3^2 - 2L_3x + x^2\right]dx + bM^P\int_{L_1}^{L_2} 2A\left[-L_3 + x\right]dx + Ww(L_3) \end{aligned} \tag{20.7}$$

which after performing the integration, can be written as

$$\begin{aligned} V = {} & 2A^2\left\{(EI)_1\left[L_3^2L_1 - L_3L_1^2 + \frac{L_1^3}{3}\right]\right. \\ & + (EI)_2\left[L_3^2(L_2 - L_1) - L_3(L_2^2 - L_1^2) + \frac{L_2^3 - L_1^3}{3}\right] \\ & \left. + (EI)_3\left[L_3^2(L - L_2) - L_3(L^2 - L_2^2) + \frac{L^3 - L_2^3}{3}\right]\right\} \\ & + 2bM^PA\left[-L_3(L_2 - L_1) + \frac{L_2^2 - L_1^2}{2}\right] + \frac{2}{3}WA\left[-L_3^3\right] \end{aligned} \tag{20.8}$$

Now setting $\delta V = 0$, the constant A is found to be

$$A = \frac{\frac{1}{6}WL_3^3 - \frac{1}{2}bM^P\left[-L_3(L_2 - L_1) + \frac{L_2^2 - L_1^2}{2}\right]}{\left\{\begin{array}{l}(EI)_1\left[L_3^2L_1 - L_3L_1^2 + \frac{L_1^3}{3}\right] \\ +(EI)_2\left[L_3^2(L_2 - L_1) - L_3(L_2^2 - L_1^2) + \frac{L_2^3 - L_1^3}{3}\right] \\ +(EI)_3\left[L_3^2(L - L_2) - L_3(L^2 - L_2^2) + \frac{L^3 - L_2^3}{3}\right]\end{array}\right\}} \tag{20.9}$$

Because in experiments performed the beam tip deflection is measured, the tip deflection is seen to be:

$$w(L) = w(L_3) + \frac{dw(L_3)}{dx}(L - L_3) \tag{20.10}$$

where $\frac{dw(L_3)}{dx} = -AL_3^2$. The final expression for $w(L)$ is

$$w(L) = \frac{\frac{1}{6}W\left(\frac{1}{3}L_3^6 - LL_3^5\right) + bM^P\left(\frac{1}{2}LL_3^2 - \frac{1}{6}L_3^3\right)\left(-L_3(L_2 - L_1) + \frac{L_2^2 - L_1^2}{2}\right)}{\left\{\begin{array}{l}(EI)_1\left[L_3^2L_1 - L_3L_1^2 + \frac{L_1^3}{3}\right] \\ +(EI)_2\left[L_3^2(L_2 - L_1) - L_3(L_2^2 - L_1^2) + \frac{L_2^3 - L_1^3}{3}\right] \\ +(EI)_3\left[L_3^2(L - L_2) - L_3(L^2 - L_2^2) + \frac{L^3 - L_2^3}{3}\right]\end{array}\right\}} \tag{20.11}$$

This is a typical problem that can be solved for a piezoelectric structure. For brevity, further examples will not be given in this text but recent literature will provided the reader with many other examples, such as those by Abramovich [20.2, 20.3] Aldraihem and Khdeir [20.4], Eisenberger and Abramovich [20.5], Azzouz, Mei, Bevan and Ro [20.6], Abramovich and Pletner [20.7], Donthireddy and Chandrashekhara [20.8], and Crawley and de Luis [18.8].

Note, if the piezoelectric couples are used to produce an in-plane extension as well as a constant moment (as in this example) then the in-plane displacement, u_0, terms in (14.2) must be retained.

If the piezoelectric patches actuation and/or the applied mechanical load are dynamic loads, then Hamilton's Principle is employed as discussed in previous chapters to obtain solutions.

20.2 References

20.1. Arters, J.T. Master's Thesis, Mechanical Engineering, University of Delaware, forthcoming.

20.2. Abramovich, H. (2003) Piezoelectric Actuation for Smart Sandwich Structures – Closed Form Solutions, *Journal of Sandwich Structures and Materials*, Vol. 5, October, pp. 377-396.

20.3. Abramovich, H. (1998) Deflection Control of Laminated Composite Beams with Piezoceramic Layers – Closed Form Solutions, *Composite Structures*, Vol. 43, pp. 217-231.

20.4 Aldraihem, O.J. and Khdeir, A.A. (2000) Smart Beams with Extension and Thickness-Shear Piezoelectric Actuators, *Smart Material Structures*, Vol. 9, pp. 1-9.

20.5. Eisenberger, M. and Abramovich, H. (1997) Shape Control of Non-Symmetric Piezolaminated Composite Beams, *Composite Structures*, Vol. 38, No. 1-4, pp. 565-571.

20.6. Azzouz, M.S., Mei, C., Bevan, J.S. and Ro, J.J. (2001) Finite Element Modeling of MFC/AFC Actuators and Performance of MFC, *Journal of Intelligent Material Systems and Structures*, Vol. 12, September, pp. 601-612.

20.7. Abramovich, H. and Pletner, B. (1997) Actuation and Sensing of Piezolaminated Sandwich Type Structures, *Composite Structures*, Vol. 38, No. 1-4, pp. 17-27.

20.8. Donthireddy, P. and Chandrashekhara, K. (1996) Modeling and Shape Control of Composite Beams with Embedded Piezoelectric Actuators, *Composite Structures*, Vol. 35, pp. 237-244.

AUTHOR INDEX

SUBJECT INDEX

Mechanics

SOLID MECHANICS AND ITS APPLICATIONS

Series Editor: G.M.L. Gladwell

Aims and Scope of the Series

The fundamental questions arising in mechanics are: *Why?*, *How?*, and *How much?* The aim of this series is to provide lucid accounts written by authoritative researchers giving vision and insight in answering these questions on the subject of mechanics as it relates to solids. The scope of the series covers the entire spectrum of solid mechanics. Thus it includes the foundation of mechanics; variational formulations; computational mechanics; statics, kinematics and dynamics of rigid and elastic bodies; vibrations of solids and structures; dynamical systems and chaos; the theories of elasticity, plasticity and viscoelasticity; composite materials; rods, beams, shells and membranes; structural control and stability; soils, rocks and geomechanics; fracture; tribology; experimental mechanics; biomechanics and machine design.

1. R.T. Haftka, Z. Gürdal and M.P. Kamat: *Elements of Structural Optimization*. 2nd rev.ed., 1990 ISBN 0-7923-0608-2
2. J.J. Kalker: *Three-Dimensional Elastic Bodies in Rolling Contact.* 1990 ISBN 0-7923-0712-7
3. P. Karasudhi: *Foundations of Solid Mechanics.* 1991 ISBN 0-7923-0772-0
4. *Not published*
5. *Not published.*
6. J.F. Doyle: *Static and Dynamic Analysis of Structures.* With an Emphasis on Mechanics and Computer Matrix Methods. 1991 ISBN 0-7923-1124-8; Pb 0-7923-1208-2
7. O.O. Ochoa and J.N. Reddy: *Finite Element Analysis of Composite Laminates.* ISBN 0-7923-1125-6
8. M.H. Aliabadi and D.P. Rooke: *Numerical Fracture Mechanics.* ISBN 0-7923-1175-2
9. J. Angeles and C.S. López-Cajún: *Optimization of Cam Mechanisms.* 1991 ISBN 0-7923-1355-0
10. D.E. Grierson, A. Franchi and P. Riva (eds.): *Progress in Structural Engineering.* 1991 ISBN 0-7923-1396-8
11. R.T. Haftka and Z. Gürdal: *Elements of Structural Optimization.* 3rd rev. and exp. ed. 1992 ISBN 0-7923-1504-9; Pb 0-7923-1505-7
12. J.R. Barber: *Elasticity.* 1992 ISBN 0-7923-1609-6; Pb 0-7923-1610-X
13. H.S. Tzou and G.L. Anderson (eds.): *Intelligent Structural Systems.* 1992 ISBN 0-7923-1920-6
14. E.E. Gdoutos: *Fracture Mechanics.* An Introduction. 1993 ISBN 0-7923-1932-X
15. J.P. Ward: *Solid Mechanics.* An Introduction. 1992 ISBN 0-7923-1949-4
16. M. Farshad: *Design and Analysis of Shell Structures.* 1992 ISBN 0-7923-1950-8
17. H.S. Tzou and T. Fukuda (eds.): *Precision Sensors, Actuators and Systems.* 1992 ISBN 0-7923-2015-8
18. J.R. Vinson: *The Behavior of Shells Composed of Isotropic and Composite Materials.* 1993 ISBN 0-7923-2113-8
19. H.S. Tzou: *Piezoelectric Shells.* Distributed Sensing and Control of Continua. 1993 ISBN 0-7923-2186-3
20. W. Schiehlen (ed.): *Advanced Multibody System Dynamics.* Simulation and Software Tools. 1993 ISBN 0-7923-2192-8
21. C.-W. Lee: *Vibration Analysis of Rotors.* 1993 ISBN 0-7923-2300-9
22. D.R. Smith: *An Introduction to Continuum Mechanics.* 1993 ISBN 0-7923-2454-4
23. G.M.L. Gladwell: *Inverse Problems in Scattering.* An Introduction. 1993 ISBN 0-7923-2478-1

Mechanics

SOLID MECHANICS AND ITS APPLICATIONS

Series Editor: G.M.L. Gladwell

24. G. Prathap: *The Finite Element Method in Structural Mechanics*. 1993 ISBN 0-7923-2492-7
25. J. Herskovits (ed.): *Advances in Structural Optimization*. 1995 ISBN 0-7923-2510-9
26. M.A. González-Palacios and J. Angeles: *Cam Synthesis*. 1993 ISBN 0-7923-2536-2
27. W.S. Hall: *The Boundary Element Method*. 1993 ISBN 0-7923-2580-X
28. J. Angeles, G. Hommel and P. Kovács (eds.): *Computational Kinematics*. 1993 ISBN 0-7923-2585-0
29. A. Curnier: *Computational Methods in Solid Mechanics*. 1994 ISBN 0-7923-2761-6
30. D.A. Hills and D. Nowell: *Mechanics of Fretting Fatigue*. 1994 ISBN 0-7923-2866-3
31. B. Tabarrok and F.P.J. Rimrott: *Variational Methods and Complementary Formulations in Dynamics*. 1994 ISBN 0-7923-2923-6
32. E.H. Dowell (ed.), E.F. Crawley, H.C. Curtiss Jr., D.A. Peters, R. H. Scanlan and F. Sisto: *A Modern Course in Aeroelasticity*. Third Revised and Enlarged Edition. 1995 ISBN 0-7923-2788-8; Pb: 0-7923-2789-6
33. A. Preumont: *Random Vibration and Spectral Analysis*. 1994 ISBN 0-7923-3036-6
34. J.N. Reddy (ed.): *Mechanics of Composite Materials*. Selected works of Nicholas J. Pagano. 1994 ISBN 0-7923-3041-2
35. A.P.S. Selvadurai (ed.): *Mechanics of Poroelastic Media*. 1996 ISBN 0-7923-3329-2
36. Z. Mróz, D. Weichert, S. Dorosz (eds.): *Inelastic Behaviour of Structures under Variable Loads*. 1995 ISBN 0-7923-3397-7
37. R. Pyrz (ed.): *IUTAM Symposium on Microstructure-Property Interactions in Composite Materials*. Proceedings of the IUTAM Symposium held in Aalborg, Denmark. 1995 ISBN 0-7923-3427-2
38. M.I. Friswell and J.E. Mottershead: *Finite Element Model Updating in Structural Dynamics*. 1995 ISBN 0-7923-3431-0
39. D.F. Parker and A.H. England (eds.): *IUTAM Symposium on Anisotropy, Inhomogeneity and Nonlinearity in Solid Mechanics*. Proceedings of the IUTAM Symposium held in Nottingham, U.K. 1995 ISBN 0-7923-3594-5
40. J.-P. Merlet and B. Ravani (eds.): *Computational Kinematics '95*. 1995 ISBN 0-7923-3673-9
41. L.P. Lebedev, I.I. Vorovich and G.M.L. Gladwell: *Functional Analysis*. Applications in Mechanics and Inverse Problems. 1996 ISBN 0-7923-3849-9
42. J. Menčik: *Mechanics of Components with Treated or Coated Surfaces*. 1996 ISBN 0-7923-3700-X
43. D. Bestle and W. Schiehlen (eds.): *IUTAM Symposium on Optimization of Mechanical Systems*. Proceedings of the IUTAM Symposium held in Stuttgart, Germany. 1996 ISBN 0-7923-3830-8
44. D.A. Hills, P.A. Kelly, D.N. Dai and A.M. Korsunsky: *Solution of Crack Problems*. The Distributed Dislocation Technique. 1996 ISBN 0-7923-3848-0
45. V.A. Squire, R.J. Hosking, A.D. Kerr and P.J. Langhorne: *Moving Loads on Ice Plates*. 1996 ISBN 0-7923-3953-3
46. A. Pineau and A. Zaoui (eds.): *IUTAM Symposium on Micromechanics of Plasticity and Damage of Multiphase Materials*. Proceedings of the IUTAM Symposium held in Sèvres, Paris, France. 1996 ISBN 0-7923-4188-0
47. A. Naess and S. Krenk (eds.): *IUTAM Symposium on Advances in Nonlinear Stochastic Mechanics*. Proceedings of the IUTAM Symposium held in Trondheim, Norway. 1996 ISBN 0-7923-4193-7
48. D. Ieşan and A. Scalia: *Thermoelastic Deformations*. 1996 ISBN 0-7923-4230-5

Mechanics

SOLID MECHANICS AND ITS APPLICATIONS

Series Editor: G.M.L. Gladwell

49. J.R. Willis (ed.): *IUTAM Symposium on Nonlinear Analysis of Fracture*. Proceedings of the IUTAM Symposium held in Cambridge, U.K. 1997 ISBN 0-7923-4378-6
50. A. Preumont: *Vibration Control of Active Structures*. An Introduction. 1997 ISBN 0-7923-4392-1
51. G.P. Cherepanov: *Methods of Fracture Mechanics: Solid Matter Physics*. 1997 ISBN 0-7923-4408-1
52. D.H. van Campen (ed.): *IUTAM Symposium on Interaction between Dynamics and Control in Advanced Mechanical Systems*. Proceedings of the IUTAM Symposium held in Eindhoven, The Netherlands. 1997 ISBN 0-7923-4429-4
53. N.A. Fleck and A.C.F. Cocks (eds.): *IUTAM Symposium on Mechanics of Granular and Porous Materials*. Proceedings of the IUTAM Symposium held in Cambridge, U.K. 1997 ISBN 0-7923-4553-3
54. J. Roorda and N.K. Srivastava (eds.): *Trends in Structural Mechanics*. Theory, Practice, Education. 1997 ISBN 0-7923-4603-3
55. Yu.A. Mitropolskii and N. Van Dao: *Applied Asymptotic Methods in Nonlinear Oscillations*. 1997 ISBN 0-7923-4605-X
56. C. Guedes Soares (ed.): *Probabilistic Methods for Structural Design*. 1997 ISBN 0-7923-4670-X
57. D. François, A. Pineau and A. Zaoui: *Mechanical Behaviour of Materials*. Volume I: Elasticity and Plasticity. 1998 ISBN 0-7923-4894-X
58. D. François, A. Pineau and A. Zaoui: *Mechanical Behaviour of Materials*. Volume II: Viscoplasticity, Damage, Fracture and Contact Mechanics. 1998 ISBN 0-7923-4895-8
59. L.T. Tenek and J. Argyris: *Finite Element Analysis for Composite Structures*. 1998 ISBN 0-7923-4899-0
60. Y.A. Bahei-El-Din and G.J. Dvorak (eds.): *IUTAM Symposium on Transformation Problems in Composite and Active Materials*. Proceedings of the IUTAM Symposium held in Cairo, Egypt. 1998 ISBN 0-7923-5122-3
61. I.G. Goryacheva: *Contact Mechanics in Tribology*. 1998 ISBN 0-7923-5257-2
62. O.T. Bruhns and E. Stein (eds.): *IUTAM Symposium on Micro- and Macrostructural Aspects of Thermoplasticity*. Proceedings of the IUTAM Symposium held in Bochum, Germany. 1999 ISBN 0-7923-5265-3
63. F.C. Moon: *IUTAM Symposium on New Applications of Nonlinear and Chaotic Dynamics in Mechanics*. Proceedings of the IUTAM Symposium held in Ithaca, NY, USA. 1998 ISBN 0-7923-5276-9
64. R. Wang: *IUTAM Symposium on Rheology of Bodies with Defects*. Proceedings of the IUTAM Symposium held in Beijing, China. 1999 ISBN 0-7923-5297-1
65. Yu.I. Dimitrienko: *Thermomechanics of Composites under High Temperatures*. 1999 ISBN 0-7923-4899-0
66. P. Argoul, M. Frémond and Q.S. Nguyen (eds.): *IUTAM Symposium on Variations of Domains and Free-Boundary Problems in Solid Mechanics*. Proceedings of the IUTAM Symposium held in Paris, France. 1999 ISBN 0-7923-5450-8
67. F.J. Fahy and W.G. Price (eds.): *IUTAM Symposium on Statistical Energy Analysis*. Proceedings of the IUTAM Symposium held in Southampton, U.K. 1999 ISBN 0-7923-5457-5
68. H.A. Mang and F.G. Rammerstorfer (eds.): *IUTAM Symposium on Discretization Methods in Structural Mechanics*. Proceedings of the IUTAM Symposium held in Vienna, Austria. 1999 ISBN 0-7923-5591-1

Mechanics

SOLID MECHANICS AND ITS APPLICATIONS

Series Editor: G.M.L. Gladwell

69. P. Pedersen and M.P. Bendsøe (eds.): *IUTAM Symposium on Synthesis in Bio Solid Mechanics*. Proceedings of the IUTAM Symposium held in Copenhagen, Denmark. 1999 ISBN 0-7923-5615-2
70. S.K. Agrawal and B.C. Fabien: *Optimization of Dynamic Systems*. 1999 ISBN 0-7923-5681-0
71. A. Carpinteri: *Nonlinear Crack Models for Nonmetallic Materials*. 1999 ISBN 0-7923-5750-7
72. F. Pfeifer (ed.): *IUTAM Symposium on Unilateral Multibody Contacts*. Proceedings of the IUTAM Symposium held in Munich, Germany. 1999 ISBN 0-7923-6030-3
73. E. Lavendelis and M. Zakrzhevsky (eds.): *IUTAM/IFToMM Symposium on Synthesis of Nonlinear Dynamical Systems*. Proceedings of the IUTAM/IFToMM Symposium held in Riga, Latvia. 2000 ISBN 0-7923-6106-7
74. J.-P. Merlet: *Parallel Robots*. 2000 ISBN 0-7923-6308-6
75. J.T. Pindera: *Techniques of Tomographic Isodyne Stress Analysis*. 2000 ISBN 0-7923-6388-4
76. G.A. Maugin, R. Drouot and F. Sidoroff (eds.): *Continuum Thermomechanics*. The Art and Science of Modelling Material Behaviour. 2000 ISBN 0-7923-6407-4
77. N. Van Dao and E.J. Kreuzer (eds.): *IUTAM Symposium on Recent Developments in Non-linear Oscillations of Mechanical Systems*. 2000 ISBN 0-7923-6470-8
78. S.D. Akbarov and A.N. Guz: *Mechanics of Curved Composites*. 2000 ISBN 0-7923-6477-5
79. M.B. Rubin: *Cosserat Theories: Shells, Rods and Points*. 2000 ISBN 0-7923-6489-9
80. S. Pellegrino and S.D. Guest (eds.): *IUTAM-IASS Symposium on Deployable Structures: Theory and Applications*. Proceedings of the IUTAM-IASS Symposium held in Cambridge, U.K., 6–9 September 1998. 2000 ISBN 0-7923-6516-X
81. A.D. Rosato and D.L. Blackmore (eds.): *IUTAM Symposium on Segregation in Granular Flows*. Proceedings of the IUTAM Symposium held in Cape May, NJ, U.S.A., June 5–10, 1999. 2000 ISBN 0-7923-6547-X
82. A. Lagarde (ed.): *IUTAM Symposium on Advanced Optical Methods and Applications in Solid Mechanics*. Proceedings of the IUTAM Symposium held in Futuroscope, Poitiers, France, August 31–September 4, 1998. 2000 ISBN 0-7923-6604-2
83. D. Weichert and G. Maier (eds.): *Inelastic Analysis of Structures under Variable Loads*. Theory and Engineering Applications. 2000 ISBN 0-7923-6645-X
84. T.-J. Chuang and J.W. Rudnicki (eds.): *Multiscale Deformation and Fracture in Materials and Structures*. The James R. Rice 60th Anniversary Volume. 2001 ISBN 0-7923-6718-9
85. S. Narayanan and R.N. Iyengar (eds.): *IUTAM Symposium on Nonlinearity and Stochastic Structural Dynamics*. Proceedings of the IUTAM Symposium held in Madras, Chennai, India, 4–8 January 1999 ISBN 0-7923-6733-2
86. S. Murakami and N. Ohno (eds.): *IUTAM Symposium on Creep in Structures*. Proceedings of the IUTAM Symposium held in Nagoya, Japan, 3-7 April 2000. 2001 ISBN 0-7923-6737-5
87. W. Ehlers (ed.): *IUTAM Symposium on Theoretical and Numerical Methods in Continuum Mechanics of Porous Materials*. Proceedings of the IUTAM Symposium held at the University of Stuttgart, Germany, September 5-10, 1999. 2001 ISBN 0-7923-6766-9
88. D. Durban, D. Givoli and J.G. Simmonds (eds.): *Advances in the Mechanis of Plates and Shells The Avinoam Libai Anniversary Volume*. 2001 ISBN 0-7923-6785-5
89. U. Gabbert and H.-S. Tzou (eds.): *IUTAM Symposium on Smart Structures and Structonic Systems*. Proceedings of the IUTAM Symposium held in Magdeburg, Germany, 26–29 September 2000. 2001 ISBN 0-7923-6968-8

Mechanics

SOLID MECHANICS AND ITS APPLICATIONS

Series Editor: G.M.L. Gladwell

90. Y. Ivanov, V. Cheshkov and M. Natova: *Polymer Composite Materials – Interface Phenomena & Processes*. 2001 ISBN 0-7923-7008-2
91. R.C. McPhedran, L.C. Botten and N.A. Nicorovici (eds.): *IUTAM Symposium on Mechanical and Electromagnetic Waves in Structured Media*. Proceedings of the IUTAM Symposium held in Sydney, NSW, Australia, 18-22 Januari 1999. 2001 ISBN 0-7923-7038-4
92. D.A. Sotiropoulos (ed.): *IUTAM Symposium on Mechanical Waves for Composite Structures Characterization*. Proceedings of the IUTAM Symposium held in Chania, Crete, Greece, June 14-17, 2000. 2001 ISBN 0-7923-7164-X
93. V.M. Alexandrov and D.A. Pozharskii: *Three-Dimensional Contact Problems*. 2001 ISBN 0-7923-7165-8
94. J.P. Dempsey and H.H. Shen (eds.): *IUTAM Symposium on Scaling Laws in Ice Mechanics and Ice Dynamics*. Proceedings of the IUTAM Symposium held in Fairbanks, Alaska, U.S.A., 13-16 June 2000. 2001 ISBN 1-4020-0171-1
95. U. Kirsch: *Design-Oriented Analysis of Structures*. A Unified Approach. 2002 ISBN 1-4020-0443-5
96. A. Preumont: *Vibration Control of Active Structures*. An Introduction (2nd Edition). 2002 ISBN 1-4020-0496-6
97. B.L. Karihaloo (ed.): *IUTAM Symposium on Analytical and Computational Fracture Mechanics of Non-Homogeneous Materials*. Proceedings of the IUTAM Symposium held in Cardiff, U.K., 18-22 June 2001. 2002 ISBN 1-4020-0510-5
98. S.M. Han and H. Benaroya: *Nonlinear and Stochastic Dynamics of Compliant Offshore Structures*. 2002 ISBN 1-4020-0573-3
99. A.M. Linkov: *Boundary Integral Equations in Elasticity Theory*. 2002 ISBN 1-4020-0574-1
100. L.P. Lebedev, I.I. Vorovich and G.M.L. Gladwell: *Functional Analysis*. Applications in Mechanics and Inverse Problems (2nd Edition). 2002 ISBN 1-4020-0667-5; Pb: 1-4020-0756-6
101. Q.P. Sun (ed.): *IUTAM Symposium on Mechanics of Martensitic Phase Transformation in Solids*. Proceedings of the IUTAM Symposium held in Hong Kong, China, 11-15 June 2001. 2002 ISBN 1-4020-0741-8
102. M.L. Munjal (ed.): *IUTAM Symposium on Designing for Quietness*. Proceedings of the IUTAM Symposium held in Bangkok, India, 12-14 December 2000. 2002 ISBN 1-4020-0765-5
103. J.A.C. Martins and M.D.P. Monteiro Marques (eds.): *Contact Mechanics*. Proceedings of the 3rd Contact Mechanics International Symposium, Praia da Consolação, Peniche, Portugal, 17-21 June 2001. 2002 ISBN 1-4020-0811-2
104. H.R. Drew and S. Pellegrino (eds.): *New Approaches to Structural Mechanics, Shells and Biological Structures*. 2002 ISBN 1-4020-0862-7
105. J.R. Vinson and R.L. Sierakowski: *The Behavior of Structures Composed of Composite Materials*. Second Edition. 2002 ISBN 1-4020-0904-6
106. Not yet published.
107. J.R. Barber: *Elasticity*. Second Edition. 2002 ISBN Hb 1-4020-0964-X; Pb 1-4020-0966-6
108. C. Miehe (ed.): *IUTAM Symposium on Computational Mechanics of Solid Materials at Large Strains*. Proceedings of the IUTAM Symposium held in Stuttgart, Germany, 20-24 August 2001. 2003 ISBN 1-4020-1170-9

Mechanics

SOLID MECHANICS AND ITS APPLICATIONS

Series Editor: G.M.L. Gladwell

109. P. Ståhle and K.G. Sundin (eds.): *IUTAM Symposium on Field Analyses for Determination of Material Parameters – Experimental and Numerical Aspects.* Proceedings of the IUTAM Symposium held in Abisko National Park, Kiruna, Sweden, July 31 – August 4, 2000. 2003 ISBN 1-4020-1283-7
110. N. Sri Namachchivaya and Y.K. Lin (eds.): *IUTAM Symposium on Nonlnear Stochastic Dynamics.* Proceedings of the IUTAM Symposium held in Monticello, IL, USA, 26 – 30 August, 2000. 2003 ISBN 1-4020-1471-6
111. H. Sobieckzky (ed.): *IUTAM Symposium Transsonicum IV.* Proceedings of the IUTAM Symposium held in Göttingen, Germany, 2–6 September 2002, 2003 ISBN 1-4020-1608-5
112. J.-C. Samin and P. Fisette: *Symbolic Modeling of Multibody Systems.* 2003 ISBN 1-4020-1629-8
113. A.B. Movchan (ed.): *IUTAM Symposium on Asymptotics, Singularities and Homogenisation in Problems of Mechanics.* Proceedings of the IUTAM Symposium held in Liverpool, United Kingdom, 8-11 July 2002. 2003 ISBN 1-4020-1780-4
114. S. Ahzi, M. Cherkaoui, M.A. Khaleel, H.M. Zbib, M.A. Zikry and B. LaMatina (eds.): *IUTAM Symposium on Multiscale Modeling and Characterization of Elastic-Inelastic Behavior of Engineering Materials.* Proceedings of the IUTAM Symposium held in Marrakech, Morocco, 20-25 October 2002. 2004 ISBN 1-4020-1861-4
115. H. Kitagawa and Y. Shibutani (eds.): *IUTAM Symposium on Mesoscopic Dynamics of Fracture Process and Materials Strength.* Proceedings of the IUTAM Symposium held in Osaka, Japan, 6-11 July 2003. Volume in celebration of Professor Kitagawa's retirement. 2004 ISBN 1-4020-2037-6
116. E.H. Dowell, R.L. Clark, D. Cox, H.C. Curtiss, Jr., K.C. Hall, D.A. Peters, R.H. Scanlan, E. Simiu, F. Sisto and D. Tang: *A Modern Course in Aeroelasticity.* 4th Edition, 2004 ISBN 1-4020-2039-2
117. T. Burczyński and A. Osyczka (eds.): *IUTAM Symposium on Evolutionary Methods in Mechanics.* Proceedings of the IUTAM Symposium held in Cracow, Poland, 24-27 September 2002. 2004 ISBN 1-4020-2266-2
118. D. Ieşan: *Thermoelastic Models of Continua.* 2004 ISBN 1-4020-2309-X
119. G.M.L. Gladwell: *Inverse Problems in Vibration.* Second Edition. 2004 ISBN 1-4020-2670-6
120. J.R. Vinson: *Plate and Panel Structures of Isotropic, Composite and Piezoelectric Materials, Including Sandwich Construction.* 2005 ISBN 1-4020-3110-6
121. *Forthcoming*
122. G. Rega and F. Vestroni (eds.): *IUTAM Symposium on Chaotic Dynamics and Control of Systems and Processes in Mechanics.* 2005 ISBN 1-4020-3267-6